T0213251

This book provides a comprehensive introduction to the theory of magnetic field line reconnection, now a major subject in plasma physics. The book focuses on the various reconnection mechanisms dominating magnetic processes under the different plasma conditions encountered in astrophysical systems and in laboratory devices. The book consists of two major parts: the first deals with the classical resistive approach, while the second presents an overview of weakly collisional or collisionless plasmas. Applications primarily concern astrophysical phenomena and dynamo theory, with emphasis on the solar dynamo and the geodynamo, as well as on magnetospheric substorms, the most spectacular reconnection events in the magnetospheric plasma. The theoretical procedures and results also apply directly to reconnection processes in laboratory plasmas, in particular the sawtooth phenomenon in tokamaks.

The book will be of value to graduate students and researchers interested in magnetic processes both in astrophysical and laboratory plasma physics.

Dieter Biskamp received his Ph.D. from the University of Munich. Following a postdoctoral period at the Max-Planck-Institute for Astrophysics he worked at the Space Research Institute in Frascati and became senior research scientist at the Max-Planck-Institute for Plasma Physics in 1972. Since 1981 he has been head of the General Theory Group and since 1995 has been head of the Nonlinear Plasma Dynamics Group. In 1979 he was visiting professor at the University of Texas and in 1995 COE visiting professor at the National Institute for Fusion Science in Nagoya. His scientific activities cover many areas of plasma physics, in particular magnetohydrodynamics and reconnection theory. He is the author of the book *Nonlinear Magnetohydrodynamics*.

CAMBRIDGE MONOGRAPHS ON PLASMA PHYSICS

General Editors: M. G. Haines, K. I. Hopcraft, I. H. Hutchinson, C. M. Surko and K. Schindler

1. D. Biskamp *Nonlinear Magnetohydrodynamics*
2. H. R. Griem *Principles of Plasma Spectroscopy*
3. D. Biskamp *Magnetic Reconnection in Plasmas*

Magnetic Reconnection
in Plasmas

Dieter Biskamp

Max-Planck-Institute for Plasma Physics, Garching

CAMBRIDGE
UNIVERSITY PRESS

CAMBRIDGE UNIVERSITY PRESS
Cambridge, New York, Melbourne, Madrid, Cape Town, Singapore, São Paulo

Cambridge University Press
The Edinburgh Building, Cambridge CB2 2RU, UK

Published in the United States of America by Cambridge University Press, New York

www.cambridge.org
Information on this title: www.cambridge.org/9780521582889

First published 2000
This digitally printed first paperback version 2005

A catalogue record for this publication is available from the British Library

Library of Congress Cataloguing in Publication data

Biskamp, D.
Magnetic reconnection in plasmas / Dieter Biskamp.
p. cm.
Includes bibliographical references and index.
ISBN 0 521 58288 1 (hb)
1. Magnetic reconnection. 2. Plasma (Ionized gases) I. Title.
QC718.5.M3 B53   2000
538'.6–dc21   99–087680   CIP

ISBN-13 978-0-521-58288-9 hardback
ISBN-10 0-521-58288-1 hardback

ISBN-13 978-0-521-02036-7 paperback
ISBN-10 0-521-02036-0 paperback

To the memory of my parents

Contents

Preface

This book is in a sense a sequel to my previous book *Nonlinear Magneto-hydrodynamics*, which contained a chapter on magnetic reconnection. Judging from many discussions it appeared that it was this chapter that was particularly appreciated. The plan to write a full monograph on this topic actually took a concrete shape during a stay at the National Institute for Fusion Science at Nagoya, where I found the time to work out the basic conception of the book. It became clear that resistive theory, to which most of the previous work was restricted, including that chapter of my previous book, covers only a particular aspect of this multifaceted subject and not even the most interesting one, in view of the various applications, both in fusion plasma devices and in astrophysical plasmas, where collisionless effects tend to dominate over resistivity.

While resistive reconnection theory had reached a certain level of maturity and completion about a decade ago (few theories are really complete before becoming obsolete), the understanding of collisionless reconnection processes has shown a rapid development during the past five years or so. The book therefore consists of two main parts, chapters 3–5 deal with resistive theory, while chapters 6–8 give an overview of the present understanding of collisionless reconnection processes. I mainly emphasize the reconnection mechanisms, which operate under the different plasma conditions, to explain the apparent paradox that formally very weak effects in Ohm's law account for the rapid dynamic time-scales suggested by the observations.

Applications concern primarily astrophysical phenomena. Chapter 5 introduces dynamo theory, considering in some detail the generation of the solar and the geomagnetic field, while chapter 8 deals with magnetospheric substorms, the most important reconnection process in the Earth's magnetosphere. Both chapters are rather autonomous and can be read independently of the remainder of the book. Concerning laboratory plasmas I resume the discussion of the sawtooth phenomenon considered in some length in *Nonlinear Magnetohydrodynamics*, giving an update of the experimental and theoretical situation. I also discuss briefly several laboratory experiments designed specifically to study magnetic reconnection physics.

It is a pleasure to express my gratitude to the many colleagues with whom I enjoyed fruitful and illuminating discussions on the topics of this book, in particular Jim Drake, who taught me the importance of collisionless reconnection, and Wolfgang Baumjohann, Michael Hesse and Manfred Scholer for introducing me to the realm of magnetospheric physics. I also acknowledge the financial support by the COE programme of Monbusho and the kind hospitality of the National Institute for Fusion Science with special thanks to Tetsuya Sato. Finally I would like to thank Brian Watts for his painstaking copy-editing of the manuscript.

Garching, January 2000   *Dieter Biskamp*

1

Introduction

Since the early 1950s, when magnetohydrodynamics – MHD in short – became an established theory and along with it the concept of a "frozen-in" magnetic field within an electrically conducting fluid, the problem of how magnetic field energy could be released in such a fluid has been generally acknowledged. In the early days the major impetus came from solar physics. Estimates readily showed that the energies associated with eruptive processes, notably flares, can only be stored in the coronal magnetic field, all other energy sources being by far too weak. On the other hand the high temperature in the corona, which makes the coronal plasma a particularly good electrical conductor, appeared to preclude any fast magnetic change involving diffusion. For a coronal electron temperature $T_e \sim 10^6$ K the magnetic diffusivity is $\eta \sim 10^4$ cm^2/s, hence field diffusion in a region of diameter $L \sim 10^4$ km as typically involved in a flare would require a time-scale $\tau_\eta = L^2/\eta \sim 10^{14}$ s, whereas the observed flash phase of a flare takes less than $\sim 10^3$ s.

It had, however, soon been realized that the discrepancy is not quite as bad as this. Contrary to magnetic diffusion in a solid conductor, a fluid is stirred into motion by the change of the magnetic field. As it carries along the frozen-in field, it may generate steep field gradients typically located in sheet-like structures, and hence lead to much shorter diffusion times. By the early 1960s the concept of resistive current sheets, or Sweet–Parker sheets named after their major protagonists, had become standard knowledge. The transverse scale of such a sheet is $\delta_\eta \sim (\tau_A \eta)^{1/2}$, where the Alfvén time τ_A is a typical MHD time-scale. Though magnetic diffusion across the sheet is very fast, $\delta_\eta^2/\eta \sim \tau_A \sim 1$ s, the overall dynamics is now limited by the rate of convective field transport toward the sheet. A simple back-of-the-envelope analysis shows that the time for magnetic energy release in the presence of a Sweet–Parker sheet is $\tau_{SP} \sim (\tau_A \tau_\eta)^{1/2} \sim 10^7$ s, which is much shorter than the global diffusion time of 10^{14} s, but still many orders of magnitude longer than observed.

A break-through occurred in 1964 at a symposium on the physics of solar flares, when Petschek presented a mechanism which appeared to allow very fast energy release, depending only logarithmically on η. The model was soon generally accepted, forming the basis for the theory during the following two decades, during which more and more refined configurations were proposed. However Petschek's model is, properly speaking, not a genuine theory of fast reconnection, but the MHD configuration set up in the *presence* of an efficient reconnection mechanism. Petschek and his successors assumed, explicitly or tacitly, that resistivity provides such a mechanism. It was only during the 1980s it became gradually clear that Petschek's configuration is not valid in resistive MHD where, instead, extended current sheets are formed. It thus appeared that theory was thrown back to the pre-Petschek period of relatively slow Sweet–Parker sheet reconnection. The situation seemed to converge with the school of thought that had developed independently in the eastern hemisphere starting from Syrovatskii's elegant theory of current sheets. A possible way to reach faster reconnection speeds was to give up the usual assumption of quasi-stationarity by allowing MHD turbulence to develop. In many applications, however, such macroscopic turbulence is not observed.

A different approach to the problem was followed in magnetospheric physics. Except for the ionosphere, the magnetospheric plasma is so dilute that Coulomb collisions are, practically speaking, absent and hence classical resistivity vanishes. Can magnetic reconnection occur in a collisionless plasma? The problem had already been realized by Dungey in the early 1960s in his theory of magnetospheric convection. The usual approach was to assume the action of small-scale turbulence excited by some microinstability, where the scattering of electrons by charge fluctuations has a similar effect as Coulomb collisions. Various candidates for such effective or anomalous resistivity were discussed, though none was found to be fully satisfactory to account for reconnection in the magnetotail.

The question of fast quasi-collisionless reconnection became also relevant for hot fusion plasmas, in particular in tokamaks, where much theoretical effort was made to explain the characteristics of the sawtooth phenomenon, an internal relaxation oscillation. It was realized that in Ohm's law nonideal terms other than resistivity play the dominant role. Combining the results from both magnetospheric and tokamak plasma theory achieved in recent years we now seem to be close to a consistent picture of collisionless reconnection. Such processes seem to be significantly more efficient than resistive diffusion and allow fast quasi-Alfvénic reconnection velocities.

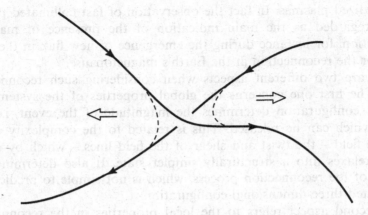

Fig. 1.1. Elementary picture of field line reconnection: relaxation of magnetic tension and plasma acceleration by a local change of field line connectivity.

The reader may have noticed that the term magnetic reconnection has slipped into our vocabulary without introduction. There is hardly a term in plasma physics which exhibits more scents, facets and also ambiguities, and which at times seems to be used even with a touch of magic. The basic picture is that of two field lines being frozen in and carried along with the fluid, until they come close to each other at some point where, due to weak nonideal effects in Ohm's law, they are cut and reconnected in a different way. (The alternative term "field line merging" with essentially the same meaning, which was popular in the 1970s, is gradually falling from use.) Simple as it may look this picture already contains important characteristic features of the reconnection process. Consider the field line configuration shown in fig. 1.1 where the upper and lower endpoints move around such as, for instance, the footpoints of coronal fields anchored in the photosphere. Physically the close encounter of the field lines implies that magnetic field gradients become locally strong, thus enhancing the formally weak nonideal process in Ohm's law. Hence reconnection is a localized process, in contrast to the slow global field diffusion and dissipation by the omnipresent, but weak, resistivity. The local enhancement of the nonideal effects determines the time-scale of the reconnection process, while the overall magnetic energy dissipation remains small owing to the localization of the process. The crucial point to note is that after local rearrangement of the magnetic connectivity, the global magnetic configuration may relax by reducing magnetic tension and accelerating the fluid in the way indicated in fig. 1.1, often called "the magnetic slingshot effect". Hence reconnection allows rapid conversion of magnetic into kinetic energy, which is the essence of impulsive events in magnetized plasmas. In fact the observation of fast collimated plasma flow is regarded as the main indication of the presence of magnetic reconnection, for instance during the emergence of new flux in the solar corona or the reconnection at the Earth's magnetopause.

There are two different aspects when considering such reconnection events. The first one concerns the global properties of the system. The magnetic configuration determines the magnitude of the event, i.e., the energy, which can be released. This is related to the complexity of the magnetic field – the twist and shear of the field lines – which by reconnection relaxes into a structurally simpler state. It also determines the location of the reconnection process, which is not simple to predict in a complicated three-dimensional configuration.

The second aspect refers to the local properties in the reconnection region – the physics of the actual reconnection mechanism – to which the major part of the book is devoted. Here one considers:

(a) Time-scale τ. It depends of course on the free magnetic energy, being the shorter the larger the latter, simply because higher velocities are generated. But τ is also expected to depend more or less strongly on the nonideal process R in Ohm's law. More specifically $\tau \sim L/v$, where L is the spatial extent of the configuration which is affected by the relaxation and v some average velocity generated during the process. According to the rule given above, we have $v \propto v_A$, where $v_A = \delta B/\sqrt{4\pi\rho}$ is the Alfvén velocity corresponding to the change δB of the magnetic field. Hence we can write

$$\tau \sim [L/v_A]f(R) = \tau_A f(R). \qquad (1.1)$$

The function $f(R)$ represents the influence of the particular reconnection mechanism, which dominates for the given plasma parameters, for instance $f = (\tau_\eta/\tau_A)^{1/2}$ for the Sweet–Parker process. Determination or at least estimation of $f(R)$ is a major topic of this book.

(b) Energy partition. The magnetic energy released by reconnection in an eruptive event is transformed into (i) bulk plasma motion often generating a strong shock, the blast wave, (ii) electron and ion heating and (iii) acceleration of a certain number of ions and electrons to high superthermal energies. Energy partition depends on the reconnection process. Until recently this point has received little attention in reconnection theory. A fundamental question concerns the efficiency of particle acceleration, in particular whether the observed high particle energies are primarily generated at the reconnection site or by acceleration at the shock.

(c) Threshold conditions. In general, a certain amount of free energy has to be accumulated before rapid relaxation sets in. A particular facet is to explain the trigger mechanism responsible for the sudden onset of energy release often observed.

The book consists of three parts: an introductory one, chapter 2, where the more formal, conceptual aspects of magnetic reconnection are discussed; a second part, chapters 3–5, which deals with resistive theory; and a third part, chapters 6–8, which treats collisionless reconnection. While the first two chapters in each of the two main parts are devoted to the basic theory, the last chapters introduce a particular application – dynamo theory and magnetospheric substorms, respectively.

Chapter 2 gives an introduction to the more formal aspects of reconnection theory. These are basic kinematic concepts referring to the topological properties of the magnetic field, where the dynamics of the embedding conducting fluid need not to be specified. Only Ohm's law and Faraday's law enter this discussion, and give rise to two conservation laws: the conservation of magnetic flux, from which derives in particular the physical meaning of field lines as infinitesimally thin flux tubes; and the conservation of magnetic helicity, which is a measure of the structural complexity of the configuration, the interconnection of the individual field lines. We then define magnetic topology and derive a criterion for topology conservation, which is related to but more general than magnetic flux conservation. A question which has evoked considerable discussion concerns the definition of reconnection; in particular, what are the criteria to decide whether, in an evolving magnetic configuration, reconnection actually takes place. While this point can easily be decided for a symmetric configuration where a separatrix is defined, general three-dimensional systems require individual consideration which refers to the local field line connectivity rather than their global topology. Finally, the problem of helicity conservation in reconnection is considered. We define the concepts of twist, knottedness and linkage and show that, while the helicities of the constituent flux tubes lose their meaning, the total helicity is in fact conserved, which is the basis of Taylor's theory of relaxation to a linear force-free state.

Chapters 3 and 4 deal with resistive reconnection theory: chapter 3 is focused on the characteristic stationary configuration, the resistive current sheet; chapter 4 considers the dynamics arising in resistive MHD, instabilities and their nonlinear evolution. We start, in chapter 3, with a cursory derivation and discussion of the MHD equations and their conservation properties and introduce a reduced set of MHD equations which, by eliminating the fast compressional effects, is particularly suited for treating the typically slower reconnection processes. Resistive recon-

nection occurs in current sheets. We analyze the self-similar process of
sheet formation and the structure of the stationary resistive sheets. The
properties of current-sheet reconnection are conveniently studied in the
form of driven reconnection, i.e., in an open system of a configuration
with reversing magnetic field and boundary conditions enforcing stationary inflow and outflow of plasma and magnetic flux. A simple and
elegant analytical approach has been developed by Syrovatskii, where current sheets appear as branch cuts of a complex function. This theory is
confirmed by numerical solution of the resistive MHD equations, which
yields the current-sheet scaling properties in terms of the resistivity and
the (enforced) reconnection rate. The simulations demonstrate that, in the
framework of resistive MHD, Petschek's slow-shock solution is not valid.
Petschek's configuration, which we briefly discuss, is only set up in the
presence of a reconnection mechanism more efficient than resistivity.

In chapter 4 we discuss in some detail the various types of instabilities
which occur in a resistive MHD system driven by gradients of current
density, pressure or velocity. For each mode we first outline the linear
stability characteristics and subsequently study its observable effects, i.e.,
nonlinear evolution and saturation. The prototype of a current-driven
resistive instability is the tearing mode, which leads to a configuration
containing magnetic islands. Though the growth rate is rather small, the
perturbation may reach a substantial amplitude or island width. Analytical
expressions are derived for the nonlinear growth and for the saturation
level. Nonlinear evolution is actually a slow magnetic diffusion process,
where no current sheet is formed. Hence not only the value of the resistivity
at the reconnection point, but also the resistivity profile in the island affect
the nonlinear properties. The neoclassical tearing mode, which may be
destabilized in a toroidal plasma at low collisionality, is due to a similar
nonlocal effect.

There are other paradigms of current-driven instabilities, which give rise
to faster dynamics and thus to current sheets. In the double tearing mode,
magnetic islands grow at neighboring resonant surfaces mutually enforcing
the reconnection process. In a plasma column we find the resistive kink
mode, a quasi-rigid helical shift of the central part of the column, which
is energetically favored. Here a single magnetic island boosts itself due
to the geometry of the configuration. Finally, two magnetic islands or
flux bundles having parallel currents attract each other and coalesce, their
magnetic fields reconnecting across a current sheet in between.

Pressure-driven modes arise due to field line curvature, which is equivalent to a gravitational force and may thus drive a Rayleigh–Taylor-type
instability. Here finite resistivity can destabilize modes by weakening the
effect of magnetic shear. Nonlinearly, however, the mode is self-focusing,
thus essentially avoiding the stabilizing shear effect. Thus reconnection

is not required, which makes the nonlinear evolution of this so-called ballooning instability particularly violent. A third source of free energy is a sheared flow driving the Kelvin–Helmholtz instability. Intense shear flows occur along current sheets, but the Kelvin–Helmholtz instability is, in general, stabilized by the parallel magnetic field adjacent to the sheet. The current sheet may, however, break up due to the tearing mode, if the sheet is sufficiently thin.

Chapter 5 gives an introduction to dynamo theory. Devoting a full chapter to this topic may at first sight appear somewhat surprising, since the dynamo effect – the amplification and sustainment of magnetic fields by motions in a conducting fluid – is usually not directly associated with a reconnection process. In particular the basic model of field amplification by stretching, twisting and folding of flux tubes does not involve reconnection. However, an important aspect of the dynamo problem is to account for large-scale fields, which have to be generated by reconnecting the naturally amplified small-scale flux tubes.

We start with kinematic dynamo theory, where the nonlinear reaction of the field through the Lorentz force is neglected. Here the convenient tool is the mean-field approximation, which allows a linear analysis in terms of dynamo modes for a phenomenological description of the magnetic fields in astrophysical objects such as stars and galaxies. We briefly review the theory of the solar dynamo, which has again become an open subject since recent observations, especially from helioseismology, have shattered what was believed to be a rather solid building.

The fundamental process of planetary and stellar dynamo action is convection in a rotating fluid shell. Numerical computations of model systems have demonstrated the basic mechanism. Though direct dynamical simulations of the Earth's liquid core are out of reach, simulations using phenomenological eddy diffusion coefficients have already revealed important properties of the geodynamo.

The convection driving dynamo action is highly turbulent in most systems of interest. Chapter 5 therefore presents also a brief overview of MHD turbulence and turbulent magneto-convection. I concentrate on recent developments obtained mainly from high-resolution three-dimensional numerical simulations, while referring for a broader introduction to chapter 7 in my book *Nonlinear Magnetohydrodynamics*.

Conventional theory has considered resistivity as the natural, if not the only possible, mechanism for magnetic reconnection. As a result, in weakly collisional plasmas such as in the solar corona or in a tokamak discharge the discrepancy between observed and predicted time-scales is particularly large caused by the inherent inefficiency of the Sweet–Parker process. In reality, however, in such systems noncollisional effects in Ohm's law are more important than resistivity, which no longer controls the reconnection dynamics. The simplest framework to deal with these effects is a two-fluid model, generalizing the one-fluid MHD theory. Since in this approach the presence of some collisions is still required to justify the use of isotropic pressures and to smooth weak flow singularities, we call reconnection in this regime, which is treated in chapter 6, noncollisional instead of collisionless.

It is useful to distinguish between high- and low-β systems. In the high-β case, ions and electrons decouple at scales below the ion inertia scale c/ω_{pi}, where the dynamics is described by the approximation of electron magnetohydrodynamics (EMHD). It is derived analytically and verified by numerical simulations, that the reconnection rate does not depend on the electron scale-length c/ω_{pe}. Also, the dependence on the ion scale-length c/ω_{pi} is found to be weak, no macroscopic current sheet being formed. In the low-β case, the strong guide field tightly couples electron and ion motions, apart from diamagnetic drifts, which give rise to the smallness parameter ρ_s, the effective ion Larmor radius. As in the high-β case, there is no macroscopic current sheet and reconnection is quasi-Alfvénic. The sawtooth oscillation in a tokamak plasma is a classical reconnection paradigm. I give a brief review of the present understanding of this phenomenon in chapter 6, again referring for a more detailed introduction to chapter 8 of my previous book.

For truly collisionless plasmas, such as in the solar wind, a kinetic description is required, which is presented in chapter 7. Since the usual approach to reconnection phenomena occurs from the fluid side, I presume that many readers are less familiar with the kinetic plasma aspects and will appreciate a somewhat broader introduction to microscopic theory. We therefore start the chapter with an outline of linear Vlasov theory and the different nonlinear effects, quasi-linear relaxation, mode coupling and particle trapping. We then discuss several microinstabilities and their nonlinear properties, which may give rise to anomalous transport effects in Ohm's law: the ion-sound instability; the lower-hybrid-drift instability; and the anisotropy-driven whistler instability. The collisionless tearing mode in a neutral sheet has played a special role in collisionless reconnection theory, in particular in the presence of a weak normal field component. After a longstanding debate it is now generally believed that the mode is stable, so that the break-up of such a collisionless configuration requires a finite perturbation generating an X-point.

Since collisionless reconnection is studied by particle simulation, we briefly describe the different numerical approaches. Simulation results indicate that, different from the resistive regime and similar to the behavior in the high-β noncollisional approximation, reconnection rates are independent of the electron physics and do not seem to depend on ion inertia, either, i.e., reconnection speeds scale with the Alfvén velocity.

Probably the most important application of collisionless reconnection is the magnetospheric substorm, which is considered in chapter 8. We first briefly describe the structure of the magnetosphere, which results from the interaction of the solar wind with the geomagnetic field. Dungey's model of magnetospheric convection gives a qualitative account of the reconnection processes at the front and in the tail and their dependence on the orientation of the interplanetary magnetic field. Observations indicate that tail reconnection occurs mainly in the form of a relaxation oscillation, a sequence of substorms. Combining observational and numerical results, the most probable substorm model is a fast reconnection of the tail magnetic field generating a plasmoid which is ejected tailward, though a nonreconnective process involving a ballooning-type instability cannot be excluded. Finally, the mechanism of particle acceleration in the magnetotail by the reconnection electric field is outlined.

Concerning notation and coordinate systems, I have tried to compromise between the demands of uniformity throughout the book and the use of notation which has become standard in a particular field. To a large extent the individual chapters are autonomous, but a certain amount of cross-referencing between chapters is certainly useful. The equations are primarily written in Gaussian units which, in my view, are more natural in collisionless theory than are SI units. But, as soon as the basic equations are specified, I introduce convenient normalizations depending on the specific problem to write the equations in dimensionless form.

2
Basic kinematic concepts

In this chapter we introduce some formal properties of the magnetic field embedded in an electrically conducting fluid, which play a fundamental role in reconnection theory. Conservation of magnetic flux leads to the concept of magnetic field lines as infinitesimally thin flux tubes, which are frozen in to the fluid. A slightly different concept is that of magnetic topology or connectivity. Since this is a property of the magnetic field alone, topology conservation is valid under more general conditions than flux conservation. We then try to give a comprehensive definition of reconnection. While, in two dimensions, reconnection can be defined globally in terms of plasma flows across a separatrix surface, such surfaces, separating regions of different global field-line topology, no longer exist in a general three-dimensional configuration. Here, reconnection can only be defined by a local change of field-line connectivity, which is due to a nonvanishing parallel electric field. Finally we consider the behavior of magnetic helicity under the influence of reconnection. Though the helicities of the individual flux tubes involved in the reconnection process lose their meaning, the total helicity of the system is conserved.

2.1 Magnetic flux and magnetic helicity

The coupling between an electrically conducting fluid and the magnetic field immersed in the fluid is described by the generalized Ohm's law

$$\mathbf{E} + \frac{\mathbf{v}}{c} \times \mathbf{B} = \mathbf{R}, \tag{2.1}$$

where $\mathbf{v}(\mathbf{x}, t)$ is the fluid velocity and $\mathbf{R}$ comprises the different so-called nonideal effects of the plasma, dissipative or nondissipative, which are discussed in more detail in section 6.1. Here, we only note that in most cases of interest $\mathbf{R}$ is formally very small. The left side of (2.1) is the

10

electric field $\mathbf{E}'$ in the rest frame of the fluid element. Neglecting $\mathbf{R}$ simply means that $\mathbf{E}' = 0$, which is called the ideal Ohm's law

$$\mathbf{E} + \frac{\mathbf{v}}{c} \times \mathbf{B} = 0. \tag{2.2}$$

(We shall see, in section 6.1, that Ohm's law is just the equation of motion of the electron fluid.) Insertion into Faraday's law

$$\partial_t \mathbf{B} = -c \nabla \times \mathbf{E} \tag{2.3}$$

gives

$$\partial_t \mathbf{B} = \nabla \times (\mathbf{v} \times \mathbf{B}), \tag{2.4}$$

which has the form of a continuity equation for the vector field $\mathbf{B}$. In integral form (2.4) states the conservation of the magnetic flux

$$\phi = \int_F \mathbf{B} \cdot d\mathbf{F} \tag{2.5}$$

through an arbitrary surface $F(t)$ bounded by a closed curve $l(t)$ moving with the fluid. Taking the surface integral of (2.4) one obtains

$$\int_F \partial_t \mathbf{B} \cdot d\mathbf{F} = \oint_l (\mathbf{v} \times \mathbf{B}) \cdot d\mathbf{l} = -\oint_l \mathbf{B} \cdot (\mathbf{v} \times d\mathbf{l}), \tag{2.6}$$

and hence

$$\frac{d\phi}{dt} = \int_F \partial_t \mathbf{B} \cdot d\mathbf{F} + \oint_l \mathbf{B} \cdot (\mathbf{v} \times d\mathbf{l}) = 0, \tag{2.7}$$

since

$$\oint \mathbf{B} \cdot (\mathbf{v} \times d\mathbf{l}) dt = \int_{dF} \mathbf{B} \cdot d\mathbf{F} \tag{2.8}$$

is the flux through the change dF of the surface F during the time interval dt. Sweeping the boundary curve l along the field lines defines a tube, called a magnetic flux tube. Because of flux conservation the picture of field lines frozen in to the fluid obtains a well-defined physical meaning as flux tubes of infinitesimal diameter. Within the validity of the ideal Ohm's law (2.2) field lines conserve their individuality, i.e., they cannot be 'broken' but only swirled around by the fluid. (As will be seen in section 2.4, field lines defined in this way also have an internal structure given by the twist of the magnetic field in the thin tube.) It should be mentioned that a similar conservation law arises in ideal incompressible hydrodynamics. The Euler equation written in terms of the vorticity $\mathbf{w} = \nabla \times \mathbf{v}$,

$$\partial_t \mathbf{w} = \nabla \times (\mathbf{v} \times \mathbf{w}), \tag{2.9}$$

has the same form as Faraday's law (2.4). Hence the vorticity flux or circulation $\int_F \mathbf{w} \cdot d\mathbf{F}$ is conserved, which gives vortex lines a physical meaning and identity. Reconnection of vortex lines or bundles is an interesting topic in hydrodynamics, which has recently received considerable attention (see for instance Pumir & Kerr, 1987; Kerr & Hussain, 1989, Zabusky *et al.*, 1991; Kida & Takaoka, 1994). Since **v** and **w** are intimately related, the reconnection process is, however, inherently different from that of a magnetic flux tube, where **v** and **B** are primarily independent dynamic variables.

Because of the frozen-in property, the structure of the magnetic field in a conducting fluid may become very complicated. A convenient measure of the complexity of a configuration is the magnetic helicity H defined by

$$H = \int_V \mathbf{A} \cdot \mathbf{B} \, dV, ^* \tag{2.10}$$

where $\mathbf{B} = \nabla \times \mathbf{A}$ and V is the volume of a flux tube. Choosing the gauge of **A** such that the scalar potential vanishes, $\mathbf{E} = -\partial_t \mathbf{A}/c$, one finds by use of (2.3)

$$\int_V \partial_t (\mathbf{A} \cdot \mathbf{B}) dV = \int_V (\mathbf{B} \cdot \partial_t \mathbf{A} + \mathbf{A} \cdot \partial_t \mathbf{B}) dV$$

$$= -2c \int_V \mathbf{E} \cdot \mathbf{B} \, dV + c \oint_F (\mathbf{A} \times \mathbf{E}) \cdot d\mathbf{F}. \tag{2.11}$$

Inserting Ohm's law (2.2), the first term on the right-hand side vanishes, while the second becomes $-\oint \mathbf{A} \cdot \mathbf{B}\mathbf{v} \cdot d\mathbf{F}$ since the normal component B_n vanishes for a flux tube. Hence the helicity of a co-moving flux tube is conserved,

$$\frac{dH}{dt} = \int_V \partial_t (\mathbf{A} \cdot \mathbf{B}) dV + \oint_F \mathbf{A} \cdot \mathbf{B}\mathbf{v} \cdot d\mathbf{F} = 0, \tag{2.12}$$

since

$$\oint \mathbf{A} \cdot \mathbf{B}\mathbf{v} \cdot d\mathbf{F} dt = \int_{dV} \mathbf{A} \cdot \mathbf{B} \, dV,$$

where dV is the change of V during dt. The condition $B_n = 0$ is also necessary to make expression (2.10) gauge invariant. Performing a gauge transformation $\mathbf{A}' = \mathbf{A} + \nabla \chi$ we find

$$H' - H = \int_V \mathbf{B} \cdot \nabla \chi dV = \oint_F \chi \mathbf{B} \cdot d\mathbf{F}. \tag{2.13}$$

The problem of gauge invariance of H is, however, more subtle. When we introduced the flux tube above, we did not specify its longitudinal extent.

---

* This expression is formally similar to the kinetic helicity $H^K = \int \mathbf{v} \cdot \mathbf{w} \, dV$, conserved in ideal hydrodynamics, which plays an important role in dynamo theory, see (5.19). Contrary to the hydrodynamic case, where the helicity density $\mathbf{v} \cdot \mathbf{w}$ has a concrete physical meaning, the corresponding magnetic helicity density is not defined, since it depends on the gauge.

The global behavior of a flux tube is usually very complicated. Defining a tube by the field lines crossing a certain smoothly bounded surface, the cross-section becomes increasingly distorted when moving along the tube because of the shear of the field lines. Simple closed flux tubes arise only in sufficiently regular magnetic configurations. Such aspects are discussed in section 2.4. Here we consider two special situations, where the helicity in the form (2.10) is not well defined.

(a) A magnetized toroidal plasma column enclosed by a perfectly conducting vessel is of particular interest in laboratory plasma physics. Since (2.4) used in the proof of the ideal invariance of H is valid only inside the plasma region, which in this case is multiply connected, the gauge function χ in (2.13) need not be single valued. Hence H is, strictly speaking, not gauge invariant,

$$H' - H = (\chi_1 - \chi_2) \int_F \mathbf{B} \cdot d\mathbf{F},$$

where $\int_F \mathbf{B} \cdot d\mathbf{F}$ is the toroidal flux and $\chi_1 - \chi_2$ the change of χ once around the torus. (For instance the toroidal angle φ is an admissible gauge function.) The following expression is, however, gauge invariant (Taylor, 1986)

$$H_1 = \int_V \mathbf{A} \cdot \mathbf{B} \, dV - \oint \mathbf{A} \cdot d\mathbf{l} \oint \mathbf{A} \cdot d\mathbf{s}, \tag{2.14}$$

where $\oint \mathbf{A} \cdot d\mathbf{l}$ and $\oint \mathbf{A} \cdot d\mathbf{s}$ are the magnetic fluxes across a horizontal cross-section (poloidal flux) and a meridional cross-section (toroidal flux), respectively, $\oint d\mathbf{l}$ and $\oint d\mathbf{s}$ denoting loop integrals the long and the short way around the toroidal surface. It is also clear that, because of the boundary condition $B_n = 0$, the fluxes and hence H_1 are conserved.

(b) Magnetic configurations of interest in astrophysics are either open with field lines extending to infinity, such as the solar wind or the Earth's magnetotail, or bounded by surfaces which are crossed by field lines, $B_n \neq 0$, such as coronal loops bounded by the photosphere. In both cases the helicity as defined by (2.10) is not gauge invariant. An alternative expression has been proposed by Finn & Antonsen (1985) (see also Berger & Field, 1984)

$$H_2 = \int_V dV (\mathbf{A} + \mathbf{A}_0) \cdot (\mathbf{B} - \mathbf{B}_0), \tag{2.15}$$

where $\mathbf{B}_0 = \nabla \times \mathbf{A}_0$ is some reference field, which can be suitably chosen. In an open infinite system $\mathbf{B}_0$ may be a static field with the same asymptotic properties as $\mathbf{B}$, while in the bounded case the normal components of $\mathbf{B}_0$ and $\mathbf{B}$ on the boundary should be equal. H_2 is apparently invariant under separate gauge transformations of $\mathbf{A}$ and $\mathbf{A}_0$ and is also conserved in time in the ideal MHD limit (2.2). Because of gauge invariance we can choose the gauge such that the scalar potential vanishes, $\mathbf{E} = -\partial_t \mathbf{A}/c$. Hence

$$
\begin{aligned}
\frac{dH_2}{dt} &= \int \Big[(\mathbf{B} - \mathbf{B}_0) \cdot \partial_t (\mathbf{A} + \mathbf{A}_0) + (\mathbf{A} + \mathbf{A}_0) \cdot \partial_t (\mathbf{B} - \mathbf{B}_0) \Big] dV \\
&= -c \int_V \Big[(\mathbf{B} - \mathbf{B}_0) \cdot (\mathbf{E} + \mathbf{E}_0) + (\mathbf{B} + \mathbf{B}_0) \cdot (\mathbf{E} - \mathbf{E}_0) \Big] dV \\
&\quad + c \oint_F (\mathbf{A} + \mathbf{A}_0) \times (\mathbf{E} - \mathbf{E}_0) \cdot d\mathbf{F} .
\end{aligned}
$$

In both cases mentioned above the surface term vanishes, either since $\mathbf{E}, \mathbf{E}_0 \to 0$ asymptotically, or since the tangential components of $\mathbf{E}, \mathbf{E}_0$ are equal at the boundary because of $B_n = B_{n0}$. Hence

$$
\frac{dH_2}{dt} = -2c \int_V dV (\mathbf{B} \cdot \mathbf{E} - \mathbf{B}_0 \cdot \mathbf{E}_0) = 0 \tag{2.16}
$$

using the ideal Ohm's law (2.2).

2.2 Conservation of magnetic topology

Conservation of magnetic flux and helicity, discussed in section 2.1, is intuitively associated with the conservation of magnetic topology, i.e., the orientation, linkage and knottedness of the field lines. Let us discuss the relationship between flux and topology conservation somewhat more in detail (Hornig & Schindler, 1996). Since magnetic topology is a property of the magnetic field alone, topology conservation can be more generally valid than flux conservation (2.7). Magnetic field lines may be described by a generating function $\mathbf{F}_B(\mathbf{x}, s, t)$ defined by the field line equation

$$
\partial_s \mathbf{F}_B = \mathbf{B}^* \quad \text{with} \quad \mathbf{F}_B(\mathbf{x}, 0; t) = \mathbf{x}. \tag{2.17}
$$

Hence $\mathbf{F}_B(\mathbf{x}, s; t)$ is a mapping of $\mathbf{x}$ along the field line parameterized by the parameter s, for instance the (oriented) distance along the line. Since field lines can move, $\mathbf{F}_B$ in general depends also on time. Consider the flow $\mathbf{u}(\mathbf{x}, t)$, which gives rise to a transport of field lines. The process can be described by a generating function $\mathbf{F}_u(\mathbf{x}, t, t_0)$, which maps the field lines at time t_0 onto those at time $t > t_0$, following the equation

$$
\partial_t \mathbf{F}_u = \mathbf{u}, \quad \text{with} \quad \mathbf{F}_u(\mathbf{x}, t_0, t_0) = \mathbf{x}. \tag{2.18}
$$

The flow is topology conserving if the points that form a field line at time t_0 remain on the same line for any $t > t_0$. Consider two points $\mathbf{x}, \mathbf{y}$ on a field line at time t_0, $\mathbf{x} = \mathbf{F}_B(\mathbf{x}, 0, t_0)$, $\mathbf{y} = \mathbf{F}_B(\mathbf{x}, s, t_0)$, which are transported

---

* To simplify notations, $\mathbf{B}$ has the dimension of a velocity, i.e., should be considered as the Alfvén velocity $v_A = \mathbf{B}/\sqrt{4\pi\rho}$, $\rho =$ mass density.

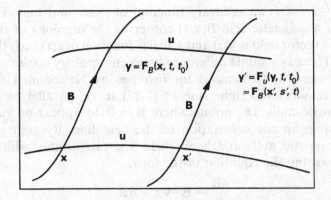

Fig. 2.1.   Field line conservation of a vector field **B** mapped by the flow function $\mathbf{F}_u$ (from Hornig & Schindler, 1996).

by **u** to $\mathbf{x}', \mathbf{y}'$ at time t, hence $\mathbf{x}' = \mathbf{F}_u(\mathbf{x}, t, t_0)$, $\mathbf{y}' = \mathbf{F}_u(\mathbf{y}, t, t_0)$ or, inserting **y**,

$$\mathbf{y}' = \mathbf{F}_u[\mathbf{F}_B(\mathbf{x}, s, t_0), t, t_0]. \tag{2.19}$$

The condition for topology conservation is that $\mathbf{x}', \mathbf{y}'$ are again connected by a field line, i.e., by the mapping $\mathbf{y}' = \mathbf{F}_B(\mathbf{x}', s', t)$ or, inserting $\mathbf{x}'$,

$$\mathbf{y}' = \mathbf{F}_B[\mathbf{F}_u(\mathbf{x}, t, t_0), s', t]. \tag{2.20}$$

(Note that s' may differ from s, since parallel flows are admitted, hence $s'(s, t)$ is some arbitrary function.) From (2.19) and (2.20) we obtain the commutation relation (see fig. 2.1)

$$\mathbf{F}_u[\mathbf{F}_B(\mathbf{x}, s, t_0), t, t_0] = \mathbf{F}_B[\mathbf{F}_u(\mathbf{x}, t, t_0), s', t]. \tag{2.21}$$

Equation (2.21) can be transformed into a more familiar differential form. Differentiation with respect to both s and t

$$\partial_s \partial_t \mathbf{F}_u[\mathbf{F}_B(\mathbf{x}, s, t_0), t, t_0] = \partial_t \partial_s \mathbf{F}_B[\mathbf{F}_u(\mathbf{x}, t, t_0), s', t] \tag{2.22}$$

taken at $s = 0, t = t_0$, and use of the definitions of $\mathbf{F}_B, \mathbf{F}_u$, (2.17) and (2.18), gives

$$\partial_s \{\mathbf{u}[\mathbf{F}_B(\mathbf{x}, s, t_0), t_0]\}|_{s=0} = \partial_t \left\{ \mathbf{B}[\mathbf{F}_u(\mathbf{x}, t, t_0), t] \partial_s s' \right\}|_{t=t_0}. \tag{2.23}$$

Carrying out the differentiations, using $\partial_s \mathbf{u}(\mathbf{F}) = \partial_s \mathbf{F} \cdot \nabla \mathbf{u} = \mathbf{B} \cdot \nabla \mathbf{u}$, yields

$$\mathbf{B} \cdot \nabla \mathbf{u} = \partial_t \mathbf{B} + \mathbf{u} \cdot \nabla \mathbf{B} + \mathbf{B} \partial_t \partial_s s'. \tag{2.24}$$

Hence we obtain the condition

$$\partial_t \mathbf{B} + \mathbf{u} \cdot \nabla \mathbf{B} - \mathbf{B} \cdot \nabla \mathbf{u} = \sigma \mathbf{B}, \tag{2.25}$$

where $\sigma = -\partial_t\partial_s s'$ is an arbitrary function of space and time. Hence the evolution of a magnetic field $\mathbf{B}(\mathbf{x}, t)$ conserves the topology of the field if there exists a vector field $\mathbf{u}(\mathbf{x}, t)$ and a scalar function $\sigma(\mathbf{x}, t)$ such that (2.25) is satisfied. There is a subtle difference between topology conservation and field line preservation discussed for instance by Newcomb (1958). The latter only requires the right side of (2.25) to be parallel to $\mathbf{B}$, which allows magnetic nulls, i.e., points where $\mathbf{B} = 0$, to appear or vanish and hence a change in the orientation of the field line. By contrast, (2.25) preserves magnetic nulls and hence field line orientation, which is seen directly by writing the equation in the form

$$\frac{d\mathbf{B}}{dt} = \mathbf{B} \cdot \nabla\mathbf{u} + \sigma\mathbf{B}. \tag{2.26}$$

Here the right side vanishes for $\mathbf{B} = 0$, such that a null cannot be generated by such flows, only the position of an existing null can be changed. The freedom in choosing σ is restricted by imposing the condition $\nabla \cdot \mathbf{B} = 0$ on the field, which is not only a basic physical requirement but is also necessary for the very existence of field lines. Equation (2.25) can also be written in the form

$$\partial_t\mathbf{B} - \nabla \times (\mathbf{u} \times \mathbf{B}) = \mathbf{B}(\sigma - \nabla \cdot \mathbf{u}) \tag{2.27}$$

where $\nabla \cdot \mathbf{B} = 0$ implies

$$\mathbf{B} \cdot \nabla(\sigma - \nabla \cdot \mathbf{u}) = 0. \tag{2.28}$$

From Faraday's law (2.3) we obtain

$$\mathbf{E} + \mathbf{u} \times \mathbf{B} = \mathbf{S}^* \tag{2.29}$$

with

$$\nabla \times \mathbf{S} = \mathbf{B}(\sigma - \nabla \cdot \mathbf{u}). \tag{2.30}$$

Hence flows $\mathbf{v}$ satisfying ideal Ohm's law also conserve magnetic topology, as can be seen by choosing $\sigma = \nabla \cdot \mathbf{v}$. However, ideal flows are not the only topology-conserving ones. The Hall term (see section 6.1), for instance, is a nonideal but topology-conserving effect in Ohm's law,

$$\mathbf{E} + \mathbf{v} \times \mathbf{B} = \frac{1}{en}\mathbf{j} \times \mathbf{B}, \tag{2.31}$$

where $\mathbf{j}$ is the current density. The physical meaning is simply that the magnetic field is convected with the electron velocity $\mathbf{v}_e = \mathbf{v} - (\mathbf{j}/en)$ instead of the plasma velocity $\mathbf{v}$, such that

$$\mathbf{E} + \mathbf{v}_e \times \mathbf{B} = 0. \tag{2.32}$$

Hence when the field is convected by $\mathbf{u} = \mathbf{v}_e$, the topology is not changed.

---

[*] Note that $\mathbf{S}$ differs from $\mathbf{R}$ in (2.1), since $\mathbf{u}$ need not be the plasma velocity.

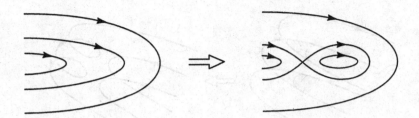

Fig. 2.2. Change of magnetic topology by formation of a plasmoid.

2.3 Conditions for magnetic reconnection

In chapter 1 we introduced the concept of reconnection in a rather loose intuitive way. Reconnection was said to produce a global change of the magnetic configuration due to a local decoupling of plasma and magnetic field, i.e., a local violation of the frozen-in condition. Over the last three decades many authors have tried to give a more rigorous and more generally valid definition. Though the discussion about the right way may sometimes sound rather semantic, a number of important aspects have been clarified in this still ongoing debate.

Originally, reconnection was considered mainly in two-dimensional (2D) systems. Here magnetic field lines, or their projection in the case of a finite field component along the symmetry axis, are given by the contour lines $\psi = const$ of a flux function ψ, since in 2D the general solution of $\nabla \cdot \mathbf{B} = 0$ can be written in the form

$$\mathbf{B} = \mathbf{h} \times \nabla\psi + \mathbf{h}f = \mathbf{B}_p + \mathbf{B}_t, \qquad (2.33)$$

where $\mathbf{h}$ is a vector in the symmetry direction (see, e.g., Biskamp, 1993a, p. 29). We propose to call the first term the poloidal field $\mathbf{B}_p$, the second the toroidal or axial field $\mathbf{B}_t$. In the 2D case Vasyliunas' definition (Vasyliunas, 1975) applies: *Magnetic reconnection or field line merging is the process whereby plasma flows across a surface separating regions of topologically different field lines.* Such a separatrix surface (or line in projection) exists in a 2D reconnecting configuration, where the different branches join in an X-type neutral line (point) or, more generally, in Y-lines (points)[*] as discussed in chapter 3. The process is illustrated in fig. 2.2, which shows the formation of a plasmoid. It can be seen that the plasma inside the plasmoid has moved across the separatrix or, more properly speaking, the separatrix has moved across the plasma inside the plasmoid.

---

[*] When explicitly discussing 2D configuration, we usually refer to separatrix lines and neutral points. It should also be noted that the terms 'neutral' point or 'neutral' line do not imply the vanishing of the magnetic field since there may be a finite field component B_t in the symmetry direction, which in many cases is actually strong, $B_t \gg B_p$.

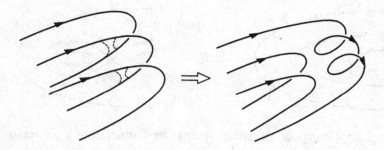

Fig. 2.3. Plasmoid formation in a weakly three-dimensional configuration.

Unfortunately this concept holds only in the strictly 2D situation, since the separatrix is structurally unstable to slight deviations from 2D geometry. Consider the generalization of the process shown in fig. 2.2 to a plasmoid of finite axial extent in the presence of a weak axial field, fig. 2.3. Obviously reconnection takes place in a way similar to that in the 2D case, but we can no longer distinguish different global field line topologies, all field lines coming from and ending up on the left part of the configuration. There is no separatrix surface anymore.

An equivalent of the 2D separatrix for the 3D case can be defined in the presence of points where $\mathbf{B} = 0$, called magnetic nulls. These isolated points are structurally stable, a slight change of the configuration just shifts them by a short amount. Magnetic nulls have therefore attracted some attention (Fukao *et al.*, 1975; Green, 1988; Lau & Finn, 1990). Let us briefly discuss the topological structure of the field in the presence of nulls. In the vicinity of a null, which we assume to be located at $\mathbf{x} = 0$, $\mathbf{B}$ has the form

$$B_i = \sum_i \alpha_{ij} x_j . \qquad (2.34)$$

Because of $\nabla \cdot \mathbf{B} = 0$ the trace of the (real) matrix α_{ij} vanishes. The matrix has three eigenvalues, which we can use to classify the different types of nulls. First, consider the case where all eigenvalues are real. Because of the trace condition either two are positive and one negative, which we call A-type, or two are negative and one positive, which we call B-type. The eigenvectors of the two equal-sign eigenvalues span a plane locally, which can be continued into a global surface Σ by following the field lines in this plane. Following the field lines along the third eigenvector defines a curve γ in space. The field lines near an A-type null are shown in fig. 2.4, the surface Σ_A dividing space into two regions not connected by field lines, all field lines in one region converging into a bundle around γ_A. A B-type null has the same topology with the field line directions reversed. A- and B-type nulls are generalizations of a two-dimensional X-point, to

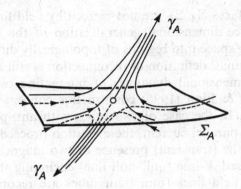

Fig. 2.4.   Field lines near an *A*-type magnetic null (after Lau & Finn, 1990).

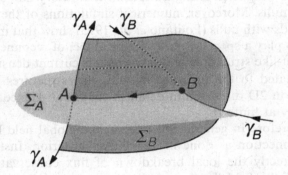

Fig. 2.5. Topology of a magnetic configuration including one *A*-type and one *B*-type null (after Lau & Finn, 1990).

which they degenerate if one of the two equal-sign eigenvalues vanishes. These nulls are therefore expected to be important in the context of reconnection. Nulls with one real and two conjugate complex eigenvalues are called *S*-type. They are generalizations of a two-dimensional *O*-point, to which they degenerate if the real eigenvalue vanishes (in this case the remaining two are purely imaginary).

A magnetic configuration with two nulls, one *A*-type and one *B*-type, is of special interest. The relative positions of the surfaces Σ_A, Σ_B and the curves γ_A, γ_B are illustrated in fig. 2.5. Since in Σ_B all field lines are directed away from *B* and are collected in the bundle around γ_A, the line γ_A bounds $\sum_B$, and conversely all field lines in Σ_A must originate in the bundle γ_B, i.e., γ_B bounds Σ_A.

Hence Σ_A, Σ_B are semi-infinite half-sheets intersecting in a line connecting *A* and *B*, called a null–null line. A topologically equivalent configuration arises when superimposing a constant field on a dipole field, for instance the Earth's magnetic field together with the interplanetary field. Since the surfaces Σ_A, Σ_B are not pierced by field lines they serve as separators, the three-dimensional generalization of the two-dimensional separatrix, dividing space into regions of topologically different field lines. In this case, Vasyliunas' definition of reconnection is still applicable. Studies of the three-dimensional dynamics of interacting twisted flux tubes performed by Lau & Finn (1996) confirm this concept of reconnection across a separator. In the case of two tubes with anti-parallel field and opposite twist, i.e., parallel current, the evolution proceeds in two phases. In the first phase, the (transient) presence of two magnetic nulls leads to formation of a closed X-line (null–null line) encircling the flux tubes. In the second phase, field lines from both tubes are reconnected, forming loops thereafter.

However, few magnetic configurations in laboratory and space do contain magnetic nulls. Moreover, numerical simulations of the dynamics of general 3D fields with nulls (Politano *et al.*, 1995) show that magnetic nulls do not seem to play a special role. The typical loci of reconnection characterized by sheet-like structures of quasi-singular current density, are found to be uncorrelated to the presence of nulls, the structures being locally similar to that in 2D discussed in detail in chapter 3. Hence reconnection occurs typically at finite B.

One can therefore in general not refer to the global field line behavior to define reconnection as done in Vasyliunas' criterion. Instead, one has to consider directly the local breakdown of flux conservation resulting in a change of the field line connection, as suggested by Axford (1984), i.e., a violation of the topology conservation discussed in section 2.2. Reconnection occurs if there are points which are originally located on the same field line but end up on different field lines. This concept has been elaborated by Schindler *et al.* (1988) and Hesse & Schindler (1988). It clearly applies also to the 3D plasmoid formation illustrated in fig. 2.3. (Since the effect of reconnection should be global, we exclude the case of a purely local change of the magnetic configuration by requiring that the points considered remain outside the small diffusion region, where **R** is important). A formal criterion for reconnection is obtained from (2.30):

$$\mathbf{B} \times (\nabla \times \mathbf{S}) \neq 0. \tag{2.35}$$

Here $\mathbf{S} = R_{\parallel}\,\mathbf{B}/B$ is the parallel component of the nonideal part in Ohm's law, since the perpendicular component can always be incorporated in the general velocity **u** in (2.29) by writing $\mathbf{R}_{\perp} = \mathbf{u}' \times \mathbf{B}$, $\mathbf{u} = \mathbf{v} - \mathbf{u}'$. Hence only $R_{\parallel} = E_{\parallel}$ can cause reconnection. But not all $R_{\parallel}$ actually do so. Clearly an electrostatic field does not, since for $\mathbf{S} = \nabla\phi$ the left side of (2.35) vanishes. Even in the standard resistive case $\mathbf{R} = \eta\mathbf{j}$ (see section 3.1), topology is conserved for special fields, for instance for a linear force-free field $\mathbf{j} = \mu\mathbf{B}$, $\mu = const$, and uniform resistivity. Practically speaking these exceptions are irrelevant because of the self-consistent dynamics of **v** in the diffusion region, as is discussed in subsequent chapters. One should also note that although $\mathbf{R}_\perp$ does not directly lead to reconnection, it may strongly increase the reconnection efficiency, for instance the Hall term in high-β noncollisional reconnection considered in section 6.2.

2.4 Conservation of magnetic helicity in reconnection

As shown in section 2.1 magnetic helicity is conserved for ideal flows, i.e., processes where the ideal Ohm's law (2.2) holds. For more general motions, H is not strictly conserved, not even in the general topology-conserving case (2.25). Straightforward evaluation gives, assuming for simplicity that the system is bounded by a perfectly conducting vessel for which the tangential electric field component vanishes, $E_t = 0$,

$$\frac{dH}{dt} = 2 \int \mathbf{A} \cdot \mathbf{B}(\nabla \cdot \mathbf{v} - \sigma)dV, \qquad (2.36)$$

which vanishes only in the ideal case $\sigma = \nabla \cdot \mathbf{v}$. However the helicity change is in general very slow, of the order of the global diffusion rate. Using (2.11), again for $E_t = 0$ on the boundary,

$$\frac{dH}{dt} = -2c \int \mathbf{E} \cdot \mathbf{B}dV, \qquad (2.37)$$

we find that dH/dt is small, since $E_\parallel = R_\parallel$ is usually small and may at most be finite in the small regions where reconnection takes place. By contrast the total energy, which is also ideally conserved, changes more rapidly. Neglecting viscosity and pressure (low-β approximation) we have

$$\frac{dW}{dt} \equiv \frac{d}{dt} \int \left(\frac{1}{8\pi}B^2 + \rho\frac{v^2}{2} \right) dV = \int \mathbf{R} \cdot \mathbf{j}dV, \qquad (2.38)$$

(where $\rho =$ the mass density), which may be larger than $O(R)$, since $\mathbf{j}$ may become large locally. (A more detailed discussion of the energy balance is given in section 3.1.)

Using the disparity of relaxation rates, Taylor (1974, 1986) has conjectured (see also Berger, 1984) that a magnetic configuration relaxes through a dynamic, generally turbulent phase to the lowest energy state consistent with the conservation of the initial value of the magnetic helicity. The final state is obtained from the variational principle

$$\delta [W - \alpha H] = \delta \left[\int \left(\frac{1}{8\pi}B^2 + \rho\frac{v^2}{2} \right) dV - \alpha \int \mathbf{A} \cdot \mathbf{B}dV \right] = 0, \qquad (2.39)$$

where α is a Lagrange multiplier. Variation with respect to **A** gives the equation

$$\nabla \times \mathbf{B} = 8\pi\alpha\mathbf{B}, \qquad (2.40)$$

while variation with respect to **v** gives **v** = 0. Hence the final state is a static constant-α force-free magnetic field, where α is determined by the value of the helicity. In general, the relaxation process requires reconnection of field lines resulting in a simpler, less complex magnetic field structure. Since magnetic helicity is a measure of the linkedness and knottedness of the magnetic field, breaking such linkages seems to involve also a change of helicity. We will show or at least make plausible that while the individual helicities of the constituent flux ropes become meaningless, since the latter are merged and reorganized and hence lose their identity, the sum of their helicities, nevertheless, is conserved.

To analyze the topological structure of a magnetic configuration (to be definite we restrict consideration to closed configurations with a perfectly conducting boundary surface) it is useful to divide the volume into small subregions bounded by flux surfaces, the simplest case being that of a toroidal configuration with nested flux surfaces or magnetic surfaces, as they are usually called. Such a surface is generated by following a field line around the torus. In general, field lines are infinitely long such that a single line will span the entire surface, which is called an irrational surface. Intermixed are special surfaces, called rational surfaces, where field lines are closed after a certain number of turns. Such surfaces are simply defined as limiting cases of the neighboring irrational ones. The subregions we consider are thin toroidal shells between two magnetic surfaces. The helicity of such a shell is

$$
\begin{aligned}
\Delta H &= \int_{\Delta V} \mathbf{A} \cdot \mathbf{B} \, dV \\
&= \int_{\Delta V} [\mathbf{A}_p \cdot \mathbf{B}_p + \mathbf{A}_t \cdot \mathbf{B}_t] \, dV \\
&= \oint \mathbf{A}_p \cdot d\mathbf{s} \int_{\Delta F_s} \mathbf{B}_p \cdot d\mathbf{F}_s + \oint \mathbf{A}_t \cdot d\mathbf{l} \int_{\Delta F_l} \mathbf{B}_t \cdot d\mathbf{F}_l \\
&= -\psi_t \Delta\psi_p + \psi_p \Delta\psi_t.
\end{aligned}
\qquad (2.41)
$$

Here $\oint d\mathbf{s}$ and $\oint d\mathbf{l}$ indicate the line integrals the short way and the long way around the torus, $d\mathbf{F}_s$ and $d\mathbf{F}_l$ are the surface elements with normals along these directions, $-\Delta\psi_p$ and $\Delta\psi_t$ are the poloidal and toroidal fluxes in the shell. $\psi_t = \oint \mathbf{A}_p \cdot d\mathbf{s}$ is the toroidal flux in the torus volume bounded by the shell and $\psi_p = \oint \mathbf{A}_t \cdot d\mathbf{l}$ is the poloidal flux threading the hole of the shell surface. The geometry is illustrated in fig. 2.6. Note that increasing the radius of a toroidal flux surface ψ_t increases, while ψ_p decreases, whence the minus sign in the first term of (2.41). There is a simple interpretation

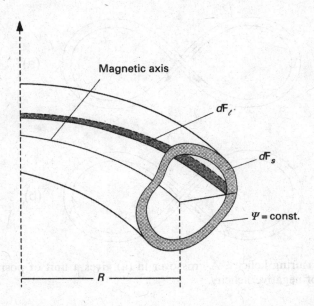

Fig. 2.6. Toroidal geometry defining the surface elements $d\mathbf{F}_s$ and $d\mathbf{F}_l$.

of the two terms in ΔH. The first term accounts for the linkage between the flux in the shell volume ΔV with the fields inside the shell surface, the second one for the linkage with the fields outside. The helicity of the torus can be written in the following way

$$H = 2\int_0^\phi T\psi_t d\psi_t \tag{2.42}$$

with $T = -d\psi_p/d\psi_t$ and ϕ = total toroidal flux, which is obtained by partial integration of the second term in (2.41)

$$\int_0^\phi \psi_p d\psi_t = \int_{\phi_p}^0 \psi_p \frac{d\psi_t}{d\psi_p} d\psi_p = -\int_{\phi_p}^0 \psi_t d\psi_p = \int_0^\phi T\psi_t d\psi_t. \tag{2.43}$$

Here the boundary contributions vanish, since at the lower boundary ψ_t vanishes, at the upper ψ_p, the field outside the torus being assumed zero. The quantity T is the (in general fractional) number of times a field line winds around the torus in the poloidal direction for one turn in the toroidal direction. In the theory of toroidal MHD equilibria T is usually called the rotational transform denoted by ι (pronounced iota), while in tokamak theory the inverse T^{-1} is called the safety factor q. For a uniformly twisted torus with twist $T = const$ we have the helicity

$$H = T\phi^2. \tag{2.44}$$

Fig. 2.7. Measuring helicity. A crossover in (a) gives a unit of positive helicity, in (b) a unit of negative helicity.

If the torus cross-section is sufficiently small, we are dealing with a thin closed flux tube or a flux rope. In this case the variation of T in the tube becomes unimportant and we can use expression (2.44) with T representing the average twist.

Let us now discuss the connection between the internal field-line twist T and the external distortion or kink K of the flux tube. A plane closed flux tube with twist $T = \pm 1$ may be kinked into a figure-8 configuration with no internal twist. To see this, unkink an untwisted figure-8 tube (as for instance shown in fig. 2.7) with a field line drawn on its upper side into a plane torus to find that this field line now winds around the torus just once corresponding to $T = \pm 1$ depending on the orientation of the field with respect to the kink. Hence we see that the helicity of a flux tube consists of an internal part H_T and an external one H_K,

$$H = H_T + H_K. \tag{2.45}$$

The decomposition is, however, not topologically invariant since, as we have just seen, one can be converted into the other. Let us assume that an arbitrarily kinked and knotted flux tube has zero twist, $H_T = 0$, if one can draw a field line on top of the tube[*] along its entire length.

There is a simple way of measuring the helicity of an untwisted tube. We have seen that untwisting a tube by kinking it into a figure-8 shape

---

[*] Strictly speaking on the plane projection of the tube. To simplify the discussion, we avoid the complicated discussion of nonplanar flux tubes, see, e.g., Moffatt & Ricca (1992).

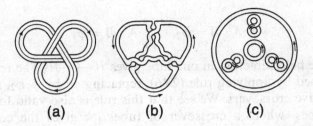

Fig. 2.8. Helicity of a trefoil knot (after Berger & Field, 1984).

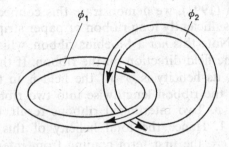

Fig. 2.9. Two linked flux tubes.

introduces a crossover. There are two types of such crossovers as indicated in fig. 2.7, which we call positive, fig. 2.7(a), and negative, fig. 2.7(b). Then the helicity of such a tube is

$$H = H_K = \phi^2(N_+ - N_-), \tag{2.46}$$

where N_+ and N_- are the numbers of positive and negative crossovers.

Consider for instance in untwisted trefoil knot, fig. 2.8(a), which according to relation (2.46) has helicity $H = 3\phi^2$. This can be proved by reconnecting the knot in the way indicated in fig. 2.8(b,c). The final state consists of three kinked tubes and two unkinked ones, all nontwisted. While the former ones contribute ϕ^2 each, the helicity contributions of the latter are, of course, zero.

However, how do we know that the reconnection does not change the helicity? We postpone the answer to this question for a while and introduce first the concept of linkedness. Consider two flux tubes with fluxes ϕ_1, ϕ_2, interlinked as shown in fig. 2.9, untwisted for simplicity. The total helicity can easily be computed as

$$
\begin{aligned}
H &= \int_{V_1} \mathbf{A} \cdot \mathbf{B}\, dV + \int_{V_2} \mathbf{A} \cdot \mathbf{B}\, dV \\
&= \phi_1 \oint_{l_1} \mathbf{A} \cdot d\mathbf{l} + \phi_2 \oint_{l_2} \mathbf{A} \cdot d\mathbf{l} \\
&= 2\phi_1 \phi_2,
\end{aligned}
\tag{2.47}
$$

since

$$\oint_{l_1} \mathbf{A} \cdot d\mathbf{l} = \phi_2, \quad \oint_{l_2} \mathbf{A} \cdot d\mathbf{l} = \phi_1. \tag{2.48}$$

Reversing the field direction in one tube gives $H \to -H$. The result (2.47), is also obtained by applying rule (2.46), replacing ϕ^2 by $\phi_1 \phi_2$, since there are two positive crossovers. We see that this rule is also valid for a system of linked tubes, where a crossover of tubes i, k gives the contribution $\pm \phi_i \phi_k$.

We now discuss the connection between twist and linkage. Following Pfister & Gekelman (1991), we demonstrate this connection by a simple experiment. Take a sufficiently long ribbon or paper strip, twist it by 360° and join the ends. (Note this *not* a Moebius ribbon, which is only twisted by 180°). Indicate the field direction on the ribbon. If this (flat) flux tube contains the flux ϕ, its helicity is ϕ^2 (if the field is in the positive twist direction). Now cut the ribbon lengthwise into two ribbons of fluxes ϕ_1 and ϕ_2, $\phi_1 + \phi_2 = \phi$. Two interlinked ribbons result, each one twisted by 360°, i.e., $T = 1$. Hence the total helicity of this configuration is $H = 2\phi_1 \phi_2 + \phi_1^2 + \phi_2^2$, the first term coming from interlinkage, the two others from the individual twists. Thus $H' = \phi^2 = H$, the helicity of the original twisted ribbon ϕ. [By the way, cutting a Moebius ($T = \frac{1}{2}$) ribbon lengthwise results in a single ribbon twisted twice ($T = 2$) with helicity $2(\phi/2)^2 = \frac{1}{2}\phi_1^2$].

This is a simple example showing that the helicity of a closed configuration can be expressed by those of the constituent closed flux tubes. If there are N constituents of flux ϕ_i, $\sum_{i=1}^{N} \phi_i = \phi$, we have

$$H = \sum_{i,j} L_{ij} \phi_i \phi_j + \sum_i H_i, \tag{2.49}$$

where L_{ij} gives the number of linkages of tubes i and j and H_i are the internal helicities of the constituents given by their twists, kinks and knots. Relation (2.49) also shows that for large N the contribution from the internal helicities becomes negligible, since $H_i \propto \phi_i^2 = O(N^{-2})$, hence $\sum_i H_i = O(N^{-1})$, while the first term, the interlinkage sum, is finite.

We now return to the question raised previously, when we computed the helicity of a trefoil knot: Is the helicity conserved in reconnection? At first sight this does not seem to be the case. If we naively reconnect two interlinked field lines, we seem to be able to reach different configurations like those in fig. 2.10(a) or (b) where, in neither process, is H conserved.

This would, however, imply an oversimplification of the concept of field lines, which in a conducting medium are to be viewed as thin flux tubes with an internal structure, the twist. One might think that in the limit of vanishing tube radius, field lines in the tube become parallel, such that the

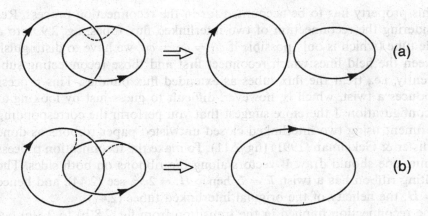

Fig. 2.10. Different ways of reconnecting an unstructured field line.

Fig. 2.11. Ribbon model to demonstrate helicity conservation in reconnection (after Pfister & Gekelman, 1991).

twist loses its meaning. The helicity of a tube does, however, not depend explicitly on the tube radius, as seen in (2.44). The helicity of a twisted tube is of the same order as that of two interlinked tubes of similar fluxes. We may say that field lines in a plasma are not "naked" but "dressed with twist".

This property has to be accounted for in the reconnection process. Reconsidering the reconnection of two interlinked flux tubes, fig. 2.9, into a single tube (which is only possible if $\phi_1 = \phi_2 = \phi$), we have to distinguish between the field lines which reconnect first and those reconnecting subsequently, i.e., treat the flux tubes as extended flux bundles. This process introduces a twist, which is, however, difficult to guess just by looking at the configuration. I therefore suggest that you perform the corresponding experiment using two interlinked closed untwisted paper ribbons as done by Pfister & Gekelman (1991) (fig. 2.11). To make the reconnection process definite, one should draw **B** vectors along the ribbons on both sides. The resulting ribbon has a twist $T = 2$, hence $H' = 2\phi^2$, see (2.44), and hence $H' = H$, the helicity of the original interlinked tubes (2.47).

The reconnection implied in the transition from fig. 2.8(b) to 2.8(c) occurs between flux tube parts which can be considered plane. Reconnection proceeds in a natural peeling manner, which does not introduce additional field line twists.

Though it seems to be difficult to give a mathematical proof we have at least made very plausible that reconnection does not change the helicity. The term on the right side of (2.16) can be made arbitrarily small by localizing the volume where $E_\parallel \neq 0$.

3

Current sheets

In an electrically conducting magnetized plasma even slow motions do not, in general, preserve a smooth magnetic field, but give rise to sheet-like tangential field discontinuities, called current sheets. These are the natural loci of magnetic reconnection. In this chapter we consider the properties of current sheets and reconnection via current sheets in the traditional framework of resistive magnetohydrodynamics (MHD). This restriction is justified, since macroscopic current sheets mostly occur, when the reconnection process **R** in Ohm's law (2.1) is dissipative. Collisionless reconnection processes, which are treated in chapters 6 and 7, usually give rise to microcurrent sheets.

The chapter starts with a brief introduction to MHD theory, discussing the basic equations, magnetostatic equilibria, and linear MHD waves. In low-β plasmas it is often convenient to eliminate the fast compressional MHD wave by using a reduced set of equations. In section 3.2 we first consider the conditions under which a current sheet arises and how it is formed by rapid thinning, which continues until finite resistivity leads to a stationary sheet configuration. Then the structure of a resistive current sheet, called a Sweet–Parker sheet (Sweet, 1958; Parker, 1963), is analyzed. While the global properties of such sheets follow from the basic conservation laws, the detailed structure requires a more specific analysis. Section 3.3 deals with the role of current sheets as centers of reconnection in a magnetic configuration. Syrovatskii (1971) has developed a simple and elegant theory of current sheets which captures many features of the fully dynamic resistive theory, in particular the complicated structure of the sheet edges, the so-called Y-points. We then discuss the scaling laws of stationary current-sheet reconnection. Since in a macroscopic sheet this is a rather slow process, Petschek's reconnection model (Petschek, 1964), predicting a much higher reconnection rate, had attracted strong interest, being considered for many years as the fundamental reconnection

29

configuration. Though it has now become clear that Petschek's model is not valid in the most interesting limit of small resistivity, it may nevertheless be valid as a phenomenological model in the presence of a more efficient reconnection process.

3.1 Resistive MHD

3.1.1 The MHD equations

The MHD equations can be obtained in a systematic way from Braginskii's two-fluid theory (see chapter 6) at sufficiently large spatial scales. Here, we introduce these equations in a simple heuristic way. The acceleration of a fluid element is due to the sum of the pressure gradient and the magnetic, gravitational and viscous forces,

$$\rho\frac{d\mathbf{v}}{dt} = -\nabla p + \frac{1}{c}\mathbf{j}\times\mathbf{B} + \rho\mathbf{g} + \rho\mu\nabla^2\mathbf{v}, \tag{3.1}$$

where $d/dt = \partial_t + \mathbf{v}\cdot\nabla$, and ρ is the mass density, which obeys the continuity equation

$$\partial_t\rho + \nabla\cdot\rho\mathbf{v} = \frac{d\rho}{dt} + \rho\nabla\cdot\mathbf{v} = 0. \tag{3.2}$$

On macroscopic scales the plasma is electrically neutral, there are just as many positive as negative charges in a fluid element, hence the charge density vanishes and there is no direct electric force in (3.1). Since the gravitational force $\rho\mathbf{g}$ is only important in certain astrophysical systems, we will usually ignore it in (3.1). The viscous term contains only a scalar kinematic viscosity μ, ignoring the anisotropy of the viscosity tensor induced by the magnetic field. In this form the viscosity is not meant to describe real momentum transfer but only to prevent singularities of the flow. The pressure tensor is assumed to be isotropic, corresponding to conditions close to local thermodynamic equilibrium. In most cases of interest, plasmas are sufficiently dilute to follow the ideal gas equation $p = (n_i + n_e)k_BT$, where $n_{i,e}$ are the particle densities of ions and electrons, k_B is the Boltzmann constant, and $\rho \simeq m_in_i$ is the mass density.

The Lorentz force in (3.1) can also be written in the following way:

$$\frac{1}{c}\mathbf{j}\times\mathbf{B} = -\nabla\frac{B^2}{8\pi} + \mathbf{bb}\cdot(\mathbf{B}\cdot\nabla\mathbf{B}) + (\mathbf{I}-\mathbf{bb})\cdot(\mathbf{B}\cdot\nabla\mathbf{B}), \tag{3.3}$$

$\mathbf{b} = \mathbf{B}/B$, where we have used Ampère's law

$$\mathbf{j} = \frac{c}{4\pi}\nabla\times\mathbf{B}, \tag{3.4}$$

i.e., the magnetic force acts as an isotropic pressure added to the thermal pressure p, a tension which strives to shorten the field lines and a stress which tends to straighten the field lines.

On the large scales over which the MHD approximation is valid, heat conduction effects are negligible, hence changes of state are adiabatic,

$$d(p/\rho^\gamma)/dt = 0,$$

or, using the continuity equation (3.2),

$$\partial_t p + \mathbf{v} \cdot \nabla p + \gamma p \nabla \cdot \mathbf{v} = 0, \tag{3.5}$$

where $\gamma = C_P/C_V$, C_P, C_V = specific heats, $\gamma = 5/3$ in the usual case.

The change of the magnetic field is given by Faraday's law (2.3) with $\mathbf{E}$ from Ohm's law (2.1),

$$\partial_t \mathbf{B} = \nabla \times (\mathbf{v} \times \mathbf{B}) - c\nabla \times \mathbf{R}. \tag{3.6}$$

The basic assumption of the resistive MHD approximation is that collisions are sufficiently frequent, such that the collisional effects in $\mathbf{R}$ dominate, of which resistivity is usually the most important one,

$$\mathbf{R} = \eta \mathbf{j}. \tag{3.7}$$

Here, too, a scalar resistivity η is assumed, which is, however, a much less drastic approximation than the assumption of a scalar viscosity, since $\eta_\perp$ and $\eta_\parallel$ differ only by a factor of order one, $\eta_\perp \simeq 2\eta_\parallel$ (see Braginskii, 1965). Inserting Ampère's law, (3.4), (3.6) becomes

$$\partial_t \mathbf{B} = \nabla \times (\mathbf{v} \times \mathbf{B}) + \nabla \cdot (\lambda \nabla \mathbf{B}). \tag{3.8}$$

The coefficient $\lambda = \eta c^2/4\pi$ has the dimensions of a diffusion coefficient and is called the magnetic diffusivity.

As an alternative to adiabatic changes of state, (3.5), one often assumes incompressible motions $\nabla \cdot \mathbf{v} = 0$. The conditions for the validity of this assumption are discussed in section 3.1.2. For $\nabla \cdot \mathbf{v} = 0$, the continuity equation reduces to the simple convection equation $d\rho/dt = 0$, such that (at least in the absence of a gravitational force) we can take $\rho = \rho_0 = const$ for simplicity. In the incompressible case the total pressure $P = p + B^2/8\pi$ is no longer an independent dynamic variable, but is determined from Poisson's equation obtained by taking the divergence of (3.1). Taking the curl of this equation eliminates the pressure. The incompressible MHD equations can thus be written in terms of the vorticity $\mathbf{w} = \nabla \times \mathbf{v}$ and $\mathbf{B}$:

$$\partial_t \mathbf{w} + \mathbf{v} \cdot \nabla \mathbf{w} - \mathbf{w} \cdot \nabla \mathbf{v} = \frac{1}{c\rho_0}(\mathbf{B} \cdot \nabla \mathbf{j} - \mathbf{j} \cdot \nabla \mathbf{B}) + \mu \nabla^2 \mathbf{w}, \tag{3.9}$$

$$\partial_t \mathbf{B} + \mathbf{v} \cdot \nabla \mathbf{B} - \mathbf{B} \cdot \nabla \mathbf{v} = \lambda \nabla^2 \mathbf{B}. \tag{3.10}$$

These equations assume a more symmetric form if written in terms of the Elsässer fields (Elsässer, 1950):

$$\mathbf{z}^{\pm} = \mathbf{v} \pm \mathbf{B}/\sqrt{4\pi\rho_0}, \qquad (3.11)$$

which gives

$$\partial_t \mathbf{z}^{\pm} + \mathbf{z}^{\mp} \cdot \nabla \mathbf{z}^{\pm} = -\frac{1}{\rho_0}\nabla P + \mu_+ \nabla^2 \mathbf{z}^{\pm} + \mu_- \nabla^2 \mathbf{z}^{\mp}, \qquad (3.12)$$

$$\mu_+ = \tfrac{1}{2}(\mu + \lambda), \quad \mu_- = \tfrac{1}{2}(\mu - \lambda).$$

3.1.2 Equilibrium and linear modes

Magnetic plasma configurations, both in laboratory and in space, vary in time because of changes of the external circuit and because of internal transport processes. These time-scales are usually long compared to the dynamics described by the MHD equation (3.1), such that the system changes quasi-statically evolving through a sequence of states which satisfy the equilibrium equation

$$\nabla p - \frac{1}{c}\mathbf{j} \times \mathbf{B} = 0, \qquad (3.13)$$

neglecting the gravitational force. Equation (3.13) can also be written in terms of the total pressure $P = p + B^2/8\pi$:

$$\nabla P - \frac{1}{4\pi}\mathbf{B} \cdot \nabla \mathbf{B} = 0. \qquad (3.14)$$

From (3.13) it follows immediately that p is constant along magnetic field lines and along current density lines,

$$\mathbf{B} \cdot \nabla p = 0, \quad \mathbf{j} \cdot \nabla p = 0.$$

Since, for a finite pressure gradient, $\mathbf{B}$ and $\mathbf{j}$ are not parallel, we can locally construct a surface $\psi(x, y, z) = const$, where p is constant, $p = p(\psi)$. In an equilibrium configuration this property must, however, hold globally, i.e., there must be surfaces of constant pressure on which a field line runs on for ever, a condition which is rather stringent for general nonsymmetric configurations. That infinitely long field lines span smooth magnetic surfaces can be shown directly for configurations with a continuous symmetry, where a coordinate system ξ, η, ζ exists such that all physical quantities including the elements of the metric tensor depend only on two coordinates ξ, η. If this condition holds, (3.13) can be reduced to an elliptic differential equation for a scalar function $\psi(\xi, \eta)$, such that surfaces $\psi = const$ have the desired properties of magnetic surfaces (Edenstrasser, 1980a). It turns out that the requirements on the coordinates ξ, η, ζ are rather restrictive. The most general case is that of helical symmetry (Edenstrasser, 1980b), of which plane and axisymmetry are special cases.

One-dimensional equilibria can easily be obtained analytically. In plane geometry, with $p(x), \mathbf{B} = (0, B_y(x), B_z(x))$, the configuration is a sheet pinch, which follows the equation

$$\frac{d}{dx}\left(p + \frac{B^2}{8\pi}\right) = 0, \tag{3.15}$$

since $\mathbf{B} \cdot \nabla \mathbf{B} = 0$. In cylindrical geometry, where we have $p(r), \mathbf{B} = (0, B_\theta(r), B_z(r))$, the most general configuration is a screw pinch,

$$\frac{dp}{dr} = \frac{1}{c}(j_\theta B_z - j_z B_\theta) = -\frac{d}{dr}\frac{B_z^2}{8\pi} - \frac{B_\theta}{4\pi r}\frac{d}{dr}rB_\theta. \tag{3.16}$$

A quantity often used in laboratory plasma physics is the safety factor q measuring the field-line twist T,

$$q(r) = T^{-1} = \frac{rB_z(r)}{RB_\theta(r)}, \tag{3.17}$$

where $2\pi R$ is the length of the plasma column. A screw pinch with $q \sim 1$ serves as a simple model of a tokamak plasma, while a reversed-field pinch has $q \ll 1$. An important parameter for the stability of a plasma column is the derivative of q,

$$\hat{s} = \frac{d \ln q}{d \ln r}, \tag{3.18}$$

the magnetic shear, which measures the variation of the field-line inclination.

Since there is only one equation for three unknown functions, boundary conditions determine the equilibrium state only up to two free functions, for instance the profiles of pressure and current density. This freedom exists also in the case of more general 2D or 3D equilibria (see, e.g., Lifshitz, 1989). The computation of explicit equilibrium configurations in these cases requires, however, sophisticated numerical techniques (see, e.g., Bauer et al., 1978).

Including gravity, but neglecting the magnetic field, leads to stratified equilibria,

$$\nabla p = \rho \mathbf{g}. \tag{3.19}$$

The resulting density profile depends on the adiabaticity coefficient γ. With $\mathbf{g} = -g\mathbf{e}_z$ we have

$$\rho(z) = \rho_0\left(1 - \frac{\gamma - 1}{\gamma}\frac{g\rho_0}{p_0}z\right)^{1/(\gamma-1)}. \tag{3.20}$$

Hence the thickness of a stratified atmosphere is in general finite, ρ vanishing for $z \geq [\gamma/(\gamma-1)]p_0/g\rho_0$. Only in the special case $\gamma = 1$, i.e., for isothermal conditions, is the extent not limited,

$$\rho(z) = \rho_0 \exp\{-g\rho_0 z/p_0\}. \tag{3.21}$$

Given a certain equilibrium it is important to know how it responds to small perturbations. If the equilibrium is stable, the perturbation generates a propagating wave, which is most conveniently discussed for a homogeneous plasma embedded in a uniform magnetic field. By linearizing the MHD equations writing $\mathbf{B} = \mathbf{B}_0 + \tilde{\mathbf{B}}$, $p = p_0 + \tilde{p}$, $\rho = \rho_0 + \tilde{\rho}$, $\mathbf{v} = \tilde{\mathbf{v}}$, the perturbations $\tilde{\mathbf{B}}, \tilde{\mathbf{v}}, \tilde{p}, \tilde{\rho}$ can be Fourier transformed in space and time, $\tilde{\mathbf{B}} \propto \exp\{i\mathbf{k} \cdot \mathbf{x} - i\omega t\}$ etc., such that the differential equations reduce to a homogeneous algebraic form, the solubility condition of which gives the dispersion relation for three types of modes $\omega_1, \omega_2, \omega_3$:

(a) The *shear Alfvén wave*

$$\omega_1^2 = k_\parallel^2 v_A^2, \tag{3.22}$$

where

$$v_A = B_0/\sqrt{4\pi\rho_0} \tag{3.23}$$

is the Alfvén velocity and $k_\parallel = \mathbf{k} \cdot \mathbf{b} = k\cos\theta$. The plasma motion $\tilde{\mathbf{v}}$ associated with this mode is incompressible $\mathbf{k} \cdot \tilde{\mathbf{v}} = 0$, i.e., the mode is transverse. $\tilde{\mathbf{v}}$ is also perpendicular to $\mathbf{B}_0$, and so is the perturbation of the magnetic field $\tilde{\mathbf{B}} = \pm\sqrt{4\pi\rho_0}\tilde{\mathbf{v}}$, which causes a bending of the field lines. Hence we see that the Elsässer fields, (3.11), describe Alfvén waves propagating in the direction of or opposite to the mean magnetic field $\mathbf{B}_0$.

(b) The *(fast) magnetosonic wave*, also called *compressional Alfvén wave*,

$$\omega_2^2 = \frac{k^2}{2}\left[v_A^2 + c_s^2 + \sqrt{(v_A^2 + c_s^2)^2 - 4v_A^2 c_s^2 \cos^2\theta}\right]. \tag{3.24}$$

(c) The *slow magnetosonic wave* or *slow mode*

$$\omega_3^2 = \frac{k^2}{2}\left[v_A^2 + c_s^2 - \sqrt{(v_A^2 + c_s^2)^2 - 4v_A^2 c_s^2 \cos^2\theta}\right]. \tag{3.25}$$

Here

$$c_s = \sqrt{\gamma p_0/\rho_0} \tag{3.26}$$

is the sound velocity. Both modes (b) and (c) are, in general, compressible. The phase velocity of the magnetosonic mode is in the range $v_A^2 + c_s^2 \geq (\omega_2/k)^2 \geq \max\{v_A, c_s\}$. Here, the left-hand equality refers to perpendicular

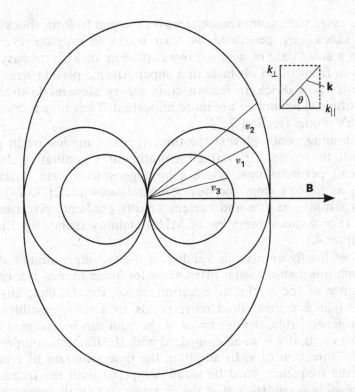

Fig. 3.1. Phase velocities $v_{1,2,3} = \omega_{1,2,3}(\theta)/k$ of the MHD modes (3.22), (3.24), (3.25).

propagation $\mathbf{k} \perp \mathbf{B}_0$, where the mode is longitudinal $\tilde{\mathbf{v}} \parallel \mathbf{k}$ leading to pure field compression. Since both B^2 and p are simultaneously compressed, the restoring force and hence the frequency are particularly high. For parallel propagation, the mode degenerates to the shear Alfvén wave, (3.22), for the usual case of $v_A > c_s$ in a magnetized plasma.

The phase velocity of the slow mode satisfies $0 \le (\omega_3/k)^2 \le \min \{v_A, c_s\}$. In the case of perpendicular propagation (left-hand equality) it is longitudinal, but since the variations of B^2 and p are now just opposite in phase, $\delta B^2/8\pi = -\delta p$, the restoring force and hence the frequency vanishes, corresponding to a quasi-static change of equilibrium. For parallel propagation (right-hand equality) the slow mode is again longitudinal, the nonmagnetic sound wave $\omega_3^2 = k^2 c_s^2$. (In the case $c_s > v_A$ the parallel modes of (b) and (c) just interchange roles.)

MHD waves are nondispersive, the phase velocity ω/k equalling the group velocity $\partial\omega/\partial k$ which depends only on the angle θ between $\mathbf{k}$ and $\mathbf{B}$. This is to be expected, since ideal MHD theory contains no intrinsic spatial scale. The frequencies $\omega_j(\mathbf{k})$ of the three modes are illustrated in fig. 3.1.

·At finite amplitude, compressional waves steepen to form shocks. Magnetosonic shocks are generated as blast waves in eruptive events, for instance in a solar flare or a supernova explosion, or as stationary stand-off shocks in front of an obstacle in a super-Alvénic plasma stream such as the Earth's bow shock. In reconnection theory, slow-mode shocks, i.e., finite amplitude slow modes, are more important. They play a crucial role in Petschek's model (section 3.3.4).

In an inhomogeneous plasma the three types of modes are in general coupled, but they remain nevertheless useful for a qualitative classification of small perturbations. While a homogeneous system is stable, ω being real, as follows from the dispersion relations (3.22), (3.24), (3.25), sufficiently strong pressure and current density gradients give rise to instabilities. (For a quick overview of MHD stability theory see Biskamp, 1993a, chapter 4.)

Finally, we briefly discuss the validity of the incompressibility assumption. Though this assumption is often made for linear modes, it is basically a consequence of the nonlinear equations since, for instance, any linear perturbation in a neutral fluid corresponds to a (compressible) sound wave. As a general rule, the dynamics of the fluid can be assumed incompressible, $\nabla \cdot \mathbf{v} = 0$, if $\mathbf{v}$ is small compared with the (fastest) compressional mode in the direction of $\mathbf{v}$. In addition, the time variation of $\mathbf{v}$ must be weak, i.e., the frequency must be small compared with the frequency of the compressible wave, such that the ∂_t terms are small compared with the advective terms $\mathbf{v} \cdot \nabla$. Consider first a neutral fluid or a high-β-plasma. From the pressure equation (3.5) we have

$$\gamma p \nabla \cdot \mathbf{v} \simeq -\mathbf{v} \cdot \nabla p.$$

Substituting ∇p from the equation of motion (3.1) for high β, $-\nabla p \simeq \rho \mathbf{v} \cdot \nabla \mathbf{v}$, we find

$$\nabla \cdot \mathbf{v} \simeq \frac{\rho}{2\gamma p} \mathbf{v} \cdot \nabla v^2 \sim \frac{v^2}{c_s^2} \frac{v}{L}, \qquad (3.27)$$

where L is a typical gradient scale-length. In a low-β plasma one has to distinguish between perpendicular (to $\mathbf{B}$) and parallel motions. While in the latter case (3.27) applies, in the former case Faraday's law (3.6) gives

$$B^2 \nabla \cdot \mathbf{v} \simeq -\mathbf{v} \cdot \nabla B^2/2.$$

From the equation of motion (3.1) for low β one has $\rho \mathbf{v} \cdot \nabla v^2/2 \simeq -\mathbf{v} \cdot \nabla B^2/8\pi$, hence

$$\nabla \cdot \mathbf{v} \simeq \frac{2\pi\rho}{B^2} \mathbf{v} \cdot \nabla v^2 \sim \frac{v^2}{v_A^2} \frac{v}{L}. \qquad (3.28)$$

As discussed in more detail in section 3.1.4, the perpendicular motion is usually incompressible or quasi-incompressible, while the parallel motion is often compressible.

In MHD turbulence theory, one often assumes full incompressibility $\nabla \cdot \mathbf{v} = 0$ (see section 5.4). While this assumption is valid in 2D, if one can rely on the presence of a strong axial field, in the general 3D case the assumption is less well founded and more a matter of convenience.

3.1.3 Conservation laws

The resistive MHD equations (3.1), (3.2), (3.5), (3.8) satisfy a number of conservation laws, which can easily be derived. The mass M in a volume $V(t)$ moving with the fluid is conserved

$$\frac{dM}{dt} \equiv \frac{d}{dt} \int_{V(t)} \rho \, dV = 0. \tag{3.29}$$

The change of momentum $\mathbf{P}$ is due to the total stress acting on the surface F of the volume,

$$\frac{d\mathbf{P}}{dt} \equiv \frac{d}{dt} \int_{V(t)} \rho \mathbf{v} dV = -\oint_F \mathscr{T} \cdot d\mathbf{F}. \tag{3.30}$$

Here $\mathscr{T} = \{T_{ij}\}$ is the stress tensor

$$T_{ij} = \left(p + \frac{B^2}{8\pi} \right) \delta_{ij} - \frac{1}{4\pi} B_i B_j + \frac{1}{2} \mu (\partial_i v_j + \partial_j v_i), \tag{3.31}$$

consisting of the thermal, the magnetic, and the viscous stresses.

The energy balance is

$$\frac{dW}{dt} \equiv \frac{d}{dt} \int_{V(t)} \left(\rho \frac{v^2}{2} + \frac{B^2}{8\pi} + \frac{p}{\gamma - 1} \right) dV$$

$$= -\oint_F (\mathbf{v} \cdot \mathscr{T}) \cdot d\mathbf{F} - \eta \int j^2 dV - \mu \int \rho \sum_{i,j} (\partial_i v_j)^2 dV. \tag{3.32}$$

In an isolated system, where $p, \mathbf{v}, \mathbf{B}$ vanish at the boundary, the momentum is conserved and the energy decays due to ohmic and viscous dissipation.[*] In most cases of interest, however, $\mathbf{B}$ is not zero at the boundary and there is an exchange of momentum and energy with the external system.

Also, the magnetic helicity $H = \int \mathbf{A} \cdot \mathbf{B} dV$ is conserved for $\eta = 0$ as discussed in section 2.1. Note that this conservation law is a property of

---

[*] The boundary terms also vanish for periodic boundary conditions, which are usually assumed in studies of homogeneous turbulence.

the frozen-in magnetic field, which is valid for any velocity $\mathbf{v}$, not only for motions satisfying the MHD equations.

In the incompressible case $\nabla \cdot \mathbf{v} = 0$, there is an additional ideal invariant of the MHD equations, the cross-helicity

$$K = \int_{V(t)} \mathbf{v} \cdot \mathbf{B} \, dV. \tag{3.33}$$

The cross-helicity of a flux tube moving with the fluid is conserved up to resistive and viscous effects,

$$\frac{dK}{dt} = -(\mu + \lambda) \int \sum_{i,j} (\partial_i v_j)(\partial_i B_j) \, dV. \tag{3.34}$$

Note that the kinetic helicity

$$H^K = \int \mathbf{v} \cdot (\nabla \times \mathbf{v}) d^3x, \tag{3.35}$$

which is an invariant of the Euler equation, is not conserved in MHD.

Finally, there is an important linear invariant, if the dynamics is restricted to helical geometry, i.e., if all physical variables depend only on r and $\theta + \alpha z$. Such motions occur in the development of a kink instability in a cylindrical plasma column, which plays an important role for instance in the sawtooth phenomenon in a tokamak plasma or the instability of a twisted coronal loop. Integrating Faraday's law (2.4) for $\eta = 0$,

$$\partial_t \mathbf{A} = \mathbf{v} \times (\nabla \times \mathbf{A}) - \nabla \phi, \tag{3.36}$$

one can easily derive that the helical flux function

$$\psi_H = \alpha r A_\theta - A_z \tag{3.37}$$

is conserved along the orbit of a fluid element:

$$\partial_t \psi_H + \mathbf{v} \cdot \nabla \psi_H = 0. \tag{3.38}$$

From (3.38) it follows that for incompressible motions the total helical flux Φ_H is conserved,

$$\frac{d\Phi_H}{dt} = \frac{d}{dt} \int \psi_H dV = 0. \tag{3.39}$$

In the limiting case of plane geometry, $\alpha = 0$, ψ_H reduces to $-A_z$.

3.1.4 Reduced MHD models

In many cases the resistive MHD equations are not considered in full generality. Instead, approximations are made concerning both physics and geometry. We often encounter the situation that the system is embedded in a strong magnetic field $\mathbf{B}_0$ generated by external currents. Such a field is called a potential field, $\nabla \times \mathbf{B}_0 = 0$. By "strong" we mean that thermal and kinetic plasma energies are small compared with the magnetic field energy:

$$p \sim \rho v^2 \ll B_0^2/8\pi,$$

i.e., the ratio of thermal to magnetic energy, called β, is small,

$$\beta = 8\pi p/B_0^2 \ll 1. \tag{3.40}$$

$\mathbf{B}_0$ is, in general, not homogeneous. The gradient of the field intensity is related to the field line curvature $\kappa = \mathbf{b} \cdot \nabla \mathbf{b}$ by the equation

$$\nabla_\perp B_0 = B_0 \kappa, \tag{3.41}$$

which is easily derived from $\mathbf{B}_0 \times (\nabla \times \mathbf{B}_0) = 0$. Since for $\beta \ll 1$ the frequencies associated with the plasma motions are much slower than the frequency of the magnetosonic mode, $\partial_t \ll v_A \nabla_\perp$, the latter decouples from the plasma dynamics and can be eliminated (Drake & Antonsen, 1984). This implies that the inertia and the viscous terms in the perpendicular part of (3.1) can be neglected, such that there is instantaneous equilibrium perpendicular to the field,

$$\nabla_\perp p - \frac{1}{c}(\mathbf{j} \times \mathbf{B})_\perp = \nabla_\perp \left(p + \frac{B^2}{8\pi} \right) - \frac{1}{4\pi} B^2 \kappa = 0 \tag{3.42}$$

or

$$\nabla_\perp \left(p + \frac{\mathbf{B}_0 \cdot \mathbf{B}_1}{4\pi} \right) \simeq 0, \tag{3.43}$$

decomposing $\mathbf{B}$ into a large potential field $\mathbf{B}_0$ and a component $\mathbf{B}_1$ generated by plasma currents, $\mathbf{B} = \mathbf{B}_0 + \mathbf{B}_1$, and using (3.41). Hence we find that the variation of the longitudinal part $B_{1\|} = \mathbf{B}_1 \cdot \mathbf{B}_0/B_0$ is small,

$$\delta B_{1\|} \simeq -\frac{4\pi}{B_0} \delta p. \tag{3.44}$$

The perpendicular velocity $\mathbf{v}_\perp$ is no longer determined directly from the force balance. Instead, the incompressible (or solenoidal) and the irrotational parts of $\mathbf{v}_\perp$ are computed separately. The incompressible part is obtained from the parallel component of the curl of (3.1)

$$\mathbf{b} \cdot \left(\nabla \times \rho \frac{d\mathbf{v}}{dt} \right) = \frac{B^2}{c} \nabla_\| \frac{j_\|}{B} + 2(\mathbf{b} \times \kappa) \cdot \nabla p + \mathbf{b} \cdot (\nabla \times \mu \nabla^2 \mathbf{v}), \tag{3.45}$$

using the force balance, (3.42), to express ∇B^2 in terms of ∇p. To derive
(3.45) use $\mathbf{j} = \mathbf{b}j_\parallel + \mathbf{j}_\perp$ and the quasi-neutrality condition $\nabla \cdot \mathbf{j} = 0$. To
compute the compressible part of $\mathbf{v}_\perp$, multiply Faraday's law (3.8) by $\mathbf{B}$
to obtain (neglecting resistive effects)

$$\tfrac{1}{2}(\partial_t + \mathbf{v}_\perp \cdot \nabla)B^2 = -B^2(\nabla \cdot \mathbf{v}_\perp + \boldsymbol{\kappa} \cdot \mathbf{v}_\perp),$$

where we have used

$$\mathbf{v}_\perp = \mathbf{v}_E = c\frac{\mathbf{E} \times \mathbf{B}}{B^2}, \tag{3.46}$$

obtained from Ohm's law (2.2). Substitution of $\nabla_\perp B^2$ from (3.42) gives

$$\tfrac{1}{2}\partial_t B^2 - 4\pi \mathbf{v}_\perp \cdot \nabla p = -B^2(\nabla \cdot \mathbf{v}_\perp + 2\boldsymbol{\kappa} \cdot \mathbf{v}_\perp). \tag{3.47}$$

In a low-β plasma the terms on the l.h.s. are proportional to p by use of
(3.44), hence we find to lowest order in β

$$\nabla \cdot \mathbf{v}_\perp = -2\boldsymbol{\kappa} \cdot \mathbf{v}_\perp. \tag{3.48}$$

We call this property quasi-incompressibility. The result can easily be
interpreted. For low β the perpendicular field $\mathbf{E}_\perp$ in $\mathbf{v}_E$ is essentially
electrostatic,

$$\mathbf{E}_\perp = -\nabla_\perp \varphi - \frac{1}{c}\partial_t \mathbf{A}_\perp \simeq -\nabla_\perp \varphi, \tag{3.49}$$

since $\partial_t \mathbf{A}_\perp \propto \delta B_\parallel$ gives a correction $O(\beta)$ to the electrostatic field, hence

$$\mathbf{v}_E = c\frac{\mathbf{B} \times \nabla \varphi}{B^2}. \tag{3.50}$$

Taking the divergence and using (3.41) it is easily seen that $\mathbf{v}_E$ satisfies
(3.48).

The parallel dynamics is determined by the parallel component of (3.1)

$$\rho \mathbf{b} \cdot \frac{d\mathbf{v}}{dt} \simeq \rho \frac{dv_\parallel}{dt} = -\nabla_\parallel p + \rho\mu\nabla^2 v_\parallel. \tag{3.51}$$

To further simplify matters we make use of the fact that $B_\perp \ll B_0$, and
that $\nabla_\parallel \simeq (\mathbf{B}_0/B_0) \cdot \nabla \ll \nabla_\perp$,

$$B_\perp/B_0 \sim \nabla_\parallel/\nabla_\perp \sim \epsilon, \tag{3.52}$$

introducing the small geometric parameter ϵ, for instance the inverse
aspect ratio of a plasma column. Consistent with this ordering we also
require that the spatial variation of B_0 is small, i.e., from (3.41) the field
line curvature is weak, $\kappa a \sim \epsilon$, where a is a typical plasma gradient scale
length. The magnetic field can now be written in the form

$$\mathbf{B} = \mathbf{B}_0 + \delta \mathbf{B}_\parallel + \mathbf{B}_\perp = B_\chi(1 + f)\nabla\chi + \nabla\chi \times \nabla\psi, \tag{3.53}$$

where $\mathbf{B}_0 = B_\chi \nabla \chi$ with $B_\chi = const$ because of $\nabla \times \mathbf{B}_0 = 0$, and $\nabla^2 \chi = 0$ because of $\nabla \cdot \mathbf{B}_0 = 0$. $\nabla \chi \times \nabla \psi = \mathbf{B}_\perp$ is the field generated by currents parallel to $\mathbf{B}_0$ and $f \nabla \chi = \delta \mathbf{B}_\parallel$ the field generated by currents perpendicular to $\mathbf{B}_0$. While $\mathbf{j}_\parallel$ may flow for arbitrarily low β, $\mathbf{j}_\perp$, the so-called diamagnetic current, is proportional to the pressure gradient, $\mathbf{j}_\perp = c\mathbf{B}_0 \times \nabla p / B_0^2$, hence $f \sim \beta$. There is a slight inconsistency in (3.53). Since $\mathbf{B}$ in this form is not divergence-free, if f varies along $\mathbf{B}_0$, it must be compensated by a correction $\delta \mathbf{B}_\perp$, such that $\nabla \cdot \mathbf{B} = \nabla_\parallel f + \nabla \cdot \delta \mathbf{B}_\perp = 0$. Since the parallel derivative $\nabla_\parallel f$ is weak according to (3.52), one has $\delta B_\perp \sim \epsilon \beta$, which is usually neglected in the dynamical equations.

The equation for ψ is obtained from the parallel component of Ohm's law $E_\parallel = \eta j_\parallel$. Multiplying the electric field by $\mathbf{b}$ gives

$$E_\parallel = -\mathbf{b} \cdot \nabla \varphi - \frac{1}{c} \partial_t A_\parallel. \tag{3.54}$$

Hence the magnetic potential $\psi = -A_\parallel$ follows the equation

$$\frac{1}{c} \partial_t \psi = \mathbf{b} \cdot \nabla \varphi + \eta j_\parallel. \tag{3.55}$$

Let us now reduce the pressure equation (3.5), in particular the compressibility term $\gamma p (\nabla_\perp \cdot \mathbf{v}_\perp + \nabla_\parallel v_\parallel)$. From (3.48) and (3.52) we find

$$\nabla \cdot \mathbf{v}_\perp \sim \kappa v_\perp \sim \epsilon v_\perp / a \tag{3.56}$$

and from (3.51)

$$\nabla_\parallel v_\parallel \sim \beta v_\perp / a. \tag{3.57}$$

Hence for $\beta \lesssim \epsilon$ the compressibility term in the pressure equation is $O(\epsilon)$ and therefore negligible to lowest order. Since now the continuity equation (3.2) degenerates to $d\rho / dt = 0$, we can assume the spatial variation of the density to be small. The l.h.s. of (3.45) can then be simplified,

$$\mathbf{b} \cdot (\nabla \times \rho \frac{d\mathbf{v}}{dt}) \simeq \rho (\partial_t w + \mathbf{v} \cdot \nabla w), \tag{3.58}$$

where $w = \mathbf{b} \cdot \nabla \times \mathbf{v}$ is the parallel component of the vorticity. In this order $\nabla \cdot \mathbf{v}_\perp = 0$, such that $\mathbf{v}_\perp$ can be expressed in terms of a stream function ϕ, $\mathbf{v}_\perp = \mathbf{b} \times \nabla \phi$, which by (3.46) is essentially the electric potential φ, $\phi = c\varphi / B$. To first order in ϵ and $\beta \simeq \epsilon$ we thus obtain a closed set of reduced equations for w, ψ, p,

$$\rho(\partial_t w + \mathbf{v} \cdot \nabla w) = \frac{1}{c} \mathbf{B} \cdot \nabla j + 2(\mathbf{b} \times \boldsymbol{\kappa}) \cdot \nabla p + \rho \mu \nabla^2 w, \tag{3.59}$$

$$\partial_t \psi = \mathbf{B} \cdot \nabla \phi + \lambda \nabla^2 \psi, \tag{3.60}$$

$$\partial_t p = -\mathbf{v} \cdot \nabla p \tag{3.61}$$

with

$$\mathbf{v}_\perp = \mathbf{b} \times \nabla\phi, \quad \mathbf{B} = \mathbf{b} \times \nabla\psi + \mathbf{b}B_0,$$

$$w = \nabla_\perp^2 \phi, \quad j = j_\parallel = \frac{c}{4\pi}\nabla^2\psi, \quad \lambda = \frac{\eta c^2}{4\pi}.$$

The parallel velocity need not be small compared with the perpendicular one, but since $v_\parallel \nabla_\parallel \ll \mathbf{v}_\perp \cdot \nabla_\perp$, the coupling of $v_\parallel$ to the other quantities w, ψ, p can be neglected. Equations (3.59)–(3.61) are called high-β reduced MHD, since in tokamak plasmas $\beta \sim \epsilon$ is the highest achievable β. The κ-term in the vorticity equation acts in a way similar to a gravitational force. As discussed in detail in section 4.5, this term plays an important role for $\beta \sim \epsilon$. At smaller β, $\beta \sim \epsilon^2$, the term is negligible and the reduced equations assume the simpler low-β form

$$\rho_0(\partial_t + \mathbf{v} \cdot \nabla)w = \frac{1}{c}\mathbf{B} \cdot \nabla j + \rho_0\mu\nabla^2 w, \tag{3.62}$$

$$\partial_t \psi = \mathbf{B} \cdot \nabla\phi + \lambda\nabla_\perp^2\psi, \tag{3.63}$$

With $\mathbf{B}_0 = B_z\mathbf{e}_z$ (3.63) can also be written in the form,

$$\partial_t\psi + \mathbf{v}_\perp \cdot \nabla\psi = B_z\partial_z\phi + \lambda\nabla_\perp^2\psi,$$

from which we see that in the 2D case, $\partial_z = 0, \psi$ is ideally conserved. Since in this case field lines are located on surfaces $\psi = const$, ψ is called the flux function. This property is, however, lost in the 3D case. The low-β reduced equations (3.62) and (3.63) have been derived by Kadomtsev & Pogutse (1974) and Strauss (1976), and a Hamiltonian formulation has been developed by Morrison & Hazeltine (1984).

It is worth pointing out that in 2D (3.62) and (3.63) are identical with the 2D nonreduced MHD equations in the incompressible limit, since in this limit the axial flow v_z and the axial field B_z decouple from the poloidal equations. The conceptual difference compared with the reduced equations is that v_z and B_z, which can be computed a posteriori from their initial values, need not be small or large, respectively.

The three-field high-β reduced equations (3.59)–(3.61) have been developed by Strauss (1977). More general reduced models, such as the four-field model by Hazeltine *et al.* (1985), are beyond the MHD framework and will be treated in the context of two-fluid theory in chapter 6.

Finally, we introduce convenient normalizations to write the resistive MHD equations in non-dimensional form. Since for $\beta \ll 1$ the important dynamics occurs in the poloidal plane, we normalize the magnetic field to a typical poloidal field intensity $B_{\perp 0}$, the velocity to the corresponding Alfvén velocity $v_{A0} = B_{\perp 0}/\sqrt{4\pi\rho_0}$, and the space coordinates to a typical poloidal scale-length a. In these units the dissipation coefficients are the inverse of the magnetic and kinetic Lundquist numbers S and U,

$$S^{-1} = \frac{\eta c^2}{4\pi v_{A0} a}, \quad U^{-1} = \frac{\mu}{v_{A0} a}, \tag{3.64}$$

which we denote again by η and μ, respectively. In the following, S is called just the Lundquist number. The lowest-order reduced equations (3.62), (3.63) now have the non-dimensional form

$$\partial_t w + \mathbf{v} \cdot \nabla w = \mathbf{B} \cdot \nabla j + \mu \nabla^2 w, \tag{3.65}$$

$$\partial_t \psi + \mathbf{v} \cdot \nabla \psi = B_z \partial_z \phi + \eta \nabla^2 \psi. \tag{3.66}$$

Here $B_z \gg 1$ is the large constant axial field component which, multiplied with the weak gradient in the axial direction, gives a term of order unity.

The ideal reduced equations conserve energy W,

$$W = \tfrac{1}{2} \int (v^2 + B^2) dV = \tfrac{1}{2} \int [(\nabla \phi)^2 + (\nabla \psi)^2] dV, \tag{3.67}$$

cross-helicity K,

$$K = \int \mathbf{v} \cdot \mathbf{B} \, dV = -\int w\psi \, dV, \tag{3.68}$$

and the linear quantity $\int \psi dV$, a special form of the helical flux, (3.39), which is also related to the magnetic helicity,

$$\int \mathbf{A} \cdot \mathbf{B} \, dV = -B_z \int (\psi - rB_\theta/2) dV = -2B_z \int \psi \, dV, \tag{3.69}$$

using the notations $\psi = -A_z$ and the gauge $A_\theta = rB_z/2, A_r = 0$.

3.2 Resistive current sheets

3.2.1 Current-sheet formation

Flows in a magnetized plasma give rise, in general, to sheet-like field discontinuities, which correspond to singularities of the current density and are hence called current sheets. An alternative terminology is tangential discontinuities since, in contrast to MHD shocks, these sheets are aligned with the magnetic field. Flows may be excited externally or by some internal instability. The location of current sheets depends on the flow pattern and the magnetic topology. In a toroidal plasma column, current sheets are generated on resonant surfaces where the field line twist or the safety factor are rational numbers.

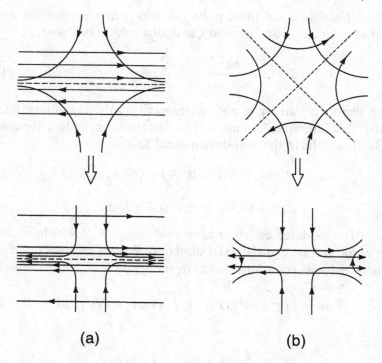

Fig. 3.2. Current-sheet formation: (a) compression of a neutral sheet configuration; (b) flattening of an X-point configuration.

The basic situation is shown in fig. 3.2, a stagnating flow in a sheared magnetic field where, in the projection perpendicular to the central field line – the "poloidal" plane – the magnetic topology is that of a neutral sheet with the poloidal field reversing direction. Note that in general: (a) flow stagnation occurs only in a particular coordinate system, *i.e.*, viewed from a moving frame of reference the stagnation property is not easily recognized; (b) the perpendicular field component – the " axial" field – does not, in general, vanish. In the low-β case the axial field is much larger than the poloidal one; (c) no X-point[*] configuration is needed initially to form a current sheet. However, if the configuration exhibits X-points, flow stagnation will preferentially occur at these, which then leads to flattening and sheet formation, as indicated in fig. 3.2(b). We therefore concentrate on this situation.

Let us discuss the process of sheet formation in more detail. Since, starting from a smooth configuration, dissipation effects are negligible – they are only important for the saturation of the sheet thinning process –

[*] When dealing with 2D configurations without reference to the third direction, we use the term neutral point, e.g., X-point; otherwise we speak of neutral line, e.g., X-line.

we restrict consideration to ideal MHD in two dimensions. In the vicinity of an X-point the MHD equations allow exact similarity solutions. First consider the compressible case, following Imshennik & Syrovatskii (1967). The equation for the flux function is

$$\partial_t \psi + \mathbf{v} \cdot \nabla \psi = 0. \tag{3.70}$$

In 2D, the equation of motion (3.1) can be written in the following way,

$$\rho(\partial_t \mathbf{v} + \mathbf{v} \cdot \nabla \mathbf{v}) = -\nabla p - j \nabla \psi, \tag{3.71}$$

where $j = \nabla^2 \psi$ is the current density in the z-direction. We assume that the pressure is a function of the density, $p = p(\rho)$, which is valid for general polytropic changes of state. Because of finite compressibility, we must also consider the continuity equation (3.2). Choosing the initial conditions

$$\psi(x, y) = \tfrac{1}{2}(x^2 - y^2),$$

$$v_x(x, y) = \gamma_0 x, \quad v_y(x, y) = \delta_0 y,$$

$$\rho(x, y) = 1,$$

the equations have the following similarity solution, as can easily be verified:

$$\psi(x, y, t) = \frac{1}{2} \left(\frac{x^2}{\xi^2(t)} - \frac{y^2}{\eta^2(t)} \right), \tag{3.72}$$

$$v_x(x, y, t) = \gamma(t)x, \quad v_y(x, y, t) = \delta(t)y, \tag{3.73}$$

$$\rho(x, y, t) = \rho(t). \tag{3.74}$$

Since the density and hence the pressure remain uniform in space, the pressure gradient in (3.71) vanishes. Upon substitution of the solution, the equations reduce to a system of ordinary differential equations for $\xi, \eta, \rho, \gamma, \delta$:

$$\dot{\xi} = \gamma \xi, \quad \dot{\eta} = \delta \eta, \tag{3.75}$$

$$\rho(\dot{\gamma} + \gamma^2) = -\frac{1}{\xi^2} \left(\frac{1}{\xi^2} - \frac{1}{\eta^2} \right), \quad \rho(\dot{\delta} + \delta^2) = \frac{1}{\eta^2} \left(\frac{1}{\xi^2} - \frac{1}{\eta^2} \right), \tag{3.76}$$

$$\dot{\rho} + \rho(\gamma + \delta) = 0. \tag{3.77}$$

From (3.75) and (3.77) we obtain

$$\frac{d}{dt} \ln(\xi \eta \rho) = 0,$$

hence

$$\xi \eta \rho = 1,$$

using the initial conditions. Elimination of γ, δ, ρ gives

$$\ddot{\xi} = -\eta \left(\frac{1}{\xi^2} - \frac{1}{\eta^2} \right), \quad \ddot{\eta} = \xi \left(\frac{1}{\xi^2} - \frac{1}{\eta^2} \right) \tag{3.78}$$

with the initial conditions $\xi_0 = \eta_0 = 1$, $\dot{\xi}_0 = \gamma_0, \dot{\eta}_0 = \delta_0$.

In general, one has $\gamma_0 \neq \delta_0$ and, to be definite, we choose $\gamma_0 < \delta_0$. From (3.75) we see that for short times $\xi < \eta$, hence $\ddot{\xi} < 0, \ddot{\eta} > 0$ from (3.78). The ratio η/ξ increases until $\xi = 0$, where the solution becomes singular, which occurs at some finite time t_0, as can be seen from the fact that $\ddot{\xi} \to -\infty$. While the full solution $\xi(t), \eta(t)$, in particular the value of t_0, must be computed numerically, the most interesting behavior in the vicinity of the singularity can easily be obtained analytically. Since $\xi \to 0$ for $t \to t_0$, while η reaches a finite value $\eta(t_0)$, as can be checked a posteriori by inserting the solution $\xi(t)$, integration of the ξ equation in (3.78), $\ddot{\xi} \simeq -\eta(t_0)/\xi^2$, gives

$$\xi \simeq \left(\frac{9}{2} \eta(t_0) \right)^{1/3} (t_0 - t)^{2/3}, \tag{3.79}$$

and hence

$$\gamma \simeq -\frac{2}{3} \frac{1}{t_0 - t}, \quad \delta \simeq const, \tag{3.80}$$

$$\rho \propto \frac{1}{(t_0 - t)^{2/3}}, \quad j \propto \frac{1}{(t_0 - t)^{4/3}}. \tag{3.81}$$

Note that, even if both γ_0 and δ_0 are positive, implying rarefactive motions in both directions initially, γ becomes negative after a finite time, leading to compression and divergence of ρ and j. Only $\gamma_0 = \delta_0 > 0$ implies unlimited expansion, while for $\gamma_0 = \delta_0 < 0$ isotropic compression gives rise to a finite-time density singularity, but in both cases j vanishes identically. Incompressible motion, i.e., $\gamma(t) = -\delta(t)$, is not possible in this class of solutions, even for initial condition $\gamma_0 = -\delta_0$.

Let us now turn to the incompressible case. Chapman and Kendall (1963) derived a solution similar to (3.72) and (3.73), where $\nabla \cdot \mathbf{v} = 0$ now requires $\gamma(t) = -\delta(t)$. The essential difference, compared with the compressible case, comes from the ∇p term in (3.71), which far from being zero plays an important part. Since fluid velocities are small compared with the Alfvén velocity, the pressure gradient nearly balances the Lorentz force and hence has the effect of retarding the formation of a current singularity. (The density can be assumed constant.) While the ψ equation (3.70) is not changed, we now use the equation of motion in the vorticity form, (3.65)

$$\partial_t w + \mathbf{v} \cdot \nabla w = \mathbf{B} \cdot \nabla j, \tag{3.82}$$

$$w = \nabla^2 \phi, \quad \mathbf{v} = \mathbf{e}_z \times \nabla \phi.$$

This system has the similarity solution

$$\psi = \frac{1}{2}\left(\frac{x^2}{\xi^2(t)} - \frac{y^2}{\eta^2(t)}\right), \tag{3.83}$$

$$\phi = \Lambda(t)xy, \tag{3.84}$$

corresponding to $j = j(t)$ and $w = 0$. (Note also that the compressible flow, (3.73), is irrotational.) From the ψ equation follows

$$\dot{\xi} = -\Lambda\xi, \quad \dot{\eta} = \Lambda\eta, \tag{3.85}$$

where $\Lambda(t)$ is a free function. Since (3.82) is satisfied identically, the solution is not completely determined, in contrast to the compressible case, (3.75)–(3.77). A special choice is $\Lambda = const$,

$$\xi = \xi_0 e^{-\Lambda t}, \quad \eta = \eta_0 e^{\Lambda t}, \tag{3.86}$$

corresponding to exponential flattening of the X-point configuration and exponential growth of the current density $j \propto e^{2\Lambda t}$. Numerical simulations (Sulem et al., 1985; Friedel et al., 1997) show that this behavior is indeed typical in 2D MHD. The result can be understood in terms of a continuous weakening of the flow nonlinearity, as the configuration becomes more and more one-dimensional, as can be seen directly from the 2D MHD equations. Taking the Laplacian of (3.70) gives

$$\partial_t j + \mathbf{v}\cdot\nabla j - \mathbf{B}\cdot\nabla w = 2\sum_i \mathbf{e}_z\cdot(\nabla\partial_i\psi\times\nabla\partial_i\phi) = 2\sum_{i,k}\epsilon_{3ik}\partial_i\mathbf{B}\cdot\partial_k\mathbf{v}. \tag{3.87}$$

Combining this equation with (3.82) and introducing the Elsässer variables $\mathbf{z}^\pm$, (3.11), and their derivatives $w^\pm = w \pm j$, we obtain

$$\partial_t w^\pm + \mathbf{z}^\mp\cdot\nabla w^\pm \equiv \dot{w}^\pm = \pm 2\sum_{i,k}\epsilon_{3ik}\partial_i\mathbf{B}\cdot\partial_k\mathbf{v}. \tag{3.88}$$

For a sheet-like structure only the cross-sheet derivative becomes large. Hence, because of the cross product, the source term in (3.88) contains only one large derivative, such that the equation can be written symbolically in the form $\dot{w} = Cw$, whence exponential growth.

   The presence or absence of a finite-time singularity (FTS) in a fluid model has been the subject of considerable debate, since this property may shed some light on the process of turbulence generation from a smooth initial state and even elucidate the character of the turbulence itself. The simplest example is the compressible flow in the 1D Burgers' equation $\partial_t u + u\partial_x u = 0$, where the presence of FTS can easily be shown. Taking the x-derivative yields

$$\partial_t w + u\partial_x w \equiv \dot{w} = -w^2, \quad w = \partial_x u, \tag{3.89}$$

which has the solution $w[x_0(t)] = (t - t_0)^{-1}$. The simplest case of incompressible flow, the 2D plane Euler flow, does not exhibit FTS, as is to be expected from the equation $\partial_t w + \mathbf{v} \cdot \nabla w \equiv \dot{w} = 0$. This is, however, no longer true for more general 2D Euler flows, since, for instance, a rotationally symmetric flow can lead to FTS (Grauer & Sideris, 1991). For nonsymmetric, i.e., three-dimensional Euler flows, FTS seems to be generic (except for high-symmetry initial conditions such as the Taylor–Green vortex, see, e.g., Brachet *et al.*, 1992), which has recently been studied numerically using the method of adaptive mesh refinement to reach the necessary high local resolution (Grauer *et al.*, 1998).

By contrast, 3D MHD flows show only exponential growth, much as in the 2D case, since narrow structures always become sheet-like eventually, even if initially the magnetic field is weak (Grauer & Marliani, 2000). The absence of a finite-time singularity seems to weaken the relevance of the similarity solution (3.86) since, for long times, spatial regions far away from the X-point, where the behavior of the solution is not universal depending on the particular configuration, may affect the behavior at this point. In practice, however, exponential growth up to the resistive saturation level is sufficiently fast that non-universal effects remain small.

Though the similarity solutions given above provide strong evidence for rapid current-sheet formation, their validity is restricted to the center region of the sheet, the immediate vicinity of the X-point, where $j \simeq const$ and $w \simeq 0$. As seen in the following sections, current sheets in their full extent are also the location of strong vorticity. In contrast to the current distribution in the sheet, which has a flat elliptical monopole character, the vorticity distribution has the structure of a flat quadrupole (hence $w = 0$ in the center). This behavior of j and w can be interpreted in terms of the variables $w^{\pm} = w \pm j$, such that $w = \frac{1}{2}(w^+ + w^-)$, $j = \frac{1}{2}(w^+ - w^-)$. In a current sheet, w^+ and w^- both form elliptical structures, slightly tilted against each other. (Since the ideal MHD equations are symmetric in the Elsässer fields, w^+ and w^- must be similar in contrast to j and w.) Hence one of the two fields, w or j, must have a quadrupole character.

Numerical simulations reveal that current sheets indeed develop in a selfsimilar way (Biskamp, 1993b). Figure 3.3 illustrates the thinning process. The system starts from the smooth state at $t = 0$ (configuration A_1 from Biskamp & Welter, 1989)

$$\phi(x, y) = \cos(x + 1.4) + \cos(y + 0.5), \qquad (3.90)$$

$$\psi(x, y) = \cos(2x + 2.3) + \cos(y + 4.1), \qquad (3.91)$$

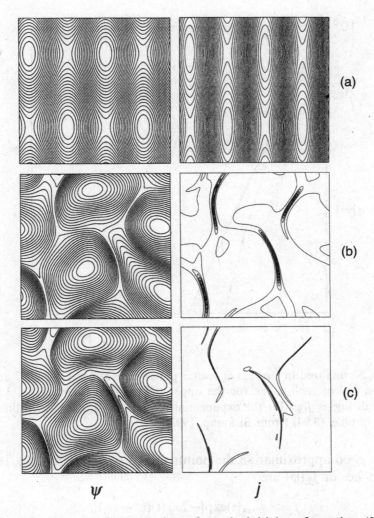

ψ j

Fig. 3.3. Development of current sheets from the initial configuration, (3.90) and (3.91): (a) $t = 0$; (b) $t = 1.0$; (c) $t = 1.3$ (from Biskamp, 1993b).

which is a generalization of the Orszag–Tang vortex (Orszag & Tang, 1979). Four current sheets develop at the X-points of this configuration. Since relaxation to equilibrium inside the narrow sheets occurs quasi-instantaneously, the magnetic field obeys the equilibrium relation $\mathbf{B} \cdot \nabla j = 0$, such that $j = j(\psi)$. In general, $j(\psi)$ is not identical on both sides of the sheet because of the asymmetry of the configuration, but has two different branches $j_\pm(\psi)$. Figure 3.4 gives scatter plots $\ln \widehat{j}$ vs $\widehat{\psi}$ for one of the sheets taken at four different times plotted on top of each other, where $\widehat{j} = j/j_{\max}(t)$, $\widehat{\psi} = \alpha(t)(\psi - \psi_0)$, and $j_{\max} = j(\psi_0)$ is the maximum current

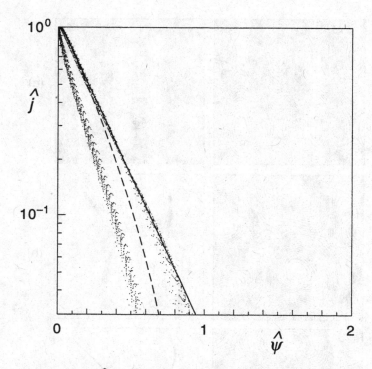

Fig. 3.4.   Normalized $\ln \widehat{j}$ versus $\widehat{\psi}$ scatter plots taken at $t = 1.0, 1.1, 1.2$ and 1.3, plotted on top of each other, for the upper-right current sheet of fig. 3.3. The straight full line is $j(\psi)$ for the exponential profile, (3.92), the dashed line for a Gaussian profile, (3.94) (from Biskamp, 1993b).

value. In good approximation the points fall on two straight lines, i.e., the two branches of $j_\pm(\psi)$ are

$$j_\pm \simeq j_{\max}(t)\exp[-2\alpha_\pm(t)(\psi - \psi_0)]. \tag{3.92}$$

The maximum current density $j_{\max}(t)$ is found to increase exponentially in time in agreement with the similarity solution (3.86), and the thickness $\delta_\pm(t) \propto \alpha_\pm(t)^{-1/2}$ decreases exponentially. In configuration space the profile $j = j_{\max} \exp(-2\alpha\psi)$ corresponds to

$$j_0(x) = j_{\max}\mathrm{sech}^2(x/\delta). \tag{3.93}$$

As seen from fig. 3.4, the numerically observed current profiles are distinctly different from a Gaussian profile, the dashed line in fig. 3.4, which is obtained from the parametric representation

$$j(x) = \frac{2}{\sqrt{\pi}}\,\mathrm{e}^{-x^2/a^2}, \quad \psi = ax\,\mathrm{erf}(x/a) + \frac{a^2}{\sqrt{\pi}}\,\mathrm{e}^{-x^2/a^2}, \tag{3.94}$$

where a is chosen such that $j(\psi)$ has the same derivative at the origin as the corresponding branch of the scatter plot.

The current profile $j = \exp(-2\psi)$ can be "derived" theoretically (Montgomery *et al.*, 1979; Taylor, 1993). Here the idea is to use statistical arguments to determine the most probable profile, which should be set up approximately under typical conditions. Using a thermodynamic approach one would calculate this state as the minimum of the free energy $\mathscr{F} = W - ST$ under certain constraints. To avoid functional integrals we consider a discrete model (Montgomery *et al.*, 1979): divide the area F containing the current sheet into M cells ΔF, $F = M\Delta F$, and decompose the current density into N filaments of current λ such that $I = N\lambda$ is the total current flowing across F with $N \gg M$. For a given current distribution there are N_i filaments in cell i. Assuming equal a priori probability of states the probability $P\{N_i\}$ to find a certain set of occupation numbers $\{N_i\}$ is proportional to the number of states with these occupation numbers, i.e.,

$$P\{N_i\} = \prod_i (N_i!)^{-1}. \tag{3.95}$$

The entropy of the system is $S = -\ln P \simeq \sum_i N_i \ln N_i$, using Stirling's formula. The energy W is

$$W = \int (\nabla\psi)^2 d^2x = -\int \psi j d^2x = -\int\int d^2x d^2x'\, j(\mathbf{x})g(\mathbf{x},\mathbf{x}')j(\mathbf{x}'),$$

where $g(\mathbf{x}, \mathbf{x}')$ is the Green's function, $\psi(\mathbf{x}) = \int g(\mathbf{x},\mathbf{x}')j(\mathbf{x}')d^2x'$. In discretized form we have $\psi_i = \lambda \sum_j g_{ij} N_i$ and therefore

$$W = -\lambda^2 \sum_{i,j} N_i g_{ij} N_j. \tag{3.96}$$

We now minimize the free energy under the constraint of constant total current by varying the occupation numbers N_i,

$$\frac{\delta}{\delta N_i}\left[\lambda^2 \sum_{j,k} N_j g_{jk} N_k + T\sum_j N_j \ln N_j + \alpha \sum_j N_j\right] = 0, \tag{3.97}$$

where α is a Lagrange multiplier. Variation gives

$$T\ln N_i + \alpha = -2\lambda^2 \sum_j g_{ij} N_j = -2\lambda\psi_i,$$

or, passing to the continuum limit,

$$j = Ce^{-2\psi/T} \tag{3.98}$$

with $C = (\lambda/\Delta F)e^{-\alpha/T}$. Hence ψ obeys the differential equation

$$\nabla^2\psi = Ce^{-2\psi/T}. \tag{3.99}$$

For the simplest case of a symmetric current sheet the solution of (3.99) is given by the expression (3.93) with $\delta = (T/C)^{1/2}$, $C = j_{\max}$. Hence the "temperature" T is a measure of the sheet width.

It is interesting to consider the Fourier transform of the profile (3.93),

$$\int_{-\infty}^{\infty} \frac{\cos kx}{\cosh^2 x} \, dx = \frac{\pi k}{\sinh(\frac{1}{2}\pi k)}. \tag{3.100}$$

Hence the magnetic energy spectrum is asymptotically exponential,

$$B^2(k) = \frac{j^2(k)}{k^2} = \frac{1}{\sinh^2(\frac{1}{2}\pi k)} \simeq 4e^{-\pi|k|}, \quad |k| \gg 1, \tag{3.101}$$

which agrees with the numerically observed energy spectrum (Frisch *et al.*, 1983). By contrast, a Gaussian profile would lead to a Gaussian Fourier spectrum.

3.2.2 Basic properties of resistive current sheets

The growth of the current density during sheet thinning saturates because finite resistivity leads to field diffusion, which balances the convective magnetic flux transport into the sheet. The result is a quasi-stationary resistive current sheet, called a Sweet–Parker sheet (Sweet, 1958; Parker, 1963). Resistivity introduces reconnection – magnetic flux is transported into the sheet, reconnected and swept out of the sheet. The configuration is illustrated in fig. 3.5. Assuming incompressible motions, the stationary quasi-one-dimensional state is characterized by six quantities, three describing the dynamics: (i) the (poloidal) magnetic field B_0 immediately outside the sheet, called the upstream field (the downstream field is negligible), (ii) the upstream flow u_0 perpendicular to the field, and (iii) the downstream flow v_0 along the field taken at the sheet edge; two geometric quantities: (iv) the sheet length Δ and (v) the width δ (in the literature one often finds a different terminology δ = thickness, Δ = width, implying a three-dimensional sheet structure with the length measured along the third direction); and, finally, (vi) the diffusion coefficient η. We will see below that the effect of the viscosity is usually weak.

These six quantities are connected by three relations derived from the continuity equation, Ohm's law, and the equation of motion, assuming stationarity. Furthermore, the analysis implies that the profiles of the dynamic quantities along and across the sheet have certain self-similar

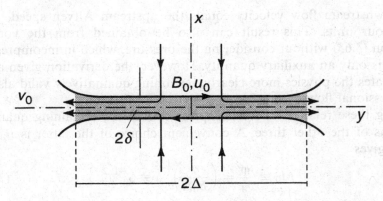

Fig. 3.5. Schematic drawing of a Sweet–Parker sheet.

properties. Integrating the continuity equation $\nabla \cdot \rho \mathbf{v} = 0$ over a quadrant of the sheet and assuming constant density gives

$$u_0 \Delta = v_0 \delta. \tag{3.102}$$

Consider Ohm's law (2.1) along the x-axis. Stationarity requires E_z to be uniform in space. In the upstream region outside the sheet, where the current density is small, the resistive term is negligible, while in the sheet center, where $j = j_{max}$, and the velocity vanishes, the resistive term dominates, which gives the relation

$$u_0 B_0 = \eta j_{max} \simeq \eta B_0 / \delta. \tag{3.103}$$

Since, usually, $u_0 \ll B_0$ as can be seen a posteriori, the inertia term is negligible in the force balance across the sheet. Hence $\partial_x(p + \frac{1}{2}B^2) = 0$, which gives

$$\tfrac{1}{2}B_0^2 = p_{max} - p_0. \tag{3.104}$$

Here p_0 is the upstream fluid pressure and p_{max} the maximum pressure in the sheet center, where the (poloidal) field vanishes. Now consider the force balance along the mid-plane of the sheet. Since B_x is negligible, so is the magnetic force, such that only the pressure force accelerates the fluid along the sheet, $v_y \partial_y v_y = -\partial_y p$. Integration between center and edge yields

$$\tfrac{1}{2}v_0^2 = p_{max} - p_0. \tag{3.105}$$

Here, the current-sheet edge $y = \Delta$ is defined by the vanishing of the pressure difference across the sheet. (In reality, the edge region may have a complicated structure, as shown in section 3.3.2.) Substituting $p_{max} - p_0$ from (3.104) we find the important result

$$v_0 = B_0, \tag{3.106}$$

the downstream flow velocity equals the upstream Alfvén speed, $B_0 = v_A$ in our units. This result can also be obtained from the vorticity equation (3.65) without considering the pressure, which in incompressible theory is only an auxiliary quantity. However, the derivation given above illuminates the physics more clearly, remaining qualitatively valid also for compressional flows.

Using these relations we can express two of the remaining quantities in terms of the other three. A convenient choice of the latter is B_0, Δ, η, which gives

$$M_0 \equiv \frac{u_0}{v_A} = \left(\frac{\eta}{B_0 \Delta} \right)^{1/2} \equiv S_0^{-1/2}, \tag{3.107}$$

$$A \equiv \frac{\Delta}{\delta} = S_0^{1/2}. \tag{3.108}$$

Here we have introduced the Mach number M_0, which has often been used as a dimensionless measure of the reconnection rate. Calling $M = u/B$ the reconnection rate can, however, lead to confusion, since properly speaking the reconnection rate is the rate of flux change at the X-point, the electric field $E_z = \partial_t \psi = \eta j$, which equals the product uB for stationary conditions. A is the aspect ratio and S_0 the Lundquist number of the current sheet. The Sweet–Parker rate $M_0 = S_0^{-1/2}$ is a characteristic quantity of a resistive current sheet. If B_0 and Δ are of the order of the global field intensity and the spatial scale of the magnetic configuration, respectively, S_0 is the global Lundquist number S. In most cases of practical interest S is very large, typically $S \sim 10^{10}$ in the solar corona, such that the Sweet-Parker rate would lead to reconnection times in a solar flare many orders of magnitude longer than observed. Obviously a single quasi-stationary current sheet with an aspect ratio $A \sim 10^5$ is a very implausible configuration in view of the tearing instability (see sections 4.1 and 4.7).

The fact that the fluid leaves the sheet at the Alfvén speed might raise concerns about the validity of the incompressibility assumption. In the case of a strong axial field component, all poloidal flows are nearly perpendicular and hence incompressible, since the fast flow along the sheet reaches only the poloidal Alfvén velocity, which is much smaller than the axial one. In the absence of a strong axial field component, the outflow is essentially parallel to the field and incompressibility would require $v \sim v_A \ll c_s \sim \beta^{1/2} v_A$, i.e., high β, which is usually not the case. But also for $\beta \sim 1$, where this condition is only marginally satisfied, numerical simulation using the full compressible equations show that the flows are practically incompressible.

Finally we consider the effect of finite viscosity μ, which changes the force balance along the sheet, $v_y \partial_y v_y = -\partial_y p + \mu \partial_x^2 v_y$, since $\nabla^2 v_y \simeq \partial_x^2 v_y$. Integration along the sheet between center and edge, using $\partial_x^2 v_y \simeq -v_y/\delta^2$

and (3.104), gives

$$v_0^2 + (\mu\Delta/\delta^2)v_0 \simeq B_0^2,$$

or, with $v_0 = \eta\Delta/\delta^2$ from (3.102) and (3.103),

$$v_0 = B_0 \left(1 + \frac{\mu}{\eta}\right)^{-1/2}, \tag{3.109}$$

and hence

$$M_0 = S_0^{-1/2} \left(1 + \frac{\mu}{\eta}\right)^{-1/4}, \tag{3.110}$$

$$A = S_0^{1/2} \left(1 + \frac{\mu}{\eta}\right)^{-1/4}, \tag{3.111}$$

generalizing (3.107) and (3.108). Thus the modifications introduced by viscosity are rather weak. While for $\mu \lesssim \eta$ the effect is negligible, $\mu \gg \eta$ results in a broader sheet with reduced inflow and outflow velocities. The cross-field collisional value of μ is $\rho_i^2 v_i$, which compared with the expression of the collisional resistivity, or rather magnetic diffusivity, $(c/\omega_{pe})^2 v_e$ gives

$$\frac{\mu}{\eta} \sim \beta \frac{m_i}{m_e} \frac{v_i}{v_e} \sim \beta \sqrt{\frac{m_i}{m_e}}.$$

Hence only for sufficiently high β may μ strongly exceed η.

3.2.3 Refined theory of the current-sheet structure

In the preceding section we have given a qualitative picture of a resistive current sheet, which will now be refined and quantified. We first consider the central sheet region near the X-point. Such a point always exists, when reconnection takes place, even if the initial ideal sheet did not have a neutral point as in fig. 3.2(a). We investigate the stationary solution of the resistive MHD equations (3.65) and (3.66) by Taylor-series-expansion of ψ and ϕ around the X-point at $(x, y) = (0, 0)$, assuming up–down and right–left symmetry,

$$\psi = \sum_{mn} \psi_{2m,2n} \frac{x^{2m} y^{2n}}{(2m)!(2n)!}, \quad \phi = \sum_{m,n} \phi_{2m+1,2n+1} \frac{x^{2m+1} y^{2n+1}}{(2m+1)!(2n+1)!}, \tag{3.112}$$

where $\psi_{2m,2n} \equiv \partial_x^{2m} \partial_y^{2n} \psi$ and similarly for ϕ. For stationary conditions, (3.65) and (3.66) read

$$\partial_x \phi \partial_y w - \partial_y \phi \partial_x w - \partial_x \psi \partial_y j + \partial_y \psi \partial_x j = \mu(\partial_x^2 w + \partial_y^2 w), \tag{3.113}$$

$$\partial_x \phi \partial_y \psi - \partial_y \phi \partial_x \psi = \eta(\partial_x^2 \psi + \partial_y^2 \psi) - E. \tag{3.114}$$

Here $E = E_z = \partial_t \psi$ is the axial electric field, which is uniform and constant in the stationary case. Since at the neutral point, which in our symmetric case is also a flow stagnation point, $\mathbf{B} = \mathbf{v} = 0$, (3.114) gives

$$\eta(\psi_{20} + \psi_{02}) = E. \tag{3.115}$$

Differentiating (3.114) twice with respect to x and twice with respect to y at the origin one obtains, respectively,

$$2\phi_{11}\psi_{20} + \eta(\psi_{40} + \psi_{22}) = 0, \tag{3.116}$$

$$2\phi_{11}\psi_{02} - \eta(\psi_{22} + \psi_{04}) = 0. \tag{3.117}$$

Differentiating (3.113) once with respect to both x and y at the origin gives

$$-\psi_{20}(\psi_{22} + \psi_{04}) + \psi_{02}(\psi_{40} + \psi_{22}) = \mu(\phi_{51} + 2\phi_{33} + \phi_{15}), \tag{3.118}$$

which by use of (3.116) and (3.117) becomes

$$-\frac{4}{\eta}\phi_{11}\,\psi_{20}\,\psi_{02} = \mu(\phi_{51} + 2\phi_{33} + \phi_{15}). \tag{3.119}$$

Cowley (1975) considered the inviscid case $\mu = 0$, where the r.h.s. of (3.119) seems to vanish. Assuming $\phi_{11} \neq 0$, i.e., stream lines to form hyperbolae, either ψ_{20} or ψ_{02} (not both because of (3.115)) must vanish. This would imply that field lines do not form hyperbolae, in particular the separatrix branches do not intersect at a finite angle but osculate, meaning that the normal magnetic field component does not increase linearly but cubically $B_x \propto y^3$. The argument is, however, not strictly valid (Uzdensky & Kulsrud, 1998), since the solution, in particular the velocity field, does not remain regular in the inviscid limit, such that the Taylor expansion (3.112) breaks down. Uzdensky & Kulsrud perform a boundary layer analysis for small viscosity $\mu \ll \eta$, where the velocity exhibits a viscous sublayer. Asymptotic matching of the solutions in the outer inviscid region and the inner viscous region yields that $\lim_{\mu \to 0} \mu\phi_{51}$ is finite, a result which they could corroborate and quantify by a fully numerical solution. It follows that, in general, both $\psi_{02} \neq 0$ and $\psi_{20} \neq 0$, their values depending on the solution of the entire current sheet with the upstream field $B_{0y}(y)$ as boundary condition. Hence in the limit $\mu \to 0$ the separatrix branches do not osculate, but intersect at a finite angle, which is of course very small in a long current sheet. For finite viscosity $\mu \sim \eta$ both ϕ and ψ are regular with converging Taylor series, (3.112). But now one cannot draw definite conclusions from relation (3.119). Numerical simulations show that, although ψ_{20} (or ψ_{02}) does not vanish, it is small, hence the separatrix branches nearly osculate.

To consider the global sheet configuration (apart from the edge regions, where a complicated behavior occurs, which is discussed separately in section 3.3.2), we can use a Taylor expansion in y along the sheet, which is valid for finite μ and η:

$$\psi(x, y) = \psi_0(x) + y^2\psi_2(x)/2 + ..., \tag{3.120}$$

$$\phi(x, y) = y\phi_1(x) + y^3\phi_3(x)/3! + ... \tag{3.121}$$

Insertion into (3.113) and (3.114) gives a set of differential equations for the coefficients $\psi_n(x), \phi_n(x)$. If $\psi_0(x)$, or the current profile $j_0(x) = \psi_0''(x)$, is given, the remaining functions $\psi_2(x), ..., \phi_1(x), ...$ can be determined successively in terms of ψ_0. Let us assume the profile

$$j_0(x) = j_{max}\text{sech}^2(x/\delta), \tag{3.122}$$

whence integration gives

$$\psi_0(x) = j_{max}\delta^2\ln[\cosh(x/\delta)]. \tag{3.123}$$

Insertion into (3.114) yields

$$\phi_1(x) = (\eta/\delta)\tanh(x/\delta),$$

and the lowest-order velocity components in the sheet become

$$v_x^{(1)} = -u_0\tanh(x/\delta), \quad v_y^{(1)} = u_0(y/\delta)\,\text{sech}^2(x/\delta). \tag{3.124}$$

Here $u_0 = \eta/\delta$ is the upstream velocity, and the downstream velocity $v_0 = B_0 = j_{max}\delta$ defines the length of the sheet Δ using mass conservation, $\Delta = (v_0/u_0)\delta$. The inverse aspect ratio of the sheet $\delta/\Delta = \eta/(j_{max}\delta^2) = S_0^{-1}$ is the smallness parameter in the expansions (3.120), (3.121). The next order terms can be obtained by straightforward but tedious calculation using both (3.113) and (3.114).

In the solution (3.120), (3.121) the zeroth-order current profile $j_0(x)$ is not determined. Less for practical reasons than for gaining more insight into the internal sheet dynamics, it is interesting to know whether the cross-sheet profiles exhibit a certain self-similarity. The exact structure requires the solution of the full 2D system for ψ and ϕ. As an approximation we could try to assume one-dimensionality for $\psi = \psi(x)$. This would, however, imply that the Lorentz force term $\mathbf{B} \cdot \nabla j$ in (3.65) vanishes and hence $w = 0$, which contradicts the fact that a current sheet is also the location of high vorticity. Instead, we may consider (3.66) only along the symmetry axis $y = 0$, where $v_y = 0$, assuming a linear velocity profile $v_x = -ux/L$ with some length-scale L. In this case Ohm's law,

$$E - \frac{u}{L}xB = \eta B', \quad B = \psi'(x),$$

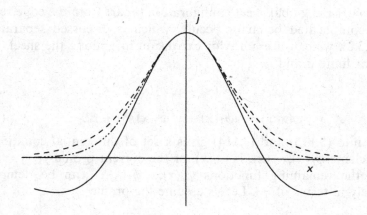

Fig. 3.6.   Theoretical current-sheet profiles: (3.125) (full), $\text{sech}^2 x$ (dashed), Gaussian (dot-dashed).

has the solution (Sonnerup & Priest, 1975):

$$B = \frac{E}{\eta} e^{-x^2/2\delta^2} \int_0^x e^{\xi^2/2\delta^2} d\xi, \quad \delta^2 = EL/u. \tag{3.125}$$

Since here $B \to 0$ for $x^2 \to \infty$, the total current vanishes and $j(x)$ has negative ("return") current parts adjacent to the central positive one, as seen in fig. 3.6. In typical resistive current sheets, however, the net current is finite and the negative contributions are small or absent, hence the solution (3.125) can, at most, model the central positive part of a current sheet. Alternative current profiles are a Gaussian e^{-x^2} or the $\text{sech}^2 x$ profile, (3.93), which are also shown in fig. 3.6. To compare these with the current profiles obtained in numerical simulations, consideration must be restricted to the central part $|x| \lesssim 2\delta$, since only here simulations exhibit a self-similar profile shape. It appears that the numerical results agree best with a Gaussian, though the difference between Gaussian and $\text{sech}^2 x$ is small.[*] More importantly, the profile does not depend on the type of resistive dissipation, a hyperresistivity η_2, obtained by the substitution $\eta \nabla^2 \psi \to -\eta_2 \nabla^4 \psi$, gives the same profile as a normal resistivity η. This shows, that the internal structure of the current sheet is primarily determined by the flow dynamics, and not by the character of the dissipation.

---

[*] This behavior seems to be at variance with fig. 3.4, where the profile (3.92), i.e., $\text{sech}^2 x$, agrees significantly better with the simulation result than a Gaussian. However, fig. 3.4 refers to the self-similar behavior in the ideal phase of current-sheet formation, while the present results deal with stationary resistive sheets. We could have chosen a Gaussian in (3.122), but the $\text{sech}^2 x$ profile is algebraically more convenient.

3.3 Driven current-sheet reconnection

In reconnection theory, one traditionally distinguishes between driven and spontaneous processes, where, however, the definition of these terms varies considerably. Driven reconnection refers to open, externally forced systems, while we call reconnection spontaneous if it arises by some internal instability or loss of equilibrium, the dynamic evolution of which depends only weakly on the external coupling. Hence in the former case the dynamics is determined by the boundary conditions, while it is rather insensitive to their choice in the latter.

The concept of driven reconnection can, however, be applied more generally. In many systems, reconnection occurs in well-defined localized regions of space. If we are primarily interested in the physics of the reconnection process, we may restrict consideration to a small region of linear dimensions L around such a location instead of the global system of scale $\Lambda \gg L$. On the other hand L should not be too small, for instance $L > \Delta$ in the case of a single current sheet.

The main advantage of the restriction to the subsystem L is that it allows us to simplify the geometry. In addition we can assume stationarity, even though the global system Λ is usually not stationary. Since the coupling of the L-system to the Λ-system occurs through the boundary conditions imposed on the L-system, which change on the global time-scale $\sim \Lambda/v_A$, while the subsystem adjusts to these changes on the much shorter time-scale $\sim L/v_A$, we may consider the latter as stationary (if a stationary state exists).

Boundary conditions are as indicated, for example, in fig. 3.2. Fluid and magnetic flux are injected from above and below, while the fluid leaves the system on both sides horizontally carrying along the reconnected field. The small region where resistivity is important is called the *diffusion region*, which is surrounded by the *ideal external region*, where resistive effects can be neglected. What can be learned from considering a driven reconnection configuration? The interpretation of such a model has led to some confusion. Since a fundamental issue of reconnection theory is to account for rapid processes in systems where the resistivity is very small, a figure of merit of a reconnection model is a weak or possibly no dependence of the reconnection rate on the resistivity. In stationary driven reconnection this point needs further specification, since the reconnection rate is determined by the boundary conditions for the inflow velocity and magnetic field intensity and hence is independent of η by definition. For the reconnection process to be independent of η one has to require instead that at fixed boundary conditions *the configuration remains unchanged if η is varied*, at least for sufficiently small values of η. A consequence of such behavior would be that the ratio of outflow energy flux to input energy flux should be independent of η (essentially unity) and so should be the ratio of the energy dissipation rate to the input energy flux (essentially zero).

3.3.1 Syrovatskii's theory of current sheets

Before we deal with the full resistive dynamics, which in general requires a numerical approach, we discuss a particularly simple theory due to Syrovatskii (1971), which deals with the ideal equations but allows sheet-like singularities. Though the theory does not describe real reconnection dynamics, it accounts for many important features thereof.

The basic equations use a vanishing pressure gradient $\nabla p = 0$ and so are somewhat different from those of two-dimensional incompressible MHD, to which the major part of this chapter is confined. Since, however, the dynamics is only computed a posteriori from the change of the magnetic field, the difference is of no importance. The main assumption is that all currents in the system are localized in isolated current sheets. Hence ψ satisfies Laplace's equation

$$\nabla^2 \psi = 0. \tag{3.126}$$

The solution is determined by the boundary conditions. If these vary in time, ψ obtains a parametric time dependence $\psi(x, y, t)$, which then determines the perpendicular component $\mathbf{v}_\perp$ of the velocity

$$\mathbf{v}_\perp = -\partial_t \psi \nabla \psi / |\nabla \psi|^2, \tag{3.127}$$

from the frozen-in condition

$$\partial_t \psi + \mathbf{v} \cdot \nabla \psi = 0,$$

while the parallel component $v_\parallel$ is calculated from the equation

$$d\mathbf{v}/dt \times \nabla \psi = 0, \tag{3.128}$$

which follows from the equation of motion (3.71) for $\nabla p = 0$. (Equation (3.128) implies that the current density and hence the Lorentz force does not vanish identically. Equation (3.126) has to be regarded as an approximation in the sense that the effect of the distributed currents is small compared to that of the sheet currents). Hence the evolution of the system is described by (3.126), i.e., a simple boundary value problem. The crucial point is that regular solutions $\psi(x, y, t)$ do not always exist. If the change of the boundary conditions leads to a change of $\psi, \partial_t \psi \neq 0$, at a neutral point $\nabla \psi = 0$, the frozen-in condition cannot be satisfied at this point, which is therefore called a singular neutral point. In the presence of such points current singularities are expected to appear. To discuss these singularities in more detail it is convenient to introduce the potential $F(z)$ in the complex plane $z = x + iy$,

$$F(z,t) = \psi(x,y,t) + i\chi(x,y,t), \qquad (3.129)$$

which is analytic in the region considered except for isolated singular points and branch cuts. The conjugate harmonic function χ is determined by the Cauchy–Riemann relations

$$\partial_x\chi = -\partial_y\psi, \qquad \partial_y\chi = \partial_x\psi. \qquad (3.130)$$

From (3.129) and the definition $\mathbf{B} = \mathbf{e}_z \times \nabla\psi$ we obtain the magnetic field in the form

$$dF/dz = B_y + iB_x, \qquad (3.131)$$

as can easily be verified by choosing a special direction of the derivative, e.g., dF/dx, since owing to (3.130) the complex derivative is independent of this choice.

If $z = z_0(t)$ is the position of a neutral point, $dF/dz|_{z=z_0} = 0$, the complex potential in the vicinity of this point has the form

$$F(z,t) = \frac{\alpha(t)}{2}[z - z_0(t)]^2 + \beta(t). \qquad (3.132)$$

Here we restrict consideration to the only practically relevant case of second-order or X-type neutral points. (Note that there is no nonsingular O-type neutral point, since this would require a nonvanishing current density $\nabla^2\psi \neq 0$.) If the change of the boundary conditions is such that $\dot{\beta} \neq 0$, the neutral point $z = z_0$ is singular and $F(z,t)$ cannot remain analytic at this point.

The natural conclusion seems to be that the electric field $\partial_t\psi = \dot{\beta}$ induces a line current I corresponding to adding a singular contribution in (3.132),

$$F(z,t) = \frac{\alpha(t)}{2}[z - z_0(t)]^2 + \beta(t) + \frac{I(t)}{2\pi}\ln[z - z_0(t)]. \qquad (3.133)$$

Here $I(t)$ is the total current induced at the neutral point, which is determined by the boundary conditions. It is, however, easy to see that this generates closed ψ lines, i.e., an O-point topology, around the neutral point $z = z_0$ and two additional X-points, as illustrated in fig. 3.7. The change of topology occurs at these secondary X-points, which is not allowed since it would require $\partial_t\psi \neq 0$ at these points, too, and hence additional singular contributions of the form (3.133) and so forth. Thus it is impossible to construct a solution with isolated singular points. The alternative is to introduce a branch cut corresponding to a current

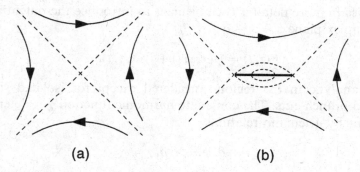

(a)   **(b)**

Fig. 3.7. Generation of a singular current at an X-point. (a) Initial nonsingular configuration; (b) effect of an induced singular current line in the original X-point, leading to a fictitious O-point and two adjacent X-points. The heavy line indicates the current sheet actually arising.

sheet carrying the induced current $I(t)$. The location of this branch cut is determined by the two fictitious X-points arising by adding the line current to the original X-point, (3.133). As indicated in fig. 3.7(b), the cut passes through this point connecting the new X-points, which drift further apart as $I(t)$ increases. Addition of the current sheet splits the original X-point into two Y-points. Let us discuss the structure of the field in the vicinity of the branch cut. At distances large compared with the length of the cut the potential $F(z,t)$ has approximately the form (3.133). The condition at the cut is that **B** does not intersect the cut, i.e., $\psi = const$ at the cut. To simplify notation, assume $z_0 = 0$ and a straight cut extending along the y-axis between the points $y = \pm\Delta$. The solution with the asymptotic form (3.133) for $|z| \gg \Delta$ and $\psi = 0$ at the cut is given by

$$F(z) = \frac{1}{2}\alpha z \sqrt{z^2 + \Delta^2} + \frac{I}{2\pi}\ln\frac{z + \sqrt{z^2 + \Delta^2}}{\Delta} \qquad (3.134)$$

with the derivative

$$\frac{dF}{dz} = B_y + iB_x = \frac{(I/2\pi) + (\alpha\Delta^2/2) + \alpha z^2}{\sqrt{z^2 + \Delta^2}}. \qquad (3.135)$$

While the magnetic potential $\psi(x, y) = Re\{F(z)\}$ is continuous, the magnetic field B_y has a jump across the cut, the line density of the current carried by the sheet,

$$J(y) \equiv B_y(0_+, y) - B_y(0_-, y) = 2\frac{(I/2\pi) + (\alpha\Delta^2/2) - \alpha y^2}{\sqrt{\Delta^2 - y^2}} \qquad (3.136)$$

with

$$\int_{-\Delta}^{\Delta} J(y)\,dy = I. \qquad (3.137)$$

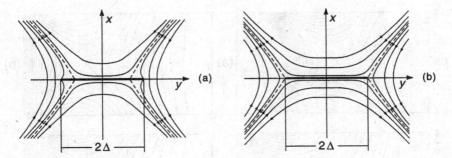

Fig. 3.8. Contours of $\psi = \text{Re}\{F\}$, where F is given by (3.134); heavy line is the current sheet, dashed line is the separatrix. (a) General case $y_0 < \Delta$ exhibiting a singularity at the current-sheet endpoints; (b) limiting regular case $y_0 = \Delta$ (from Syrovatskii, 1971).

The current distribution, (3.136), shows an interesting feature. It is positive, i.e., in the direction of the total current I, in the center part $|y| < y_0$, and negative in the outer parts $|y| > y_0$, where $y_0^2 = (I/2\pi\alpha) + \Delta^2/2$. The points $z_\pm = (0, \pm y_0)$ are neutral points of magnetic field, $dF/dz = 0$, where the separatrix branches off the y-axis, while the current sheet continues along the y-axis, fig. 3.8(a). At the end points $|y| = \Delta$ the current density $J(y)$ becomes singular giving rise to infinite magnetic field intensity. Only in the special case $I = \pi\alpha\Delta^2$, where the neutral points coincide with the current sheet endpoints, the singularity vanishes, fig. 3.8(b). This value is in fact an upper limit of the current I (or a lower limit of the sheet width Δ), since for larger values the points $z_\pm$ would form isolated neutral points connected by a separatrix similar to the dashed line drawn in fig. 3.7(b), which is topologically not possible because of the frozen-in property. The characteristic features of the Syrovatskii current sheet, (3.136), agree well with those of dynamic current sheets in a fully resistive theory, as discussed in section 3.3.2.

The velocity field $\mathbf{v}$ corresponding to the magnetic configuration (3.135), (3.127), (3.128), cannot be given in simple analytical form, but must be computed numerically, even in the stationary case $\partial_t\psi = const$. The qualitative behavior close to the current sheet is, however, easily understood. It follows from (3.127) that there is a plasma flux into the sheet, $v_x(0_+, y) = -v_x(0_-, y) \neq 0$. Mass conservation then requires the plasma to flow along the sheet, leaving the sheet at high speed close to the neutral points $z_\pm$, while the plasma flow vanishes at the singular sheet endpoints. The downstream velocity in the cone formed by the two branches of the separatrix is of the order of the upstream flow and is not related to the cone angle.

The technique of conformal mapping used by Syrovatskii is restricted

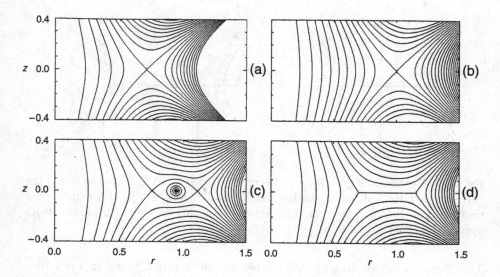

Fig. 3.9.   Construction of a current sheet in an axisymmetric configuration. (a) Initial flux distribution (3.138); (b) modified flux distribution resulting from a change of the parameters ψ_0 and a. Since this increases the flux inside the separatrix, a current must be induced at the X-point to undo this change. This is achieved in (c) by adding a single current ring, while the superposition of 16 current rings in (d) already gives a good approximation to a continuous current sheet (from Longcope & Cowley, 1996).

to 2D plane geometry. In axisymmetry for instance, $\partial_\phi = 0$ in cylindrical coordinates r, z, ϕ, an X-line is a closed line with coordinates $r = r_0, z = z_0$. If an electric field E_ϕ acts along this line, a current sheet will be generated, say in the plane $z = z_0$ (the orientation of the sheet depends on the boundary conditions), which extends between the closed lines through the points r', z_0 and r'', z_0 forming a toroidal ribbon. Longcope & Cowley (1996) devised an algorithm to compute the current density in the sheet, approximating the latter by a number n of linear ring currents, which produce a chain of magnetic islands, the width w_I of which shrinks to zero rapidly, $w_I \propto n^{-2}$ as $n \to \infty$. As a particular example the authors consider the purely poloidal field $\nabla\phi \times \nabla\psi$,

$$\psi = \frac{\psi_0}{2\pi a^4}\left[4r^2z^2 - (r^2 - 2a^2)r^2\right]. \tag{3.138}$$

Changing the parameters ψ_0 and a corresponds to a change of the external field configuration, which gives rise to change of ψ at the X-line and hence to a current sheet, an example being given in fig. 3.9.

Let us briefly discuss the case of a fully three-dimensional magnetic field. Imagine a configuration characterized by two magnetic nulls, one

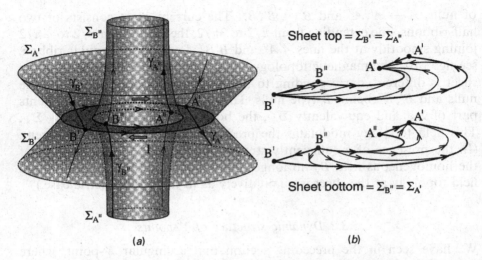

(a) (b)

Fig. 3.10. (a) Schematic plot of the field, (3.139), including the current ribbon, inside which the current flows in a clockwise direction indicated by the wide gray arrows. (b) The magnetic field lines at the top and the bottom of the current sheet, shown separately (only half of the ribbon is plotted). The current in the sheet arises from the difference in angle between the field lines in the two surfaces (from Longcope & Cowley, 1996).

A-type and one B-type (see fig. 2.5). If there is a nonvanishing electric field E along the null–null line AB, a current sheet is expected to form. It appears all we have to do is to split the null–null line thus creating pairs of nulls A', A'' and B', B'', which now span a ribbon of finite length, consisting of pieces of Σ_A and Σ_B glued together. However, because of the boundaries of the sheet, it cannot carry a net current in the direction of the applied field E, while in reality there will be such a current, which continues to flow in a diffuse way outside the sheet, a feature not allowed in Syrovatskii's theory. Hence it does not seem possible in general to extend the theory to three-dimensional configurations.

Longcope & Cowley considered the special case of a *closed* sheet forming an annular ribbon, generalizing the axisymmetric case, (3.138), by superimposing a weak homogeneous field in the y-direction,

$$\mathbf{B} = \nabla\phi \times \nabla\psi + b\mathbf{e}_y. \tag{3.139}$$

This field has two nulls A, B located at

$$r_{A,B} = (1 + \epsilon^2/8)a, \quad \phi_{A,B} = \pm\pi/2, \quad z_{A,B} = \mp\epsilon a/2,$$

where $\epsilon = \pi a^2 b/\psi_0$. Now, a current sheet may form in a way similar to the axisymmetric case, leading to a splitting of the nulls into pairs of nulls, $A \to A', A''$ and $B \to B', B''$. The current sheet consists of two half-ribbons, one extending from $\pi/2$ to $3\pi/2$, the other from $\pi/2$ to $-\pi/2$ joining smoothly at the lines $A'A''$ and $B'B''$, forming an annular ribbon, see fig. 3.10. The magnetic topology on the top and on the bottom of the sheet is different, corresponding to the properties of A', A'' being A-type nulls and B', B'' being B-type nulls. Thus the top of the sheet represents part of $\Sigma_{B'}$ and equivalently $\Sigma_{A''}$, the bottom $\Sigma_{B''}$ and equivalently $\Sigma_{A'}$. This field topology modulates the predominantly toroidal sheet current. (In the presence of a sufficiently strong toroidal field B_ϕ the addition of the homogeneous field B_y no longer gives rise to magnetic nulls and the field topology at the sheet is qualitatively as in the axisymmetric case.)

3.3.2 Dynamic structure of Y-points

We have seen in the preceding section that a singular X-point, where the induced electric field $\partial_t \psi \neq 0$, is transformed into a current sheet. Topologically, the separatrix is pulled apart, forming two Y-type neutral points connected by the sheet. Syrovatskii's theory, however, shows that only in a special case are these regular Y-points coinciding with the endpoints of the current sheet. In general, the sheet continues beyond these neutral points with 'reversed current direction, extending up to the singular endpoints, where the magnetic field becomes infinite. It is interesting to compare these predictions with the structure of the edge regions in a dynamic resistive current sheet. Since the Taylor expansion along the sheet, (3.120) and (3.121), breaks down at the sheet edge $y \sim \Delta$, the only reliable information is obtained from numerical simulations.

Previous analytical treatments of the diffusion region (see, e.g., Vasyliunas, 1975) assume a smooth transition to the ideal external region in the downstream cone, the plasma continuing to flow at the upstream Alfvén velocity, to which it has been accelerated along the sheet. Such highly super-Alfvénic flow (the local Alfvén velocity in the downstream cone is much smaller than that in the upstream region) should be prone to shock formation, which would increase the field intensity and slow down the plasma to sub-Alfvénic speed. In fact, simulations show that the flow in the downstream region is clearly sub-Alfvénic and apparently unrelated to the high flow speed inside the diffusion region.

Closer inspection of the edge of the diffusion region reveals a complicated structure, as is illustrated in figs. 3.11 and 3.12. Figure 3.11 contains stereographic plots of the current distribution viewed from (a) the upstream side and (b) the downstream side, which show the main features of the configuration, the diffusion layer represented by the central current sheet along the y-axis and a weaker sheet current along the separatrix both joining in a region of rather complex behavior. As seen in fig. 3.11(b),

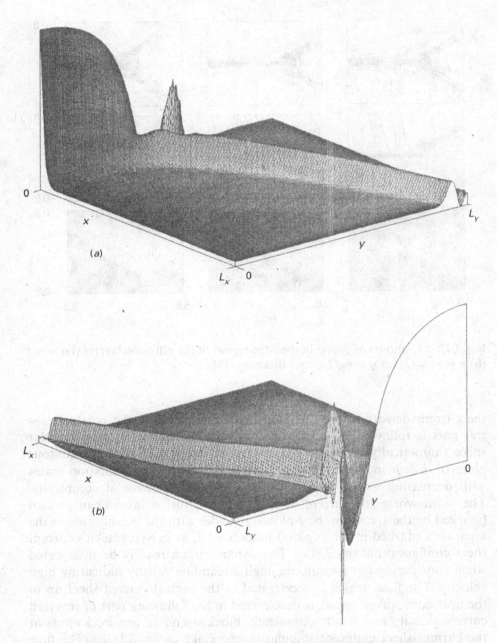

Fig. 3.11. Stereographic plots of the current distribution of a stationary reconnection configuration, viewed from (a) the upstream side, and (b) the downstream side (from Biskamp, 1986).

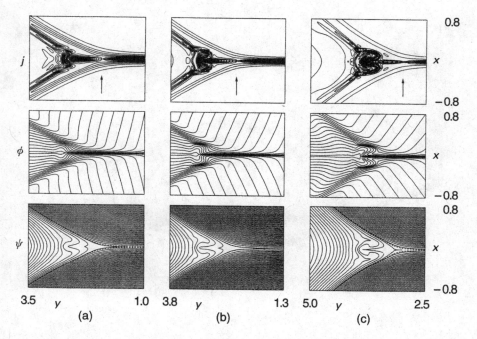

Fig. 3.12.   Contours of j, ϕ, ψ in the edge region of the diffusion layer. (a) $\eta = \eta_0$; (b) $\eta = \eta_0/\sqrt{2}$; (c) $\eta = \eta_0/2$ (from Biskamp, 1986).

the current density in the diffusion layer changes sign, the positive central part is followed by a negative part, which ends in a quasi-singular spike (numerically well resolved, though). Figure 3.12 contains contour plots of j, ϕ, ψ in the edge region of three stationary simulation states with decreasing values of η, showing the rapid increase of complexity. The point where the current density of the diffusion layer changes sign (marked by the arrows in the j-plots) coincides with the location where the separatrix (dashed in the ψ-plots) branches off, as in Syrovatskii's current sheet configuration, fig. 3.8(a). The dynamics can readily be interpreted when considering the ϕ-contours, high streamline density indicating high velocity. The flow, which is accelerated in the central current sheet up to the upstream Alfvén speed, is decelerated in the following part of reversed current density and finally completely blocked and turned backwards at the current-sheet endpoint singularity, the spike in fig. 3.11(b). The flow is subsequently again accelerated, forming two secondary current sheets adjacent to the primary one, with the same characteristics, $j > 0$ and $j < 0$ parts and a flow-blocking shock-like structure. In the limit $\eta \to 0$ a hierarchy of higher-order current sheets is generated with a self-similar scaling behavior, as drawn schematically in fig. 3.13.

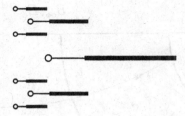

Fig. 3.13. Schematic drawing of the self-similar hierarchy of current sheets in the diffusion layer: thick line = positive j; thin line = negative j; open circle = quasi-singularity at sheet endpoint.

These properties of the diffusion layer are consistent with Syrovatskii's current-sheet model, the multi-current-sheet edge behavior representing the dynamically resolved singularity predicted in Syrovatskii's quasi-static theory. The picture indicated in fig. 3.13, however, rests on two assumptions, *viz.*, symmetry and stationarity. As will be discussed in section 4.7, a dynamic current sheet, though more stable than a static one, becomes unstable if the aspect ratio $A = \Delta/\delta$ is sufficiently large, which gives rise to nonstationary behavior. In a less symmetric configuration such non-stationary behavior will be very complex and finally lead to fully developed turbulence consisting of a statistical distribution of microcurrent sheets, see section 5.4.

3.3.3 Scaling laws of stationary current-sheet reconnection

In Syrovatskii's theory, discussed in section 3.3.1, the system is completely determined by the boundary conditions on ψ. The corresponding plasma flows are passive, i.e., they do not affect the magnetic configuration, and the physical reconnection mechanism, in particular the value of the resistivity, does not enter the theory. By contrast, numerical solutions of the resistive MHD equations exhibit a strong influence of the magnitude of η on the entire system. Hence, to determine the dependence of reconnection on η the dynamic equations must be solved. More specifically we consider an open system, where reconnection is driven by imposing suitable inflow and outflow boundary conditions. Unfortunately the problem appears to be too complicated to permit an analytical treatment without severe approximations. In particular, the usual boundary layer approach of matching the solution of the resistive region to that of the ideal external region is practically impossible, since the complicated shape of the resistive layer (section 3.3.2) makes the problem non-separable, i.e., fully two-dimensional.

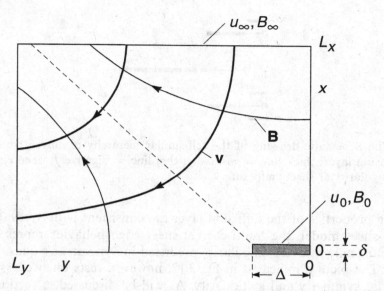

Fig. 3.14. Computational system as used in numerical simulations of driven reconnection.

Hence the problem has to be treated numerically. The simplest and also most reliable method is that of dynamical simulation, i.e., to follow the system from some initial state with stationary boundary conditions until a steady configuration is reached, which automatically eliminates unstable solutions. Accurate numerical solutions in the relevant parameter regime are state of the art, and since there are only two essential parameters their scaling laws can be determined from a limited number of computer runs. (In practice, we do not have to restart the system from the initial state for each parameter value. Instead, the run is simply continued, changing, for instance, the resistivity in not too large steps, which requires only a short time for the system to relax to the new steady state, provided such state exists.)

Since primary interest is in understanding the qualitative behavior, we can choose the simplest possible geometry with up–down and right–left symmetry, such that only one quadrant must actually be computed. The computational system is indicated in fig. 3.14. The basic equations to be solved are the 2D incompressible MHD equations (3.65) and (3.66). Boundary conditions have to be assigned to ψ, ϕ, j and w. While at the internal boundaries, the x-axis and the y-axis, these conditions follow from the imposed symmetry, ψ and j being symmetric, ϕ and w antisymmetric, at the upper ($x = L_x$) and the left-hand ($y = L_y$) boundaries they correspond to the inflow and outflow conditions, respectively. These should be chosen in a way conforming with the concept of an open system, such that boundaries, in particular the outflow boundary, do not obstruct the flow. Open boundaries are well defined for linear waves, requiring that waves are not reflected at the boundary, or, in mathematical terms, that for a hyperbolic system of differential equations all characteristics are outgoing at the boundary, which guarantees that perturbations arising due to the presence of the boundary do not propagate into the system. However, (3.65) and (3.66) of incompressible MHD are of mixed hyperbolic–elliptic type. Moreover, the equations are nonlinear and two-dimensional, and to date there is no rigorous method to determine whether a particular set of boundary conditions is admissible. (A discussion of the boundary conditions for different MHD systems has been given by Forbes & Priest, 1987.)

Let us therefore apply a more practical procedure. While it seems to be sufficient to require continuity of w and j at the boundaries in (3.65), $\partial_n w = \partial_n j = 0$, the system is more sensitive to the boundary conditions for the potentials ψ in (3.66) and ϕ in Poisson's equation $\nabla^2 \phi = w$, since an inappropriate choice may lead to singularities in w and j, which would show up in the form of slow-mode shocks in the vicinity of the boundaries. Any choice of the boundary conditions that does not give rise to such effects should be regarded as acceptable, and numerical experience indicates that there is considerable freedom in this choice (Biskamp, 1986). One may, for instance, specify $\phi(y)$ and $\partial_x \psi$, i.e., $v_x(y)$ and $B_y(y)$ at the ingoing boundary $x = L_x$, and $\phi(x)$, i.e., $v_y(x)$ at the outgoing boundary $y = L_y$, and iterate $\partial_y \psi$, i.e., $B_x(x)$ at the latter, until the outflow configuration becomes smooth.

In order to describe the dependence of the configuration on the inflow conditions it is convenient to parameterize the inflow boundary functions, writing $v_x(y) = u_\infty f(y), B_y(y) = B_\infty g(y)$. Then a stationary configuration is characterized by the inflow parameters u_∞, B_∞ or simply $u_\infty = E$ using the normalization $B_\infty = 1$, and by the parameters u_0, B_0, Δ, δ describing the diffusion layer, the internal current sheet, as illustrated in fig. 3.14. While E is prescribed, the current-sheet parameters depend on the internal dynamics and are functions of η, in particular. A homogeneous resistivity distribution is chosen to avoid additional complications arising from resistivity gradient effects. Reconnection is said to be independent of η if, for fixed boundary conditions, the configuration does not depend on η (for sufficiently small η).

First consider the case of weak driving, $E = \partial_t \psi \lesssim \eta$. We obtain an approximate stationary solution by expanding (3.65) and (3.66) in E. To lowest order the inertia term is negligible,

$$\mathbf{B} \cdot \nabla j = 0, \qquad (3.140)$$

$$\mathbf{v} \cdot \nabla \psi = -E + \eta j. \qquad (3.141)$$

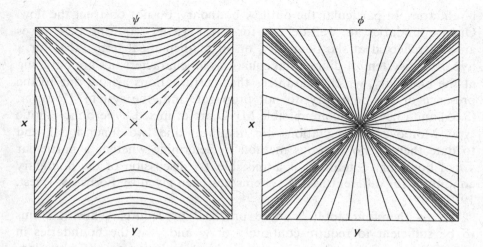

Fig. 3.15.   Contours of (a) ψ and (b) ϕ of the similarity solution (3.142), (3.143) (for clarity, the singularity of ϕ has been smoothed by $(x\pm y)^2 \to (x\pm y)^2+\delta^2, \delta = 0.1$).

If j vanishes for $|\mathbf{x}| \gg 1$, (3.140) implies $j = 0$ everywhere in a configuration with open field lines. In this case there is a simple similarity solution of (3.140), (3.141):

$$\psi = \tfrac{1}{2}(x^2 - y^2), \qquad (3.142)$$

$$\phi = \tfrac{1}{2}E \ln \left| \frac{x+y}{x-y} \right|, \qquad (3.143)$$

illustrated in fig. 3.15. The solution is, however, not valid on the separatrix $x = \pm y$, where ϕ and, a forteriori, w are singular. Hence in the vicinity of the X-point, where the velocity becomes large, the inertia term and the current density cannot be neglected. Let us estimate the distance from the X-point where the similarity solution, (3.142) and (3.143), breaks down. From the vorticity equation $\mathbf{v} \cdot \nabla w = \mathbf{B} \cdot \nabla j$, we obtain

$$j \sim \frac{E^2}{x^4}, \qquad (3.144)$$

using $v \sim E/x$, $w \sim E/x^2$, $B \sim x$. On the other hand, the current density is controlled by the resistivity close to the X-point

$$j \sim \frac{E}{\eta}. \qquad (3.145)$$

Combining both relations gives the distance δ, where resistivity and inertia become important,

$$\delta \sim (E\eta)^{1/4}. \qquad (3.146)$$

For sufficiently small E the presence of this diffusion region does not affect the configuration, (3.142) and (3.143), outside this region. This is true, if the local Lundquist number S_0, (3.107), is $S_0 \lesssim 1$, where inertia effects remain small and, according to (3.108), no current sheet is formed. Hence we have $S_0 = B\delta/\eta \sim \delta^2/\eta \sim (E/\eta)^{1/2}$ upon substitution of (3.146). $S_0 \lesssim 1$ implies

$$E \lesssim \eta. \tag{3.147}$$

At higher $E \gg \eta$ we have $S_0 \gg 1$, which means that inertia effects now give rise to a current sheet of length Δ, which modifies the configuration on a scale $\Delta \gg \delta$. The properties of this current sheet B_0, u_0, δ, Δ indicated in fig. 3.14, are determined by the asymptotic conditions on the configuration, in particular the imposed reconnection rate E, and by the resistivity. We can rewrite (3.107) and (3.108) such as to express B_0, u_0, δ in terms of E, η and the sheet length Δ,

$$B_0 \sim (E^2 \Delta/\eta)^{1/3}, \tag{3.148}$$

$$u_0 \sim (E\eta/\Delta)^{1/3}, \tag{3.149}$$

$$\delta \sim (\eta^2 \Delta/E)^{1/3}. \tag{3.150}$$

The length Δ, which depends on the specific dynamics at the sheet, cannot be determined in a simple way. A series of simulation runs has been performed (Biskamp, 1986) for identical boundary profile functions varying only the parameters $E(= u_\infty)$ and η, which yield the scaling relation

$$\Delta \sim E^4/\eta^2. \tag{3.151}$$

Inserted into relations (3.148)–(3.150) we obtain the current-sheet scaling laws in a stationary driven reconnection process in terms of the two free parameters E, η:

$$B_0 \sim E^2/\eta, \tag{3.152}$$

$$u_0 \sim \eta/E, \tag{3.153}$$

$$\delta \sim E. \tag{3.154}$$

Hence, increasing the driving E increases the field B_0 in front of the sheet (flux pile-up) and correspondingly lowers the velocity u_0 because of $u_0 B_0 = E$. The most conspicuous feature is the increase of the sheet length Δ, when η is decreased. This is illustrated in fig. 3.16, which gives the flow patterns $\phi(x, y)$ and the magnetic configurations $\psi(x, y)$ for three cases differing only in the value of η. The figure clearly shows the increase of Δ, the flux pile-up, and the slowing down of the inflow velocity.

The scaling laws (3.151)–(3.154) can be related to the physics in the current sheet, which is dominated by inertia, i.e., the acceleration along

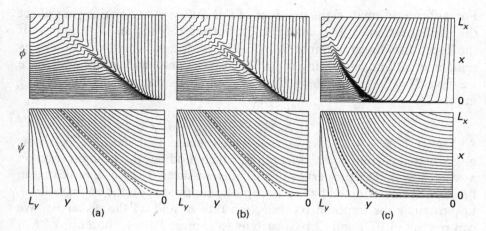

Fig. 3.16. Contours of ϕ and ψ for three stationary configurations of current-sheet reconnection. (a) $\eta = \eta_0$; (b) $\eta = \eta_0/2$; (c) $\eta = \eta_0/4$ (from Biskamp, 1986).

the sheet. As shown in section 3.2.3 the velocity v_y increases linearly, $v_y \simeq B_0 y/\Delta$, such that the average force along the sheet is

$$\overline{v_y \partial_y v_y} \simeq B_0^2/2\Delta \simeq const \tag{3.155}$$

using (3.151) and (3.152). Hence the scaling relations imply that the force along the sheet is independent of E and η, in particular remains finite for $\eta \to 0$.

In Syrovatskii's theory, the current-sheet length Δ is independent of the reconnection dynamics inside the sheet, in contrast to the result, (3.151), obtained for a fully dynamic system, which is due to the fact that in Syrovatskii's approach the reaction of the flow dynamics on the magnetic field is ignored, the flow being passive, determined only a posteriori.

The question of a finite reconnection rate in the limit $\eta \to 0$ has also been considered by Grad *et al.* (1975). Instead of the full MHD equations these authors treat a system of equations suitable for slow diffusive processes, in which plasma inertia is omitted, which reduces the dynamics to a sequence of MHD equilibria changing quasi-statically because of changes in the boundary conditions and magnetic diffusion. By solving these equations numerically it is found that reconnection always occurs at a finite-angle X-point and that the reconnection rate seems to remain finite for $\eta \to 0$. No indication of current-sheet formation is observed. The crucial objection, however, concerns the omission of the inertia term, which is only justified if fluid velocities remain much smaller than the Alfvén velocity. As we have seen, this is not true for the flow along the current sheet, which limits the validity of Grad's reduced equations to reconnection rates $E \lesssim \eta$ and thus invalidates the claim of the existence of η-independent fast reconnection.

Let us also discuss the scaling of the energy dissipation rate ϵ, which for $\mu \lesssim \eta$ is mainly due to Ohmic dissipation $\epsilon \simeq \epsilon_\eta = \int \eta j^2 dV$, by comparing ϵ_η with the input power ϵ_{in}, the flux of magnetic energy across the inflow system boundary, $\epsilon_{in} \simeq \frac{1}{2} \int uB^2 dF$. Using the scaling relations (3.151)–(3.154) we obtain

$$\frac{\epsilon_\eta}{\epsilon_{in}} \sim \eta \frac{B_0^2}{\delta^2} \Delta \delta \bigg/ u_\infty B_\infty^2 L_y \sim E^6/\eta^3 \qquad (3.156)$$

assuming that the dissipation is concentrated in the current sheet. ϵ_η equals the magnetic energy flux into the sheet, $\epsilon_\eta \simeq u_0 B_0^2 \Delta$. Relation (3.156) indicates that the fraction of the input power which is dissipated strongly increases with decreasing η.

Relation (3.151) is only valid if the current sheet is shorter than the macroscopic scale length L, which in our driven reconnection configuration is just the system size. For $\Delta \sim L$ (3.151) gives the well-known relation for the reconnection rate of a macroscopic Sweet–Parker sheet,

$$E \sim \eta^{1/2}. \qquad (3.157)$$

By inserting $\Delta = L$ in (3.148)–(3.150) we obtain the scaling relation for reconnection in the presence of a macroscopic current sheet. Examples of such processes are the nonlinear evolution of the resistive kink mode and the coalescence of two flux bundles, which are treated in sections 4.3 and 4.4, respectively.

As a consequence of the current-sheet scaling, the value of the flux ψ of a fluid element is conserved in reconnection. Consider the change $\delta\psi$ during passage across the current sheet:

$$\delta\psi \sim \eta j \delta t \sim EL/v_A \sim O(\eta^{1/2}). \qquad (3.158)$$

Since the change of ψ outside the diffusion layer is, at most, of the same order, ψ is conserved along the fluid orbit in the limit of small η. Using this property one can calculate the configuration resulting from reconnection without knowledge of the dynamics in the diffusion layer. An example, the evolution of the resistive kink mode, is considered in section 4.3.3.

3.3.4 Petschek's slow-shock model

Reconnection via macroscopic current sheets, (3.157), is obviously too slow to explain typical time-scales of eruptive processes both in laboratory and especially in space, where $\eta = S^{-1}$, the inverse Lundquist number, is truly small. Therefore Petschek's idea, which was presented at a symposium on solar flares (Petschek, 1964), soon received general acceptance as a major breakthrough in the theory of reconnection, serving as the basic model for the following two decades. However, it later became evident that the model does not stand up to its claim as a fundamental theory explaining fast reconnection in the limit of small resistivity, where the configuration is expected to develop a macroscopic current sheet, as we have shown in the preceding sections. But Petschek's configuration could be set up, if the dominant reconnection mechanism is more efficient than resistivity. Such processes will be discussed in chapter 6, where we also give a general criterion to decide about the reconnection efficiency of the different effects in Ohm's law, section 6.4.4. Here, we only briefly outline Petschek's model *assuming* the presence of a fast reconnection process without further specification.

The basic current-sheet relations (3.107) or (3.149) indicate that the inflow velocity u_0 and the reconnection rate are the higher the shorter the sheet length Δ. The fundamental ingredient of Petschek's theory is the assumption that the size of the diffusion region remains microscopic $\Delta \sim \eta$, i.e., considered on macroscopic scales one is dealing with an X-point configuration. The driving mechanism of the reconnection is the relaxation of the tension of the reconnected field lines, which like a slingshot pushes the plasma along the downstream cone away from the X-point. In response, the upstream plasma is sucked in toward the X-point together with the upstream field. Since this field is expanding in the upstream flow toward the X-point, the velocity increases because of $uB = const$, which corresponds to converging streamlines. In the downstream cone the field intensity increases away from the X-point and the velocity decreases – the streamlines are diverging. Hence there is no continuous transition of the velocity between these regions. The situation is qualitatively similar to the solution (3.143), fig. 3.15(b). The essential new element in Petschek's theory is the introduction of a shock mediating the upstream–downstream transition. There are two pairs of shocks standing back to back against the upstream flow, as shown in fig. 3.17. Since the flow is sub-Alfvénic, these shocks have the characteristic properties of the slow mode (section 3.1.2).

The slow mode is essentially (for perpendicular propagation exactly) a longitudinal mode, but since pressure and field variations are opposite in phase, $\delta p \simeq -\delta(B^2/2)$, the mode survives in the incompressible limit with finite phase velocity, $\omega^2 = k_\parallel^2 v_A^2 = k^2 B_n^2$, $B_n =$ component normal to the wave front, as is seen from (3.25) in the most interesting nearly perpendicular case $k_\parallel \ll k$. Hence by adjusting its inclination with respect to the field, the shock can be stationary against an arbitrarily small flow velocity.

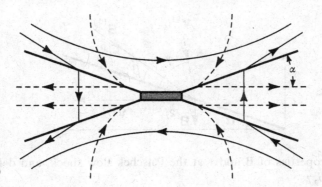

Fig. 3.17.   Schematic drawing of Petschek's reconnection configuration.

In the incompressible limit, the jump conditions at the shock can easily be derived. Equations $\nabla \cdot \mathbf{v} = \nabla \cdot \mathbf{B} = 0$ give the continuity of the normal components

$$[v_n] = [B_n] = 0, \qquad (3.159)$$

where the brackets indicate the change across the shock. From the normal component of the equation of motion (3.1) it follows that the continuity of the total pressure is

$$[p + B^2/2] = 0, \qquad (3.160)$$

while the tangential component of (3.1) gives

$$v_n[v_t] = B_n[B_t]. \qquad (3.161)$$

Ohm's law for steady state implies the continuity of $(\mathbf{v} \times \mathbf{B})_z$,

$$[v_n B_t - v_t B_n] = 0. \qquad (3.162)$$

Combining the last two relations we finally obtain

$$v_n^2 = B_n^2. \qquad (3.163)$$

Consider, for instance, the shock configuration in fig. 3.18 where, for given upstream velocity $\mathbf{v}^{(1)}$ and field $\mathbf{B}^{(1)}$, downstream velocity and field are as indicated in the figure, $v_x^{(2)} = B_y^{(2)} = 0$. This determines the shock angle α, which obeys the relation

$$\sin 2\alpha \sin(\gamma - \beta) = 2 \sin\beta \sin\gamma, \qquad (3.164)$$

as can be verified by some straightforward algebra. For small Mach number $M_0 = v^{(1)}/B^{(1)} \ll 1$ we have $\beta \ll \gamma$, the downstream cone is narrow, $\alpha \ll 1$. In this case we find $\alpha = \beta$ and $v^{(2)} = B^{(1)}$, i.e., the

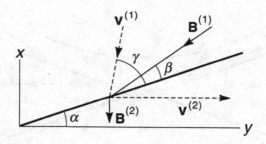

Fig. 3.18. Properties of **B** and **v** at the Petschek slow shock, and definitions of the angles α, β, γ.

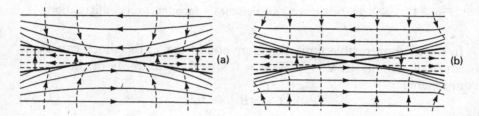

Fig. 3.19. Petschek-type configurations: (a) Petschek's original configuration; (b) Sonnerup-type configuration.

outflowing plasma moves at the upstream Alfvén velocity generalizing the result (3.106) for a quasi one-dimensional sheet configuration.

Petschek also assumed that current and vorticity are concentrated in the shocks while, in the external region, **B** and **v** are irrotational, $\nabla \times \mathbf{B} = \nabla \times \mathbf{v} = 0$, which together with $\nabla \cdot \mathbf{B} = \nabla \cdot \mathbf{v} = 0$ results in $\nabla^2 \psi = \nabla^2 \phi = 0$.[*] Solving these equations, subject to boundary conditions and the jump conditions at the shock, determines the configuration, an example being given in fig. 3.19(a). Petschek derives an approximate analytical solution for $\alpha \ll 1$, and estimates the maximum achievable Mach number $M_{\max} \sim (\ln S)^{-1}$ for a downstream cone $\alpha \sim 1$. (The weak S- or η-dependence results from the fact that M refers to the asymptotic field, assumed homogeneous, instead of the slightly reduced field in front of the diffusion layer.) A detailed discussion of Petschek's theory has been given by Vasyliunas (1975).

More general solutions can be obtained by relaxing the conditions of irrotational field and velocity, which allows, for instance, uniform upstream states as illustrated in fig. 3.19(b). Special similarity solutions of this type have been presented by Sonnerup (1970) and Yeh & Axford

---

[*] This is similar to the assumptions in Syrovatskii's theory (section 3.3.1), though the velocity equation is different.

(1970) consisting of three regions (per quadrant) of uniform velocity and field separated by two singular lines, the Petschek slow-mode shock confining the downstream cone and a second line inside the upstream region, where current and vorticity in this region are concentrated. A wider class of analytical solutions containing Petschek's and Sonnerup's solutions as special cases has been derived by Priest & Forbes (1986).

Petschek's solution (and its variants just mentioned) can be regarded as a generalization of the similarity solution (3.142) and (3.143), replacing the flow singularity on the separatrix by a shock located slightly on the downstream side thereof. Common to all these solutions, however, is the fact that the singularity at the X-point, or rather the diffusion region at finite resistivity, is ignored. It is true that Petschek's theory, in particular as discussed by Vasyliunas (1975), also considers the diffusion region. But, in view of the actual complex structure of this region, this treatment can only be regarded as a simple interpolation between the origin and the external ideal region.

Petschek's model is based on the gas dynamic analog of two supersonic jets hitting head on and being deflected sideways by a pair of shocks. The difference arises in the physics at the X-point. While this is an ordinary stagnation point in a neutral fluid, it is the location of intense dissipation in a reconnecting magnetized plasma. In the latter case the flow is supersonic only with respect to the slow mode but subsonic with respect to the magnetosonic mode, the phase velocity of which is, in fact, infinite in the incompressible approximation. If magnetic diffusion is reduced by decreasing η, the field is compressed in front of the diffusion layer, which process is communicated upstream, modifying the entire upstream configuration, in contrast to a supersonic gas stream, where no signal can propagate upstream.

A consistent theory requires the treatment of the boundary layer problem by matching the inner resistive solution to the external ideal one. Because of the complicated structure of the diffusion region, which makes the system non-separable, the problem must be treated numerically. As simulation studies have shown (section 3.3.3), the size of the diffusion layer, far from shrinking with η as assumed in Petschek's theory, stretches rapidly, (3.151), affecting the entire configuration.

The fact that numerical simulations of driven reconnection do not produce a Petschek-like configuration for small η, is due to the fundamental physics of a current sheet and not to an inappropriate choice of the boundary conditions, as has occasionally been claimed. The freedom in choosing open boundary conditions has been discussed in section 3.3.3. An even more convincing argument against this objection is provided by the various simulations of self-consistent closed systems (such as the nonlinear kink mode, see section 4.3), where no internal boundary conditions are imposed and which all develop macroscopic current sheets at small η.

Petschek's solution of the ideal external region is correct and stable, but it does not match to the diffusion region for small η. The simplest way to eliminate this problem is to assume a locally enhanced resistivity in the vicinity of the X-point ("anomalous resistivity"), such that the effective Lundquist number becomes $Lv_A/\eta_{eff} \sim 1$. In this case a Petschek–like configuration is found in numerical simulations (Sato & Hayashi, 1979; Sato, 1979; see also work by e.g., Ugai, 1995). These studies illustrate the basic merit of Petschek's model as a powerful energy converter, transforming magnetic energy into plasma flow energy. The high-speed flow is in fact a characteristic feature of fast reconnection, which is used in observations to identify the presence of such processes, for instance in the solar corona (Shibata et al., 1996), at the Earth's magnetopause (Paschmann et al., 1979), or in laboratory experiments (e.g., Ono et al., 1996). While these observations are claimed to be evidence of a Petschek-type behavior, we should keep in mind, that such a process requires the presence of an efficient reconnection mechanism.[*] In this sense Petschek does not give a self-consistent theory of magnetic reconnection, but only a phenomenological model.

---

[*] A Petschek-like configuration is, for instance, found in high-β noncollisional reconnection, see section 6.2.5.

4

Resistive instabilities

Many dynamical processes observed in fluids can be attributed to the effect of some instability of a stationary system. Instabilities may give rise to rather coherent dynamics such as the kinking of a plasma column, but more often, especially at high Reynolds number, turbulence will be generated, for instance in the case of shear-flow instability. In many systems several instability mechanisms may become active, e.g., the kink instability leading to the formation of a current sheet which subsequently breaks up due to the tearing instability, or an unstably stratified system, where the Rayleigh–Taylor instability gives rise to convection which then becomes turbulent due to the Kelvin–Helmholtz instability of the sheared flows associated with the convective eddies. This chapter gives an introduction to the resistive MHD instabilities which play a role in magnetic reconnection processes (microscopic or kinetic instabilities are treated in chapter 7). Besides the linear theory, we also discuss the characteristics of nonlinear evolution.

A useful classification of MHD instabilities can be obtained from the reduced momentum equation (3.59)

$$\rho(\partial_t w + \mathbf{v} \cdot \nabla w) = \frac{1}{c} \mathbf{B} \cdot \nabla j + 2(\mathbf{b} \times \boldsymbol{\kappa}) \cdot \nabla p, \qquad (4.1)$$

by considering small perturbations of the equilibrium. The magnetic term has two contributions,

$$\delta(\mathbf{B} \cdot \nabla j) = \mathbf{B}_0 \cdot \nabla \delta j + \delta \mathbf{B} \cdot \nabla j_0. \qquad (4.2)$$

The first term on the r.h.s. is associated with the Alfvén wave, (3.22). It involves a bending of field lines and is hence stabilizing, which is the origin of the damping effect caused by magnetic shear. The second term on the r.h.s. in (4.2) represents the free energy associated with the parallel

81

current density j_0, which drives large-scale magnetic instabilities, the most important resistive modes being the tearing mode (sections 4.1 and 4.2), and the resistive kink mode (section 4.3). The coalescence of magnetic islands or flux bundles is also caused by the parallel current distribution (section 4.4).

The second term on the r.h.s. of (4.1) gives rise to pressure-driven modes, which tend to be localized around a field line in order to minimize the magnetic shear damping. They correspond essentially to an interchange of flux tubes and are therefore called (resistive) interchanges or, if more strongly localized *along* the field line, ballooning modes (section 4.5).

Modes can also be excited by an equilibrium flow, the inertia term in (4.1),

$$\delta(\mathbf{v} \cdot \nabla w) = \mathbf{v}_0 \cdot \nabla \delta w + \delta \mathbf{v} \cdot \nabla w_0. \tag{4.3}$$

While the first term on the r.h.s. only leads to advection of the perturbation, the second $\propto w_0' = v_0''$ gives rise to shear flow-driven instabilities. Since these may be excited by the strong localized flow along a current sheet or along magnetic boundaries such as the magnetopause, they are treated in some detail in section 4.6, where we consider, in particular, the stabilizing effect of a parallel magnetic field.

Section 4.7 deals with the stability of a resistive current sheet. Here the dominant instability is the tearing mode, though it has a considerably higher threshold than in a static sheet pinch. The tearing mode leads to the generation of plasmoids, a conspicuous feature in many reconnecting systems.

4.1 The resistive tearing mode

The term "tearing instability" or "tearing mode" (implying the unstable mode) refers to a spontaneous reconnection process, which may occur in any sheared magnetic configuration. The simplest case is that of a plane sheared field, as shown schematically in fig. 4.1 and described by (3.15), which in projection corresponds to a sheet pinch with a neutral line. In a low-β plasma the pressure is negligible, such that (3.15) is simply $(d/dx)(B_y^2 + B_z^2) = 0$. In a high-$\beta$ plasma the field at the neutral line is reduced, e.g., $(d/dx)(p + \frac{1}{2}B_y^2) = 0$ in the special case of $B_z = 0$, which corresponds to the situation in the geomagnetic tail. (The collisionless tearing mode and its relevance for the magnetic activity in the magnetotail are considered in chapters 7 and 8.) The tearing mode in a sheet pinch is illustrated in fig. 4.2. Ignoring a weak variation in z-direction the field in the xy-plane $\mathbf{B}_\perp$ can be described by the flux function ψ, $\mathbf{B}_\perp = \mathbf{e}_z \times \nabla \psi$, which for a sinusoidal perturbation in the vicinity of the neutral line has

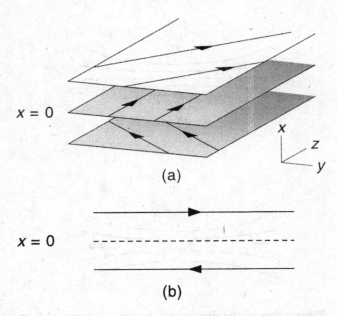

$x = 0$

(a)

$x = 0$

(b)

Fig. 4.1. Schematic drawing of a neutral sheet configuration (a), *xy* projection (b).

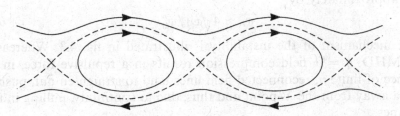

Fig. 4.2. Topology of the tearing mode.

the form

$$\psi(x, y) = \psi_0(x) + \delta\psi(x, y) \simeq \tfrac{1}{2}\psi_0'' x^2 + \psi_1 \cos ky, \qquad (4.4)$$

where ψ_0'', ψ_1 are chosen positive. The line $\psi = \psi_1$ is the separatrix between reconnected and merely deformed field lines. There is a sequence of neutral points, where the field intensity $B_\perp = |\nabla\psi|$ vanishes, alternately X-type and O-type neutral points or simply X-points and O-points.* The separatrix forms a chain of magnetic islands, where the island width, defined as the distance between the separatrix branches at an O-point, is

---

* Properly speaking, these are neutral lines along z, but following convention we call them neutral points, as long as the geometry is strictly two-dimensional.

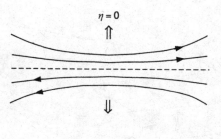

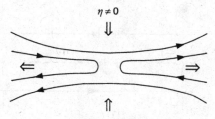

Fig. 4.3.   Illustration of the driving mechanism of a resistive instability.

given approximately by

$$w_I \simeq 4\sqrt{\psi_1/\psi_0''}. \tag{4.5}$$

The mechanism of the instability is illustrated in fig. 4.3. Whereas in ideal MHD, $\eta = 0$, field compression results in a repulsive force, in the presence of finite η, reconnected field lines tend to straighten out, pushing plasma away from the X-point and thus, due to continuity, pulling in new field lines.

Since resistive instabilities are slow compared with the Alfvén time, the associated motion is nearly incompressible, at least in the perpendicular plane. Hence in the lowest-order approximation one considers the low-β reduced equations (3.62), (3.63) for the flux function ψ and the vorticity $w = \nabla^2\phi$, where ϕ is the stream function, $\mathbf{v} = \mathbf{e}_z \times \nabla\phi$,

$$\partial_t\psi - \mathbf{B}\cdot\nabla\phi = \eta\nabla^2\psi, \tag{4.6}$$

$$\partial_t w + \mathbf{v}\cdot\nabla w - \mathbf{B}\cdot\nabla j = 0, \tag{4.7}$$

neglecting viscosity, which in typical resistive processes has only a small effect (see the discussion in section 3.2.2). Linearizing these equations about a static equilibrium $\mathbf{B}_0, j_0$, and using the ansatz $\phi_1, \psi_1 \propto e^{\gamma t}$ gives

$$\gamma\psi_1 = \mathbf{B}_0\cdot\nabla\phi_1 + \eta\nabla^2\psi_1, \tag{4.8}$$

$$\gamma\nabla^2\phi_1 = \mathbf{B}_0\cdot\nabla j_1 + \mathbf{B}_1\cdot\nabla j_0. \tag{4.9}$$

Since the growth rate is found to be small, $\gamma = O(\eta^\nu)$, $0 < \nu < 1$, inertia and resistivity are only important in a narrow layer δ around the neutral line, where $\mathbf{k} \cdot \mathbf{B}_0 = 0$. Because of this resonance property this line, or surface including the z-direction, is called a resonant surface. Thus the stability analysis leads to a boundary layer problem. One calculates separately the solution in the resistive layer using simplified geometry and in the ideal external region using simplified physics. The dispersion relation for the eigenvalue γ is obtained by matching both solutions asymptotically. It can be shown that for the system (4.8) and (4.9), γ is real in the case of instability, i.e., there is no overstability (Furth *et al.*, 1963).

Since the properties of ideal MHD instabilities are only slightly changed by finite resistivity, we restrict consideration to ideally stable configurations. It is, however, useful to distinguish between configurations which are deeply ideally stable and those close to marginal ideal stability. In the first case the instability, the tearing mode proper, grows sufficiently slowly such that the magnetic perturbation ψ_1 can diffuse across the resistive layer δ, attaining a finite value $\psi_1(x_s)$ at the resonant surface $x = x_s$. The derivative is finite, such that $\psi_1' \ll \psi_1(x_s)/\delta$, but it exhibits a jump characterized by the quantity Δ',

$$\Delta' = \lim_{\delta \to 0} [\psi_1'(x_s + \delta) - \psi_1'(x_s - \delta)]/\psi_1(x_s). \qquad (4.10)$$

This allows us to simplify the analysis in the resistive layer considerably by assuming $\psi_1 = const$.

In the second case the resistive mode is driven by the free energy of the almost unstable ideal mode, which gives rise to larger growth rates with weaker dependence on the resistivity. Here $\psi_1(x_s)$ is either small as in the double tearing mode, section 4.2, and the resistive kink mode, section 4.3, or ψ_1 is localized around $x \simeq x_s$ as in the interchange mode, section 4.5.1. In both cases $\psi_1' \sim \psi_1/\delta$, which corresponds formally to $\Delta' \to \infty$.

4.1.1 Linear tearing instability

Writing $\psi_1(x, y) = \psi_1(x)e^{iky}$, (4.8) and (4.9) in the resistive layer become

$$\gamma \psi_1 = ixkB_0'\phi_1 + \eta \psi_1'', \qquad (4.11)$$

$$\gamma \phi_1'' = ixkB_0'\psi_1'' - ikj_0'\psi_1, \qquad (4.12)$$

where we make the approximations $\nabla^2 \phi_1 \simeq \phi_1''$, $\nabla^2 \psi_1 \simeq \psi_1''$, $B_{y0}(x) \simeq xB_0'$, the constant-ψ assumption $\psi_1 \simeq \psi_1(0)$, and the coordinate system such that $x_s = 0$. The individual terms in each equation are a priori of the same order. The j_0' term in (4.12) can, however, be neglected, since it would only lead to a small correction of the stability threshold replacing Δ' by

$\Delta' + O(\eta^{2/5})$ (Bertin, 1982). The functions $\phi_1(x), \psi_1(x)$ can then be chosen with definite parities, ϕ_1 odd and ψ_1 even (or the reverse, but since we are only interested in the reconnecting mode with $\psi_1(0) \neq 0$, the former parity is the relevant one). This parity choice makes the following simple analysis semi-quantitative. With the approximations $\phi_1'' \simeq -\phi_1/\delta^2, \psi_1'' \simeq \Delta'\psi_1/\delta$, (4.12) and the two relations obtained from (4.11) by equating the l.h.s. with either the first or the second term on the r.h.s. give γ and δ as functions of Δ',

$$\gamma \simeq \eta^{3/5}(\Delta')^{4/5}(kB_0')^{2/5}, \tag{4.13}$$

$$\delta \simeq \eta\Delta'/\gamma \simeq \eta^{2/5}(\Delta')^{1/5}(kB_0')^{-2/5}. \tag{4.14}$$

(From a rigorous treatment one obtains a numerical factor of 0.55 in the expression for γ, see section 4.2.) In the analysis we have tacitly assumed that $\Delta' > 0$. It can in fact be shown (Furth *et al.*, 1963) that, in the framework of incompressible MHD, *the tearing mode is unstable if, and only if, $\Delta' > 0$.* Relations (4.13), (4.14) are consistent with the constant-ψ assumption, which requires that the skin time of the layer $\tau_\delta \sim \delta^2/\eta$ is shorter than the growth time γ^{-1}. We find from (4.14)

$$\tau_\delta \sim \eta/\gamma^2 = O(\eta^{-1/5}) \ll \gamma^{-1} = O(\eta^{-3/5}). \tag{4.15}$$

The quantity Δ' is obtained from the solution in the external region. Since the instability is weak, $\gamma \sim \tau_\eta^{-3/5}\tau_A^{-2/5} \ll \tau_A^{-1}$, where $\tau_\eta = a^2/\eta$ is the global resistive diffusion time, the inertia term in (4.9) is negligible, such that the equation reduces to the perturbed equilibrium equation

$$\mathbf{B}_0 \cdot \nabla j_1 + \mathbf{B}_1 \cdot \nabla j_0 = 0. \tag{4.16}$$

In the vicinity of the resonant surface, (4.16) assumes the form

$$\psi_1'' - \frac{\kappa}{x}\psi_1 = 0, \tag{4.17}$$

where $\kappa = j_0'/B_0'$. The solution has the form

$$\psi_1(x)/\psi_1(0) = f_1(x) + Cf_2(x), \tag{4.18}$$

with

$$f_1 = 1 + \kappa x \log|x| + O(x^2), \quad f_2 = x + O(x^2), \tag{4.19}$$

where C may have different values $C_\pm$ on both sides of the singularity, which are determined by the outer boundary conditions. To obtain an overall smooth behavior the resistive layer solution must be matched to the external solution by identifying the logarithmic derivative of ψ_1 of the former for $|x| \to \infty$, Δ', with that of the latter for $x \to 0$,

$$\Delta' = C_+ - C_-. \tag{4.20}$$

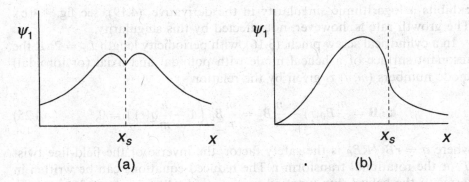

Fig. 4.4. Schematic plots of the eigenfuction ψ_1 for (a) $j_0'(x_s) = 0$ as in the symmetric sheet pinch; (b) $j_0'(x_s) \neq 0$ as in the cylindrical pinch.

Note that the logarithmic singularity of ψ_1', which is proportional to j_0', cancels in Δ'.

In the case of a plane sheet pinch, (4.13) becomes

$$\psi_1'' - \left(k^2 + \frac{j_0'(x)}{B_0(x)}\right)\psi_1 = 0. \tag{4.21}$$

For the standard symmetric sheet pinch $B_0(x) = B_0 \tanh(x/a), j_0 = (B_0/a)\mathrm{sech}^2(x/a)$, and boundary conditions $\psi_1 = 0$ for $|x| \to \infty$, (4.21) can be solved analytically (see fig. 4.4(a))

$$\psi_1 = e^{-k|x|}\left(1 + \frac{1}{ka}\tanh\frac{x}{a}\right), \tag{4.22}$$

hence

$$\Delta' = \frac{2}{a}\left(\frac{1}{ka} - ka\right). \tag{4.23}$$

We thus find that the tearing mode is unstable, $\Delta' > 0$, for long wavelength $ka < 1$ and stable, $\Delta' < 0$, for short wavelength $ka > 1$. These properties remain qualitatively valid for rather general $j_0(x)$ profiles, in particular the behavior $\Delta' \propto k^{-1}$ for small k. The constant-ψ property breaks down for $k \sim \delta/a^2$, where $\Delta' \sim \delta^{-1}$. A more general analysis without the constant-ψ assumption shows that the growth rate, which increases with decreasing k, $\gamma \sim (\Delta')^{4/5}k^{2/5} \sim k^{-2/5}$, reaches a maximum

$$\gamma_{\max} = \gamma(k_0) \sim \eta^{1/2}, \; k_0 \sim \eta^{1/4}. \tag{4.24}$$

For still smaller k, γ decreases again, as can be seen also from (4.13) and (4.14) by inserting $\Delta' \sim \delta^{-1}$.

In the general nonsymmetric case, where $j_0'(x_s) \neq 0$, for instance in the presence of a mean field $\bar{B}$, $B_0(x) = \bar{B} + B_0 \tanh(x/a)$, the eigenfunction exhibits a logarithmic singularity in the derivative, (4.19), see fig. 4.4(b). The growth rate is, however, not affected by this singularity.

In a cylindrical screw pinch, (3.16), with periodicity length $L_z = 2\pi R$ the resonant surface of a helical mode with poloidal and axial (or toroidal) mode numbers (m, n) is given by the relation

$$\mathbf{k} \cdot \mathbf{B} = \frac{m}{r} B_\theta(r) - \frac{n}{R} B_z = \frac{m}{r} B_\theta \left(1 - \frac{n}{m} q(r)\right) = 0, ^* \qquad (4.25)$$

where $q = rB_z/RB_\theta$ is the safety factor, the inverse of the field-line twist T or the rotational transform ι. The reduced equations can be written in terms of the helical flux function $\psi_* = (n/m)(r/R)A_\theta - A_z$, (3.37),[†]

$$\psi_* = \psi - \frac{n}{m}\frac{r^2}{2R}B_z, \quad j = \nabla^2 \psi = \nabla^2 \psi_* + 2\frac{n}{m}\frac{B_z}{R}, \qquad (4.26)$$

since, in the framework of reduced MHD, $B_z = const$, hence $A_\theta = rB_z/2$ and $A_z = -\psi$. Equations (4.6), (4.7) thus become

$$(\partial_t + \mathbf{v} \cdot \nabla)\psi_* = \eta j, \qquad (4.27)$$

$$(\partial_t + \mathbf{v} \cdot \nabla)w = \mathbf{B}_* \cdot \nabla j, \qquad (4.28)$$

where $\mathbf{B}_* = \mathbf{e}_z \times \nabla\psi_*$ is called the helical magnetic field. The equation for the external solution (4.16) becomes

$$\frac{d^2\psi_1}{dr^2} + \frac{1}{r}\frac{d\psi_1}{dr} - \left(\frac{m^2}{r^2} + \frac{j_0'(r)}{B_\theta(r)[1 - nq(r)/m]}\right)\psi_1 = 0. \qquad (4.29)$$

The short-wavelength stability of the tearing mode, $\Delta' < 0$ for $ka \gtrsim 1$, has important consequences in a cylindrical plasma. For equilibrium profiles with gradient scale a, tearing instability arises only for $ma/r \lesssim 1$, hence instability is in general limited to low mode number $m = O(1)$. High m-number modes are primarily driven by the plasma pressure gradient instead of the parallel current, see section 4.5.

4.1.2 Small-amplitude nonlinear behavior

The tearing instability is determined by the parameter $\Delta'(k)$, a measure of the free energy of the magnetic configuration which can be released by the tearing mode of wavenumber k. Multiplying (4.21) by ψ_1^* and integrating

---

[*] Note the sign convention of n corresponding to the mode ansatz $\exp\{im\theta - inz/R\}$.
[†] In the context of helical instabilities, it is customary to use the notation ψ_* instead of ψ_H.

over the configuration excluding the singularity at the resonant surface gives, after integration by parts,

$$\lim_{\epsilon \to 0} \left(\int_{-\infty}^{x_s-\epsilon} + \int_{x_s+\epsilon}^{\infty} \right) \left[\left| \frac{d\psi_1}{dx} \right|^2 + \left(k^2 + \frac{j_0'(x)}{B_0(x)} \right) |\psi_1|^2 \right] dx = -\Delta' |\psi_1(x_s)|^2,$$

(4.30)

using the definition of Δ', (4.10). The l.h.s. is the usual expression for the magnetic energy δW_M associated with the perturbation ψ_1. In the energy conservation relation of the linear tearing mode the magnetic energy is balanced by the kinetic energy and the ohmically dissipated energy, $\delta W_M + \delta W_K + \delta W_\Omega = 0$ (Adler et al., 1980; Wesson, 1991). Hence, since the latter two are positive, the magnetic energy has to be negative, i.e., $\Delta' > 0$ for instability.

Δ' not only determines the linear growth rate, but also the nonlinear dynamics. Contrary to the standard picture of instability evolution, where the perturbation grows at essentially the linear growth rate until nonlinear effects lead to saturation, the growth rate of the tearing mode is modified already at amplitudes much below the saturation level. Rutherford (1973) has shown that inertia effects are switched off as soon as the island width w_I exceeds the resistive layer width δ. The physics of the process can readily be understood. The dominant nonlinear effect is the diffusive broadening of the current perturbation j_1. While, in the linear phase of the instability, one has $j_1 = \psi_1'' \simeq \psi_1 \Delta'/\delta$, the layer width is replaced by the island width when the latter exceeds the former, so then $j_1 \simeq \psi_1 \Delta'/w_I$. Using the definitions of w_I, (4.5), and δ, (4.14), this cross-over occurs at amplitude $\psi_1 \sim \eta^{4/5}$. Further growth is a pure resistive diffusion process, independent of inertia,

$$\dot{\psi}_1 = \eta j_1 \simeq \eta \frac{\Delta'}{w_I} \psi_1$$

(4.31)

or in terms of the island width,

$$\dot{w}_I \simeq \eta \Delta',$$

(4.32)

i.e., the island no longer grows exponentially but linearly in time, $w_I \propto t$.

Let us now outline the rigorous derivation of the result (4.32) (Rutherford, 1973). Since inertia is negligible, the vorticity equation (4.7) degenerates to the equilibrium equation $\mathbf{B} \cdot \nabla j = 0$, hence $j = j(\psi, t)$, which satisfies Ampère's law

$$\nabla^2 \psi = j(\psi, t).$$

(4.33)

The time evolution is obtained by averaging the ψ-equation (4.6) over a flux surface

$$\langle \partial_t \psi \rangle_\psi = \eta j(\psi),$$

(4.34)

since $\langle \mathbf{B} \cdot \nabla \phi \rangle_\psi = 0$, where

$$\langle f \rangle_\psi = \oint f \frac{dl}{|\nabla \psi|} \Big/ \oint \frac{dl}{|\nabla \psi|}.$$

In general, (4.33) and (4.34) must be solved numerically, an elegant method having been developed by Grad *et al.* (1975). In the small-amplitude limit we can, however, obtain an analytic solution. Here ψ assumes the approximate form (4.4)

$$\psi(x, y, t) = \tfrac{1}{2} \psi_0'' x^2 + \psi_1(t) \cos y, \tag{4.35}$$

where $ky \to y$ to simplify the notation and the x-dependence in ψ_1 is neglected, which reflects the constant-ψ property valid for small island width. Equation (4.34) is solved by matching the solution to the linear external solution for $|x| \gg w_I$, which remains valid as long as w_I is still small,

$$\frac{1}{\pi} \int_{-\infty}^{\infty} dx \int_{-\pi}^{\pi} dy \, j(\psi) \cos y = d\psi_1/dx|_{-\infty}^{\infty} = \Delta' \psi_1. \tag{4.36}$$

The fact that higher-order harmonics of ψ are neglected does not imply that these are also negligible in j, i.e., j cannot be computed from the approximate form of ψ. But, since the higher harmonics decay faster than the first with distance from the resonant surface, the first harmonic is sufficient in the asymptotic matching. To evaluate integrals of the form

$$\int_{-\infty}^{\infty} dx \int_{-\pi}^{\pi} dy f = \int_{\psi_{min}}^{\infty} d\psi \oint f \frac{dl}{|\nabla \psi|}, \tag{4.37}$$

use is made of the small-amplitude approximation, (4.35),

$$|\nabla \psi| \simeq |x| \psi_0'' = \sqrt{2\psi_0''} \sqrt{\psi - \psi_1 \cos y}, \tag{4.38}$$

and $dl \simeq dy$, so that, for instance,

$$\oint \frac{\cos y \, dl}{\sqrt{\psi - \psi_1 \cos y}} \simeq 4 \int_0^{y_{max}} \frac{\cos y \, dy}{\sqrt{\psi - \psi_1 \cos y}} \tag{4.39}$$

with

$$y_{max} = \begin{cases} \pi & u > 1 \\ \arccos u & 1 > u > -1, \end{cases}$$

$$u = \psi/\psi_1.$$

Inserting $j(\psi)$ from (4.34) into expression (4.36) and using (4.35) gives

$$\frac{4}{\pi} A \sqrt{\frac{\psi_1}{2\psi_0''}} \dot\psi_1 = \eta \Delta' \psi_1, \tag{4.40}$$

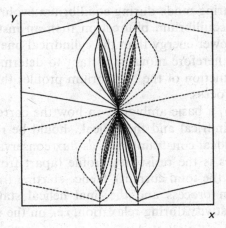

Fig. 4.5. Flow pattern around the X-point in the tearing mode evolution.

where

$$A = \int_{-1}^{\infty} du \left(\int_0^{y_{max}} \frac{\cos y \, dy}{\sqrt{u - \cos y}} \right)^2 \bigg/ \int_0^{y_{max}} \frac{dy}{\sqrt{u - \cos y}} = 1.827. \quad (4.41)$$

Hence we obtain

$$\dot{w}_I = 1.22 \eta \Delta'. \quad (4.42)$$

Rutherford's theory is exact for small island size. This does, however, not imply that inertia effects are completely negligible, since on the separatrix j need not be a flux function, $\mathbf{B}$ vanishing at the X-point. But the detailed behavior in this region does not influence the global evolution, which is determined by (4.34). From (4.42) one finds the reconnection rate

$$E = \dot{\psi}_1 \simeq \Delta'^2 \psi_0'' \eta^2 t \lesssim O(\eta) \quad \text{for} \quad t \lesssim a^2/\eta. \quad (4.43)$$

As discussed in section 3.3.3, this implies that reconnection is slow and no current sheet is formed, as can also be seen in the flow pattern in fig. 4.5, which agrees with the similarity solution, (3.143), in the vicinity of the X-point.

4.1.3 Saturation of the tearing mode

Although the nonlinear evolution of the tearing mode given by (4.42) occurs formally on the global resistive time-scale $\tau_\eta = \eta/a^2$, the actual saturation time $\sim \tau_\eta w_I^2/a^2$ is considerably shorter, since the island width is typically $w_I \sim 0.1a$. Hence the saturated islands adjust quasi-instantaneously to the slow changes of the equilibrium profiles. In fact the

appearance of a tearing mode during equilibrium evolution should rather
be regarded as an equilibrium bifurcation than an instability, where the
helical state has a lower energy than the cylindrical one. From a practical
point of view it is therefore more important to determine the saturation
island width as a function of the equilibrium profiles than to describe the
actual relaxation process.

There is, however, a basic ambiguity in how the corresponding equilibrium states, the cylindrical and the helical, should be related, because of
the lack of a strict local constraint such as flux conservation. The only tie
between these states is the resistivity profile (apart from possible global
constraints, such as the total current). Hence starting from the cylindrical
state, the relaxation process and the final helical state depend on the
behavior of the resistivity during relaxation, i.e., on the thermal transport,
assuming a classical resistivity behavior, where $\eta \propto T_e^{-3/2}$ is a function of
the electron temperature (see, e.g., Braginskii, 1965). Only for sufficiently
small island width is this transport effect weak, while for large island
width it is expected to play an important role.

A theory of the saturation of the tearing instability has been developed by White *et al.* (1977), which extends Rutherford's theory into the
saturation regime,

$$\dot{w}_I = 1.22\eta\left[\Delta'(w_I) - \alpha w_I\right]. \tag{4.44}$$

$\Delta'(w_I)$ is the finite-amplitude generalization of the linear quantity $\Delta' = \Delta'(0)$,

$$\Delta'(w_I) = [\psi_1'(x_s + w_I/2) - \psi_1(x_s - w_I/2)]/\psi_1(x_s), \tag{4.45}$$

where ψ_1 is the solution of the linear equation (4.21) or (4.29) using
the initial current profile.[*] The coefficient α in (4.44) is a rather complicated expression depending primarily on the gradients dj_0/dx and $d\eta/dx$.
Saturation occurs for

$$\Delta'(w_I) - \alpha w_I = 0. \tag{4.46}$$

Let us first consider the standard sheet pinch $B_y(x) = \tanh(x/a)$. From
the linear solution, (4.22), one obtains

$$\Delta'(w_I) = \frac{2}{a}e^{-kw_I}\left[\frac{1}{ka}\,\mathrm{sech}^2(w_I/2a) - ka - \tanh(w_I/2a)\right], \tag{4.47}$$

and $\alpha = (ka)^2 - 0.78$. The agreement with the numerical solution of
the dynamic equations (4.6), (4.7) is reasonably good for $w_I/a < 2$, w_I
increasing roughly linearly with $[(ka)^{-1} - 1]$ (note that $ka = 1$ is the
marginally stable wavenumber).

---

[*] An alternative definition closer to the boundary layer concept of matching logarithmic
derivatives is $\Delta_1'(w) = \psi_1'/\psi_1|_{x_s+w_I/2} - \psi_1'/\psi_1|_{x_s-w_I/2}$. In practice the difference is, however,
found to be small.

We now turn to the practically more important case of helical tearing modes in a cylindrical pinch. Since here the equilibrium scale-height is of the order of the resonant radius r_s, the geometry enforces $w_I/r_s < 1$. Hence one expects that the term αw_I in (4.46) is small and the saturation island width is, to good approximation, given by

$$\Delta'(w_I) = 0, \tag{4.48}$$

which is used in most practical applications (see, e.g., Carreras *et al.*, 1979). Relation (4.48) has a simple physical interpretation. The dominant nonlinear effect is the modification of the equilibrium current profile corresponding to a flattening of j_0 across the island width, while j_0 remains essentially unchanged outside the island region. Considering the external region, (4.29), the flattening of j_0 implies that the solution ψ_1 is approximately determined by integrating the equation only up to the separatrix. $\Delta'(w_I)$ can be interpreted as the free energy of the tearing mode remaining at the amplitude given by w_I. In general $\Delta'(w_I)$ is a decreasing function of w_I, as can be judged by the shape of the eigenfunction ψ_1, fig. 4.4.

In a cylindrical pinch, the stability characteristics of a tearing mode (m, n) and its saturation island width w_{mn} depend on the safety factor profile $q(r)$ or, equivalently, the current profile j_0 and the position of the resonant surface r_s, $q(r_s) = m/n$. We assume a fixed resistivity profile $\eta(r) = \eta_0/j_0(r, t = 0)$, ignoring for the moment the effect of a dynamic resistivity response. Consider the following class of profiles introduced by Furth *et al.* (1973):

$$q(r) = q_0 \left(1 + \left(\frac{r}{r_0}\right)^{2v}\right)^{1/v}, \tag{4.49}$$

corresponding to the current density

$$j_0(r) = j_0 \left(1 + \left(\frac{r}{r_0}\right)^{2v}\right)^{-(1+1/v)}, \tag{4.50}$$

where r_0 is a characteristic profile width.

By solving (4.29) numerically, w_{mn} can be computed from the condition (4.48) for any position r_s. Results are plotted in fig. 4.6, which shows $j_0(r)$ with $r_0 = 0.5$ and the saturated island width of the (2,1) mode normalized to its resonant radius, $w_{21}(r_s)/r_s$, as function of r_s for $v = 1, 2, 4$. The main feature is that w_{21}/r_s is roughly constant, $w_{21} \propto r_s$, for r_s inside the current channel and then falls off the more abruptly the steeper the decay of the current profile. The maximum value of w_{21} is reached for r_s just outside the maximum gradient of the current density. Since this point is located

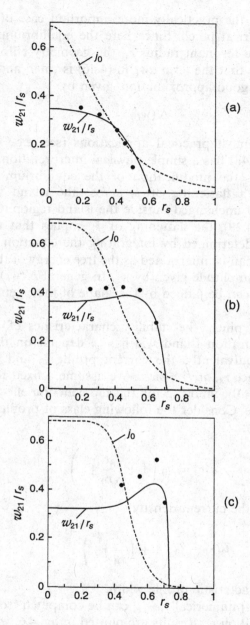

Fig. 4.6.   Normalized island width $w_{21}(r_s)/r_s$ for (a) the "peaked profile" $v = 1$; (b) the "rounded profile" $v = 2$; and (c) the "flat profile" $v = 4$. Quasi-linear result $\Delta'(w_I) = 0$ (full lines), simulation result (full circles), normalized current profiles (dashed lines).

at larger radii for more square-shaped profiles, w_{21} reaches larger values for a square-shaped than for a peaked profile.

As seen in fig. 4.6, the quasi-linear prediction $\Delta'(w_{21}) = 0$ agrees well with the exact numerical results, even for what appear to be large island widths. A discrepancy is found, however, for small r_s in the flat central part of the current profiles, where quasi-linear theory predicts constant normalized island width $w_{21}/r_s \simeq 0.35$, while numerical simulations show that the islands do not saturate but lead to full reconnection of the helical flux inside the resonant radius, i.e., islands grow toward the inside down to the center $r = 0$, which results in $w_{21}/r_s \to 1$. This behavior is related to the vacuum bubbles in a shearless plasma column (Kadomtsev & Pogutse, 1974).

The stability of the tearing mode depends rather sensitively on the current-density gradient in the vicinity of the resonant surface. A local steepening may destabilize the mode, while a flattening leads to stabilization. The latter effect is indeed the dominant self-stabilization process of an unstable tearing mode when growing to finite amplitude. The process may, however, destabilize other modes owing to the current-profile steepening in the region adjacent to the island, which appears to play an important role in the major disruption in tokamak plasmas (see, for instance, Biskamp, 1993a, chapter 8). A particular shape of the q-profile close to $q = 1$ can give rise to cascades of high-n tearing modes with n up to 20 (Hallatschek *et al.*, 1998). By current-profile tailoring around the major resonant surfaces completely tearing mode-stable profiles can be constructed (Glasser *et al.*, 1977).

4.1.4 *Effect of dynamic resistivity*

In the linear theory of the tearing mode, η is usually assumed homogeneous and constant in time (resistivity gradient effects give rise to a different type of instability, the rippling mode, see Furth *et al.*, 1963). A homogeneous resistivity distribution violates, strictly speaking, the equilibrium condition $\eta j = const$, but in linear theory this effect is negligible, since the instability time-scale is faster than the resistive diffusion time of the equilibrium. In the nonlinear regime, however, the behavior of the resistivity becomes important, in particular for large island width. Let us therefore assume that the resistivity adjusts dynamically such that $\eta(x, y) = \eta(T_e) = \eta(\psi)$ owing to fast parallel electron heat transport, where the ψ-dependence is determined by slow perpendicular heat conduction and local heat sources and sinks. Consider the difference of (4.34) taken at the X-point and the O-point, where $\nabla\psi = 0$, so that the advection term vanishes:

$$\frac{d}{dt}(\psi_X - \psi_O) = \eta_X j_X - \eta_O j_O. \tag{4.51}$$

Since $\psi_X - \psi_O = 2\psi_1$, $j_X - j_O \simeq 2\psi_1\Delta'/w_I$ and $\eta_X - \eta_O \simeq 2\psi_1 d\eta/d\psi$, one obtains approximately

$$\dot{w}_I \simeq \eta\Delta' + j\frac{d\eta}{d\psi}w_I. \tag{4.52}$$

The $d\eta/d\psi$-term is related to the α-term in (4.44).

We first discuss the tearing mode in a sheet pinch. The physical meaning of (4.51) can easily be understood. Because of the choice $\psi_0'' > 0$ the magnetic flux in the island is positive, $\psi_X - \psi_O > 0$. The variation of this flux is given by the difference between the injection rate $\eta_X j_X$ due to reconnection at the X-point and the dissipation rate $\eta_O j_O$ due to resistive dissipation in the island. Lowering η_O with respect to η_X, $d\eta/d\psi > 0$ ("island heating"), leads to enhanced growth and larger saturation width, while increasing η_O, $d\eta/d\psi < 0$ ("island cooling"), slows down the growth and reduces the final width.

In a cylindrical tokamak-like configuration, the effect of a dynamic resistivity on helical tearing modes is just the opposite of that in a sheet pinch. Since ψ is now replaced by the helical flux function $\psi_* = \psi - (n/m)r^2 B_z/2R$, (4.26), $\psi_*'(r) = B_\theta(r)[1 - nq(r)/m]$, we have $\psi_*''(r_s) < 0$ for the usual case of radially increasing $q(r)$ and $j_0 > 0$. This implies that ψ_* has its maximum at the O-point, in particular $\psi_{*X} - \psi_{*O} < 0$, such that $(d/dt)(\psi_{*X} - \psi_{*O}) > 0$ corresponds to island shrinking. Hence it follows from (4.52) that island heating $\eta_O < \eta_X$, meaning $d\eta/d\psi_* < 0$, leads to smaller islands, while island cooling leads to larger ones.

Numerical simulations illustrating this effect are shown in fig. 4.7. The initial current profile is the rounded profile ($\nu = 2$) with $r_0 = 0.5$, $r_s = 0.65$, and the resistivity $\eta(\psi_*)$ is defined such that the profile along the radius through the X-point remains constant, $\eta(r, \theta_X) = \eta_0/j_0(r, t = 0)$. The resistivity in the island is still free. For $t < 10^3\tau_A$ we assume a flat distribution $\eta_I(\psi_*) = \eta_X$, the value at the X-point. For $t > 10^3\tau_A$ the island resistivity is shaped,

$$\eta_I(\psi_*) = \eta_X\left[1 + \epsilon(\psi_* - \psi_{*X})\right].$$

Choosing $\epsilon > 0$, i.e., $\eta_O > \eta_X$ corresponding to island cooling, gives rise to larger islands (the upper branch in fig. 4.7(a)), whereas $\epsilon < 0$, i.e., $\eta_O < \eta_X$ corresponding to island heating, leads to smaller islands (the lower, dashed branch). The corresponding η-profiles are plotted in fig. 4.7(b). One can see that a relatively small change of η (~ 10 percent) has a dramatic effect on the island width.

4.1.5 Neoclassical tearing mode

The main influence of the island resistivity profile $\eta_I(\psi)$ on tearing mode evolution occurs through its effect on the current profile in the island. Since

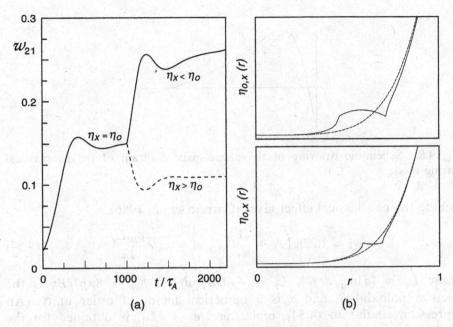

Fig. 4.7. Effect of dynamic resistivity. (a) Island width w_{21} for three island resistivity profiles; (b) radial η-profiles across the O-point (full line) and the X-point (dashed line) for the final states in (a).

in equilibrium $\eta_I(\psi)j_I(\psi) = const$, a decrease of η_I implies an increase of j_I and vice versa. Hence if j_I is reduced compared to the ambient value, the tearing mode is destabilized leading to larger saturated islands. In hot tokamak plasmas the parallel ($\simeq$ toroidal) current density is increased compared with the resistive value $E_\parallel / \eta_\parallel$ by the bootstrap current, which is proportional to the pressure gradient (see, e.g., Kikuchi & Azumi, 1995),

$$j_\parallel \simeq \frac{E_\parallel}{\eta_\parallel} - \sqrt{\epsilon}\,\frac{c}{B_\theta}\frac{dp}{dr}. \tag{4.53}$$

The last term arises from the friction between electrons circulating around the torus and those trapped on the low-field outer side. The rigorous calculation of these collision processes is the subject of neoclassical transport theory (see, e.g., Hinton & Hazeltine, 1976). Since in the island the pressure profile $p \simeq p(\psi)$ is essentially flat, there is no bootstrap-current contribution from the island. Hence the current density in the island is reduced compared to the value outside the island, which has a destabilizing effect on the tearing mode. Extending Rutherford's theory, (4.42), to

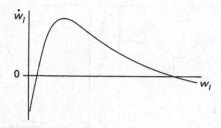

Fig. 4.8. Schematic drawing of the phase-space diagram of the neoclassical tearing mode.

include the neoclassical effect gives (Carrera *et al.*, 1986),

$$\dot{w}_I = 1.22\eta\left(\Delta' + \frac{A}{w_I}\right), \quad A = \alpha\sqrt{\epsilon}\frac{\beta_{pe}L_q}{L_p}, \tag{4.54}$$

where $L_q = (d\ln q/dr)^{-1}$, $L_p = -(d\ln p/dr)^{-1}$, $\beta_{pe} = 8\pi p_e/B_\theta^2$ is the electron poloidal β, and α is a numerical factor of order unity. (An expression similar to (4.54), replacing $\sqrt{\epsilon} \rightarrow \kappa L_q$, is obtained for the nonlinear tearing mode under the influence of magnetic field-line curvature κ, see Kotschenreuther *et al.*, 1985).

It follows from (4.54) that for small island width the growth of the tearing mode is dominated by the neoclassical term, $w_I \sim t^{1/2}$. The large-amplitude behavior and saturation depend on the sign of Δ'. For $\Delta' > 0$ these are determined by the nonlinear modification of Δ', $\Delta'(w_I) \rightarrow 0$, as discussed before. For $\Delta' < 0$, however, (4.54) indicates saturation at $w_I = A/|\Delta'|$, which seems to imply that the neoclassical tearing mode grows to finite amplitude for any negative Δ'. Under real conditions, however, the form of the neoclassical term in (4.54) is only correct for not too small island width, since the flattening of the pressure profile in the island is counteracted by cross-field diffusion. One can introduce this effect phenomenologically by the following substitution in (4.54) (Fitzpatrick, 1995):

$$\frac{1}{w_I} \rightarrow \frac{w_I}{w_I^2 + w_d^2} \tag{4.55}$$

with $w_d \propto (\kappa_\perp/\kappa_\parallel)^{1/4}$, where $\kappa_\perp$, $\kappa_\parallel$ are the perpendicular and parallel heat diffusivities. Hence there is, in general, a critical island width

$$w_{I,\text{crit}} = |\Delta'|w_d^2/A,$$

which must be exceeded, before the mode starts growing. (A detailed discussion of the threshold conditions is given by Wilson *et al.*, 1996.) Growth may be triggered by toroidal coupling to modes of different helicities, which are excited in a tokamak discharge, for instance by the $m = 1$ mode associated with the sawtooth oscillation (see section 6.5). The phase-space diagram of the neoclassical tearing mode is shown in fig. 4.8.

4.2 The double tearing mode

The basic assumption in the theory of the tearing instability, given in the preceding section, is the constant-ψ approximation, the validity of which requires that the skin time of the resistive layer is shorter than the growth time, $\tau_\delta < \gamma^{-1}$. There are, however, circumstances when this condition is not satisfied. Formally this occurs when $\Delta' \to \delta^{-1}$, which we have already encountered in the long-wavelength range $k \sim \eta^{1/4}$ at the maximum tearing mode growth rate, where $\tau_\delta \sim \gamma^{-1} \sim \eta^{-1/2}$. While in this case the constant-ψ property breaks down, because the layer width and hence τ_δ become large, in the case of quasi-ideal instability the break-down is due to a large growth rate.

Unstable motions require that the stabilizing field-line bending is minimal. This is the origin of the short-wavelength stability threshold $ka \simeq 1$ for the tearing mode. A more efficient way to avoid field-line bending involves spatially extended perturbations corresponding to a rigid displacement of the plasma, which results in faster growth. Here the most important paradigms are the double tearing mode, which is discussed in this section, and the resistive kink mode considered in section 4.3.

The double tearing mode is characterized by the presence of two (or more) resonant surfaces, the coupling of which may give rise to faster dynamics than in the standard tearing mode. It is useful at this point to introduce the linear displacement vector $\boldsymbol{\xi}$, $\partial_t \boldsymbol{\xi} = \gamma \boldsymbol{\xi} = \mathbf{v}_1$, which is generally used in ideal MHD stability theory, and more specifically the component ξ_x in the direction of the equilibrium gradient, $\xi_x = -ik_y \phi_1/\gamma$. In terms of the displacement, the linearized equations (4.8), (4.9) become

$$\gamma \psi_1 = -B_0 \gamma \xi + \eta(\psi_1'' - k^2 \psi_1), \tag{4.56}$$

$$\gamma^2(\xi'' - k^2 \xi) = B_0 k^2(\psi_1'' - k^2 \psi_1) - k^2 B_0'' \psi_1, \tag{4.57}$$

where we have simplified the notation by replacing $\xi_x \to \xi$, $k_y \to k$, $B_{0y} \to B_0$. In the external region, resistivity and inertia can again be neglected. Hence from (4.56) one has $\psi_1 = -B_0 \xi$, which on substitution in

$$B_0(\psi_1'' - k^2 \psi_1) = B_0'' \psi_1 \tag{4.58}$$

gives

$$\frac{d}{dx}\left(B_0^2 \frac{d\xi}{dx}\right) = k^2 B_0^2 \xi. \tag{4.59}$$

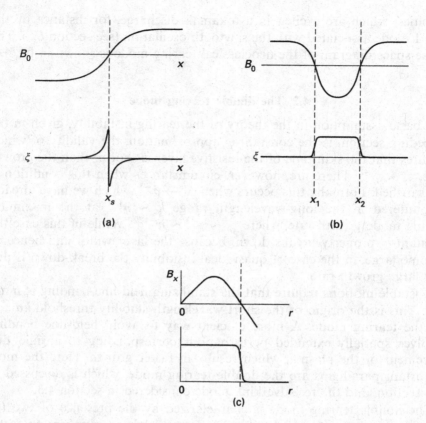

Fig. 4.9. Schematic plots of the equilibrium field B_0 and the eigenfunction ξ for (a) the standard tearing mode, (b) the double tearing mode, and (c) the resistive kink mode.

To be definite, consider a symmetric equilibrium field $B_0(x)$ with just two resonant surfaces $B_0(\pm x_s) = 0$ (Pritchett *et al.*, 1980). For long wavelength $kx_s \ll 1$ the factor on the r.h.s. is small, such that we can solve (4.59) in terms of a power series expansion in the parameter $k^2 x_s^2$, $\xi = \xi^{(0)} + \xi^{(1)} + \cdots$. To lowest order, $d\xi^{(0)}/dx = 0$ except at the resonant surfaces $x = \pm x_s$. Since for an isolated system ξ must vanish for $|x| \to \infty$, we have

$$\xi^{(0)} = \begin{cases} \xi_0 = const, & |x| < x_s, \\ 0, & |x| > x_s. \end{cases} \tag{4.60}$$

While in the case of a single resonant surface, boundary conditions enforce $\xi_0 = 0$, such that ξ is localized at the resonant surface, fig. 4.9(a), in the case of two neighboring resonant surfaces, $\xi^{(0)}$ may be finite in between, fig. 4.9(b). The first-order term is easily obtained:

$$\frac{1}{\xi_0}\frac{d\xi^{(1)}}{dx} = \begin{cases} \dfrac{k^2}{B_0^2}\displaystyle\int_0^x B_0^2 dx, & |x| < x_s, \\[2mm] \dfrac{k^2}{B_0^2}\displaystyle\int_0^{x_s} B_0^2 dx & |x| > x_s. \end{cases} \tag{4.61}$$

In the vicinity of $x = x_s$, one has $B_0(x) \simeq (x - x_s)B_0'$, such that (4.61) can be written in the form

$$\frac{1}{\xi_0}\frac{d\xi^{(1)}}{dx} = -\frac{1}{\pi}\frac{\lambda_H}{(x - x_s)^2}, \tag{4.62}$$

where

$$\lambda_H = -\pi\frac{k^2}{B_0'^2}\int_0^{x_s} B_0^2 dx \tag{4.63}$$

is a measure of the free magnetic energy of the mode. In the resistive layer (4.11) and (4.12) have to be solved which, when written in terms of ξ, read

$$\gamma^2\xi'' = (x - x_s)k^2 B_0'\psi_1'', \tag{4.64}$$

$$\gamma\psi_1 = -(x - x_s)\gamma B_0'\xi + \eta\psi_1''. \tag{4.65}$$

Asymptotic matching of the logarithmic derivatives of the solution to the external solution, (4.62), yields the growth rate. The problem can be treated analytically (Coppi *et al.*, 1976; Ara *et al.*, 1978), which gives the following expression for the normalized growth rate $\widehat{\gamma}$:

$$\frac{\widehat{\gamma}^{5/4}\widehat{\lambda}_H\Gamma[(\widehat{\gamma}^{3/2} - 1)/4]}{\Gamma[(\widehat{\gamma}^{3/2} + 5)/4]} = 8. \tag{4.66}$$

Here $\widehat{\gamma} = \gamma/(\eta k^2 B_0'^2)^{1/3} = \gamma(\tau_\eta\tau_A^2)^{1/3}$ and $\widehat{\lambda}_H = \lambda_H(kB_0'/\eta)^{1/3} = \lambda_H k(\tau_\eta/\tau_A)^{1/3}$, $\tau_A^{-1} = B_0'$, $\tau_\eta^{-1} = \eta k^2$. From the general expression (4.66), $\widehat{\gamma}$ can be obtained explicitly in three special cases:

(a) For strong ideal instability $\widehat{\lambda}_H \gg 1$ use of the asymptotic properties of the Γ-function, $\Gamma(z + a)/\Gamma(z + b) \simeq z^{a-b}$, gives the ideal growth rate $\widehat{\gamma} = \widehat{\lambda}_H$, i.e.,

$$\gamma = \lambda_H kB_0'; \tag{4.67}$$

(b) Marginal instability $\widehat{\lambda}_H = 0$ corresponds to the first pole of $\Gamma[(\widehat{\gamma}^{3/2} - 1)/4]$, which is $\widehat{\gamma} = 1$ or

$$\gamma = (\eta k^2 B_0'^2)^{1/3}; \tag{4.68}$$

(c) For deeply ideally stable MHD conditions $\widehat{\lambda}_H < 0$, $|\widehat{\lambda}_H| \gg 1$, corresponding to small growth rate $\widehat{\gamma} \ll 1$, (4.66) gives

$$\gamma = \left(\frac{\Gamma(\frac{1}{4})}{2\Gamma(\frac{3}{4})}\right)^{4/5} \eta^{3/5}(kB_0')^{2/5}|\lambda_H|^{-4/5}. \tag{4.69}$$

Hence in the MHD-stable regime the mode reduces, as expected, to the standard tearing mode. In fact λ_H is directly related to Δ'. From (4.62) one obtains

$$\xi = \begin{cases} \xi_0\left(1 + \dfrac{\lambda_H}{\pi(x - x_s)}\right), & x - x_s \to 0_- \\[2mm] \xi_0 \dfrac{\lambda_H}{\pi(x - x_s)}, & x - x_s \to 0_+ \end{cases} \tag{4.70}$$

or, using the ideal relation $\psi_1 = -(x - x_s)B_0'\xi$,

$$\psi_1 = \begin{cases} -\xi_0 B_0'[x - x_s + (\lambda_H/\pi)], & x - x_s \to 0_-, \\ -\xi_0 B_0'\lambda_H/\pi, & x - x_s \to 0_+, \end{cases} \tag{4.71}$$

whence

$$\Delta' \equiv \psi_1'/\psi_1|_{0_+} - \psi_1'/\psi_1|_{0_-} = -\pi/\lambda_H \tag{4.72}$$

(This analysis is only valid for $\Delta' > 0$, since for $\Delta' < 0$ the displacement ξ differs from expression (4.70), see below.) Evaluating the coefficient in (4.69) the growth rate thus becomes

$$\gamma = 0.55\eta^{3/5}\Delta'^{4/5}(kB_0')^{2/5}. \tag{4.73}$$

Equation (4.63) indicates that λ_H is always negative for the double tearing mode. (Because of the assumed symmetry it is sufficient to consider only one resonant surface $x = x_s$.) In fact, in a plane configuration all modes are ideally stable in the framework of reduced MHD. For long wavelength $kx_s \ll 1$, however, λ_H is small, $k\lambda_H \sim (kx_s)^3$. If η is not too small such that $|\widehat{\lambda}_H| \ll 1$, or $|\lambda_H| \ll (\eta/kB_0')^{1/3}$, the growth rate is given by (4.68), $\gamma \propto \eta^{1/3}$, which is much faster than for the standard tearing mode $\gamma \propto \eta^{3/5}$.

The eigenfunction ξ corresponds to a rigid displacement of the plasma between the resonant surfaces, as indicated in fig. 4.9(b). Note that the condition $|\widehat{\lambda}_H| \ll 1$, which implies $kx_s < O(\eta^{1/9})$, can easily be satisfied for only moderately long wavelength, even if η is small. If, on the other hand, the wavelength is short such that $|\lambda_H| \gg (\eta/kB_0')^{1/3}$, the coupling of the resonant surfaces becomes weak and the modes evolve essentially as individual tearing modes with eigenfunction ξ localized around the resonant surfaces and growth rate given by (4.73).

Equation (4.72) suggests that $\Delta' \propto k^{-2}$, hence $\Delta' > 0$ for arbitrarily large kx_s, contrary to the tearing mode result $\Delta' < 0$ for $kx_s \gtrsim 1$, see (4.23). However, the ansatz (4.60), (4.61) is no longer valid for $kx_s \sim 1$. A more accurate treatment shows that there is, in fact, a finite stability threshold k_c with $\Delta'(k_c) = 0$. As an example, consider the periodic equilibrium profile

$$B_0(x) = \cos x, \tag{4.74}$$

which gives rise to growth of a periodic sequence of double tearing modes with resonant surfaces at $x = (n + \frac{1}{2})\pi$. Here the external solution can be obtained analytically. Since $B_0'' = -B_0$, (4.58) yields immediately, for $k^2 < 1$,

$$\psi_1 = \cos(\sqrt{1 - k^2}\, x), \quad \xi = \cos(\sqrt{1 - k^2}\, x)/\cos x, \tag{4.75}$$

whence

$$\Delta' = 2\sqrt{1 - k^2} \tan(\sqrt{1 - k^2}\, \pi/2). \tag{4.76}$$

For $0 < 1 - k^2 \ll 1$ we have $\Delta' \simeq \pi(1 - k^2)$, hence the stability threshold is $k = 1$, while $\Delta' \to \infty$ in the limit $k \to 0$.

The nonlinear evolution of the double tearing mode depends on the linear properties determined by k, i.e., the length $2\pi L_y$ of the system, since the longest possible wavelength dominates the nonlinear dynamics. For ideally deeply stable conditions $kx_s \sim 1$, the different resonant surfaces decouple, such that on each surface the amplitude grows slowly as a single tearing mode saturating at island width $w_I < x_s$, see fig. 4.10(a). More interesting is the ideally weakly stable case $kx_s \ll 1$, where the eigenmode has the shape illustrated in fig. 4.9(b) indicating a strong coupling of both surfaces. Here the nonlinear dynamics is much faster, both tearing modes drive each other mutually with reconnection occurring at a macroscopic current sheet. The nonlinear behavior is illustrated in fig. 4.10(b). The analytic theory, which shows the necessity of current-sheet formation and the self-similar evolution, is developed in section 4.3.2 in conjunction with the resistive kink mode.

4.3 The resistive kink mode

The magnetic field in a flux tube is, in general, twisted about the tube axis corresponding to a current flowing along the tube. If the twist T exceeds unity or, in the tokamak terminology, the safety factor $q = T^{-1}$ falls below unity, i.e., if field lines wind around the axis more than once over the length of the tube, the system becomes unstable, the field lines tending to straighten by kinking the tube. The topological equivalence of kink and twist is discussed in section 2.4.

In a plasma column, where the unstably twisted field is embedded in a less strongly twisted stable field, the kink motion leads to reconnection

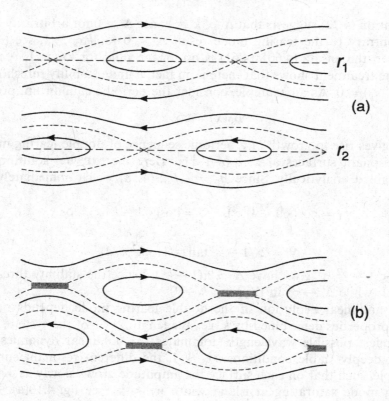

Fig. 4.10. Nonlinear behavior of the double tearing mode. (a) Saturation for weak island interaction $kx_s \lesssim 1$; (b) current-sheet formation for strong island interaction $kx_s \ll 1$.

of inner field lines with $q < 1$ with outer ones with $q > 1$, such that the resulting twist is stable. Hence resistivity plays a crucial role. The resistive kink instability has attracted much interest as a model of the sawtooth phenomenon in a tokamak plasma (see section 6.5). Here we restrict consideration to a low-β slender cylindrical pinch, (3.16) with periodic ends, the dynamics of which is described by (4.27) and (4.28).

4.3.1 Resistive kink instability

The equilibrium profiles are illustrated in fig. 4.11. The $m = 1$ mode corresponds to a rigid shift of the plasma inside the resonant radius r_s, $q(r_s) = 1$, see fig. 4.9(c). The analysis of the double tearing mode treated in the previous section remains valid with the appropriate modifications

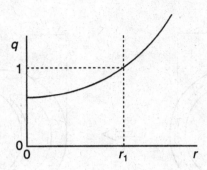

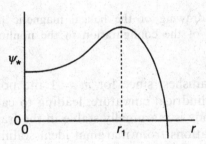

Fig. 4.11. Profiles of q and the helical flux function ψ_* for a kink-unstable cylindrical pinch.

from plane to cylindrical geometry:

$$B_{0y}(x) \rightarrow B_*(r) = B_\theta(r)[1 - q(r)], \quad j_0' \rightarrow \frac{d}{dr}\left(\frac{1}{r}\frac{d(rB_\theta)}{dr}\right), \qquad (4.77)$$

$$k = k_y \rightarrow k_\theta = \frac{1}{r}, \quad kB_0'|_{x_s} \rightarrow -F'|_{r_s} = \frac{B_\theta}{r^2}\frac{d\ln q}{d\ln r}\bigg|_{r_s} \equiv \frac{B_\theta}{r^2}\widehat{s}\bigg|_{r_s}, \qquad (4.78)$$

$$\xi \rightarrow \xi_r = -i\frac{1}{r}\frac{\phi_1}{\gamma}, \quad \psi_1 \rightarrow -rF\xi, \qquad (4.79)$$

where

$$F = \mathbf{k} \cdot \mathbf{B} = \frac{B_\theta}{r}[1 - q(r)].$$

With these substitutions the equation for the external solution (4.29) becomes

$$\frac{d}{dr}r^3F\frac{d\xi}{dr} = 0, \qquad (4.80)$$

while in the resistive layer curvature effects are negligible, hence (4.64) and (4.65) remain unchanged. Contrary to the double tearing mode, (4.59),

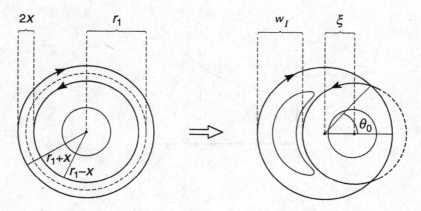

Fig. 4.12. Schematic drawing of the helical magnetic field $\mathbf{B}_* = \mathbf{e}_z \times \nabla\psi_*$, illustrating the change of the configuration in the nonlinear evolution of the resistive kink mode.

the r.h.s. of (4.80) vanishes, since for $m = 1$ the poloidal wavenumber coincides with the cylindrical curvature, leading to cancellation of terms, i.e., the ideal kink mode is marginally stable in the framework of reduced MHD, $\lambda_H = 0$. Deviations from marginal ideal stability, both stabilizing and destabilizing, arise from finite pressure (see, e.g., Rosenbluth *et al.*, 1973) and toroidicity (Bussac *et al.*, 1975). For η values typical for many laboratory plasmas these effects are, however, small, such that $|\hat{\lambda}_H| \ll 1$, and the growth rate is approximately given by (4.68),

$$\gamma = (\eta F'^2)^{1/3}. \tag{4.81}$$

For still smaller η, as in many astrophysical plasmas, resistivity is usually negligible superseded by more efficient collisionless reconnection processes, as discussed in chapters 6 and 7.

4.3.2 Small-amplitude nonlinear evolution

For island width exceeding the resistive layer width $\delta \sim \eta^{1/3}$, a current sheet of macroscopic length is generated. Reconnection at this sheet governs the nonlinear evolution. The configuration is illustrated in fig. 4.12. If reconnection would not oppose and slow down the inflow motion, the surfaces would remain circular, the shift of the core ξ_0 would equal the difference between the two radii, $\xi_0 = 2x$, and the separatrix would consist of two osculating circles. Since, however, resistive reconnection is relatively slow, the field on both sides becomes aligned over a macroscopic arc θ_0 corresponding to a current sheet, where both branches of the separatrix coincide. Hence the internal shift is somewhat larger than the difference of the two radii, $\xi_0 > 2x$. The island width, defined as the distance between the separatrix branches at the O-point, is

$$w_I = \xi_0 + 2x. \tag{4.82}$$

The configuration, in particular $x(\xi_0)$ and θ_0, can be determined quantitatively for small island width (Waelbroeck, 1989), generalizing the non-linear theory of the ideal kink mode by Rosenbluth *et al.* (1973). Since the reconnection time-scale is much longer than the Alfvén time, the evolution of the resistive kink instability proceeds through a sequence of helical MHD equilibria. From the vorticity equation (4.28) we find the equilibrium condition $j = j(\psi_*)$,

$$\nabla^2 \psi_* = j(\psi_*) - 2B_z/R. \tag{4.83}$$

For small shift ξ_0 only the radial derivative is retained,

$$\nabla^2 \psi_* \simeq \partial_r^2 \psi_*.$$

Multiplying (4.83) by $\partial_r \psi_*$ and integrating over r gives

$$\tfrac{1}{2}(\partial_r \psi_*)^2 = F(\psi_*) - G(\theta), \tag{4.84}$$

where

$$F(\psi_*) = \int_{\psi_0}^{\psi_*} [j(\psi) - 2B_z/R]d\psi \tag{4.85}$$

is essentially the total current between the surfaces ψ_0 and ψ_*, and $G(\theta)$ is a constant of integration. Solving (4.84) for $\partial_r \psi_*$,

$$\partial_r \psi_* = \pm\sqrt{2[F(\psi_*) - G(\theta)]}, \tag{4.86}$$

we find that flux surfaces ψ_* with $F(\psi_*) < \max[G(\theta)]$ will be limited to a certain θ-range between the mirror points θ_m, $G(\theta_m) = F(\psi_*)$ corresponding to flux surfaces within the island. Surfaces with $F(\psi_*) > \max[G(\theta)]$ are not restricted in azimuth and hence encircle the plasma core.

We want to determine the displacement of the flux surfaces from the initial unperturbed position. In the vicinity of the resonant surface the unperturbed helical flux function can be expanded

$$\psi_* = \tfrac{1}{2}\psi_0'' x^2, \quad x = r - r_1. \tag{4.87}$$

Since there are two surfaces with the same value of ψ_*, it is more practical to label surfaces by their initial position x. The radial displacement from this position is

$$\xi(x, \theta) = r(x, \theta) - r_1 - x, \tag{4.88}$$

where $r(r, \theta)$ is the actual position of the surface. Writing $(\partial_r \psi_*)^{-1} = \partial r/\partial \psi_* = \partial_x r/\partial_x \psi_*$, $\partial_x \psi_* = \psi_0'' x$, and taking into account the appropriate sign of the square root, (4.86) becomes

$$\partial_x r(x, \theta) = \frac{|x|}{\sqrt{f(x) - g(\theta)}}, \tag{4.89}$$

valid for both signs of x, with $f = F/\frac{1}{2}(\psi_0'')^2$, $g = G/\frac{1}{2}(\psi_0'')^2$. Integrating (4.89) and using relation (4.88) gives

$$\xi(x, \theta) = \xi[x_m(\theta), \theta] + \int_{x_m(\theta)}^{x} dx \left(\frac{|x|}{\sqrt{f(x) - g(\theta)}} - 1 \right), \tag{4.90}$$

where $x_m(\theta)$ is the x value of the flux surface with the mirror point at the angle θ, $f(x_m) = g(\theta)$. For large $|x|$, i.e., far away from the separatrix, $\xi(x, \theta)$ has the form of the linear kink mode. To match this behavior to the nonlinear solution in the vicinity of the separatrix we form the limits $\lim_{x \to \pm \infty} \xi(x, \theta) = \xi_{\pm}(\theta)$:

$$\xi_{\pm}(\theta) = \xi(\pm |x_m|, \theta) \pm \int_{|x_m|}^{\infty} dx \left(\frac{x}{\sqrt{f(x) - g(\theta)}} - 1 \right), \tag{4.91}$$

using the asymptotic behavior of $f(x) \propto \psi_* \propto x^2$. From the linear mode properties, it follows that

$$\xi_+(\theta) = 0, \quad \xi_-(\theta) = \xi_0 \cos \theta, \tag{4.92}$$

choosing the phase such that the O-point is located at $\theta = \pi$. To eliminate the unknown functions $\xi(\pm x_m, \theta)$ we use the property that the combined shift from outside and from inside to the mirror point just spans the original distance between the surfaces $\pm x_m$:

$$\xi(-x_m, \theta) - \xi(x_m, \theta) = 2x_m. \tag{4.93}$$

We then obtain the final result

$$\xi_0 \cos \theta = 2x_m(\theta) - 2 \int_{x_m(\theta)}^{\infty} dx \left(\frac{x}{\sqrt{f(x) - g(\theta)}} - 1 \right), \tag{4.94}$$

which essentially determines $g(\theta)$. To calculate $f(x)$, we use the property that reconnection is sufficiently fast such that the value of ψ_* of a fluid element is conserved, (4.27). Since the motion is incompressible, the volume between neighboring surfaces ψ_* and $\psi_* + d\psi$ is conserved, $dV_+ + dV_- = dV$. By use of relation (4.89), this property can be expressed in the form:

$$\int_{\theta_m}^{2\pi - \theta_m} \frac{d\theta}{2\pi} \frac{1}{\sqrt{f(x) - g(\theta)}} = \frac{1}{|x|}. \tag{4.95}$$

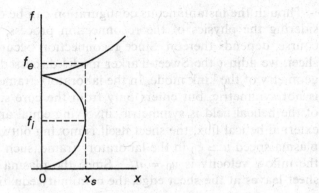

Fig. 4.13. Qualitative behavior of the function $f(x)$.

Evaluation of this equation yields $f(x)$. Without solving the equation explicitly, one can demonstrate the existence of a current sheet connecting the tips of the island. The integral is positive and diverges for $f = 0$ and for $f = g_{max}$, i.e., $x = 0$ at these points. Hence the function $f(x)$ consists of three branches, as indicated in fig. 4.13. Since f is the (normalized) current flowing in the region between the surfaces $x = 0$ and x, the physically correct function $f(x)$ must be single valued, which implies a jump Δf at the separatrix $x = x_s$ corresponding to a current sheet. The last flux surface in the island has the value f_i, which is smaller than g_{max}. Hence the island is limited in azimuthal extent to the range $\theta > \theta_0$, where $g(\theta_0) = f_i$. By contrast, the last non-reconnected flux surface has the value f_e and extends all the way around the axis, since $f_e > g_{max}$. For $-\theta_0 < \theta < \theta_0$, where the surfaces labeled $x = \pm x_s$ touch, the flux is continuous since $\psi_*(-x) = \psi_*(x)$, but the derivative $\partial_r \psi_*$, the helical field B_*, exhibits a symmetric θ-dependent discontinuity from $-\sqrt{f_e - g(\theta)}$ on the inside ($x = -x_s$) to $+\sqrt{f_e - g(\theta)}$ on the outside ($x = x_s$). Note that the current singularity is not limited to the strip between the Y-points, $-\theta_0 < \theta < \theta_0$, the current sheet proper, but extends around the island along the separatrix, since $\sqrt{f_e - g(\theta)} - \sqrt{f_i - g(\theta)} \neq 0$, though its value is smaller. This property is not captured by Syrovatskii's theory, but clearly appears in the driven reconnection simulations, fig. 3.11. On the other hand, the present theory misses Syrovatskii's singular behavior at the sheet edge, (3.136).

Equation (4.94), resulting from the equilibrium condition, and the conservation of the helical flux, (4.95), have been solved numerically for $g(\theta)$ and $f(x)$, which determines the configuration in terms of the core displacement ξ_0 (Zakharov et al., 1993). One finds in particular

$$\theta_0 = 60°, \quad w_I = 1.84\xi_0, \quad x_s = 0.4\xi_0. \tag{4.96}$$

($x_s = 0.5\xi_0$ would correspond to $\theta_0 = 0$, i.e., osculating circular surfaces.)

Though the instantaneous configuration can be calculated without considering the physics of the reconnection process, the time evolution, of course, depends thereon. Since reconnection occurs in a resistive current sheet, we adjust the Sweet–Parker model discussed in section 3.2.2 to the geometry of the kink mode. In the laboratory frame the flow into the sheet is not symmetric, but enters only from the core side. Since reconnection of the helical field is symmetric, involving equal amounts of internal and external helical flux, the sheet itself is moving outward radially at half the plasma speed $u \simeq \dot{\xi}_0$ in the laboratory frame, such that in the sheet frame the inflow velocity is $u_0 = u/2$. Since the plasma entering all along the sheet leaves at the sheet edge, the continuity equation gives the relation

$$\int_0^{\theta_0} u_{0n} d\theta = u_0 r_1 \sin \theta_0 = v_0 \delta(\theta_0), \tag{4.97}$$

inserting the normal component $u_{0n} = u_0 \cos \theta$. The outflow velocity equals the Alfvén velocity of the upstream helical field

$$v_0 = B_* = |\partial_r \psi_*| = |\psi_0''| x_s \simeq |\psi_0''| \xi_0/2. \tag{4.98}$$

(Note that in our normalization $B_* = v_A$.) In contrast to the case of a plane current sheet with a homogeneous inflow the sheet width is not constant but varies along the sheet $\delta = \delta(\theta)$, which is determined by Ohm's law, (4.27),

$$u_{0n} \simeq \eta/\delta(\theta),$$

hence

$$\delta(\theta_0) = \eta/(u_0 \cos \theta_0). \tag{4.99}$$

Combining these relations, the change of the island width $w_I \simeq 2\xi_0$ is given by the equation

$$\dot{w}_I \simeq 2 \left(\frac{\eta |\psi_0''|}{r_1 \cos \theta_0 \sin \theta_0} w_I \right)^{1/2}, \tag{4.100}$$

which yields the island width growth law

$$w_I = \frac{\eta |\psi_0''|}{r_1 \cos \theta_0 \sin \theta_0} (t - t_0)^2 \tag{4.101}$$

or

$$u = \frac{\eta |\psi_0''|}{r_1 \cos \theta_0 \sin \theta_0} (t - t_0). \tag{4.102}$$

The time-scale of the resistive kink mode evolution, defined by $w_I \sim r_1$, is $\tau = O(\eta^{1/2})$. Though the velocity into the sheet is small, $u_0 = O(\eta^{1/2})$, inertia is not negligible in the dynamics, since the v_0 velocity along the

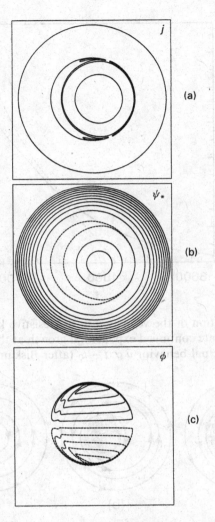

Fig. 4.14. Contour plots of (a) current density j, (b) helical flux ψ_*, and (c) stream function ϕ, from a numerical simulation of the nonlinear resistive kink mode with $\eta = 10^{-7}$ (from Biskamp, 1991).

sheet is independent of η. Equation (4.102) is confirmed by numerical simulations. Figure 4.14 illustrates the configuration during the early nonlinear phase of the instability, while fig. 4.15 shows the velocity of the plasma center (continuous curve), which clearly approaches a linear law. For comparison, the upper dashed curve indicates the continuation of the exponential growth of the linear kink instability, hence the algebraic growth of the current-sheet-controlled reconnection process is substantially slower. Numerically one finds $\dot{u} \simeq 4 \times 10^{-7}$, while inserting the equilibrium

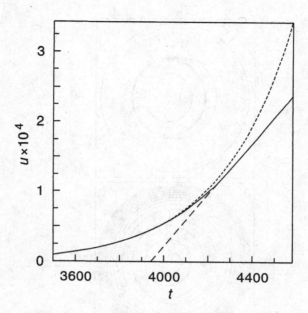

Fig. 4.15.　Time evolution of the velocity u in the resistive kink mode (full line). The small dashes indicate continued exponential growth at the linear growth rate, the large dashes the actual behavior $u \propto t - t_0$ (after Biskamp, 1991).

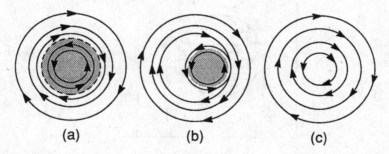

(a)　　　　　　　　(b)　　　　　　　　(c)

Fig. 4.16.　Schematic drawing of the evolution (a) → (b) → (c) of the helical magnetic field $\mathbf{B}_*$ in the resistive kink mode.

quantities ψ_0'', r_1 used in the simulation, (4.102) gives $\dot{u} \simeq 2.5 \times 10^{-7}$, a good agreement in view of the approximative character of the theory.

4.3.3 Final state of the resistive kink mode

The nonlinear evolution of the resistive kink mode continues until the entire helical flux contained inside the original resonant surface is reconnected, as illustrated schematically in fig. 4.16. The algebraic growth, (4.101), is, however, only valid for sufficiently small amplitude $w_I \ll r_1$.

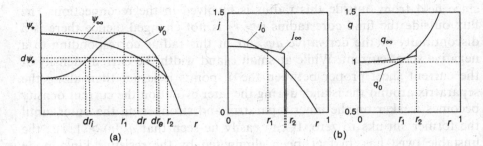

Fig. 4.17. (a) Calculation of the final state ψ_∞ of the resistive kink instability from the initial state ψ_0. (b) Initial and final current density and safety factor profiles in the resistive kink mode.

At finite island width the coefficients ψ_0'' and r_1 cannot be considered constant, and the solution loses its self-similar character. Numerical simulations show that, at large amplitude, the dynamics accelerates, becoming effectively exponential again (Aydemir, 1997). The most important effect is the shrinking of the length of the current sheet, which is proportional to the radius of the remaining core plasma $r_c = r_1 - \xi_0/2 \simeq r_1 - w_I/4$. Insertion into (4.100) shows that, in the deep nonlinear phase, island growth becomes faster than t^2.

The final state is again cylindrically symmetric, where the island containing the reconnected helical flux replaces the original core plasma, which is called complete reconnection. Conservation of ψ_* determines the final state uniquely, if one furthermore assumes that the $q = 1$ resonance has been eliminated, i.e., the final helical flux distribution ψ_∞ is a monotonically decreasing function (Kadomtsev, 1975). The construction of the final state is indicated in fig. 4.17(a). We use the conservation of area

$$r_i|dr_i| + r_e|dr_e| = r|dr| \tag{4.103}$$

and of helical flux

$$\left.\frac{d\psi_0}{dr}\right|_i dr_i = \left.\frac{d\psi_0}{dr}\right|_e dr_e = \frac{d\psi_\infty}{dr} dr. \tag{4.104}$$

Note that for $dr > 0$ one has $dr_e > 0$, $dr_i < 0$. Integration of (4.103) gives

$$r_e^2 - r_i^2 = r^2. \tag{4.105}$$

Using this relation the first part of (4.104) gives an equation for dr_i/dr, which determines r_i, r_e as functions of r. Then the second part yields the final helical flux function ψ_∞.

Figure 4.17(b) illustrates the qualitative features of the final state. The radius of the reconnected core is larger than the initial resonant radius r_1,

since field from outside this radius is involved in the reconnection. The
flux outside the final core radius $r > r_2$ is not changed, hence there is a
discontinuity of the derivative $d\psi_\infty/dr$ at this radius corresponding to a
negative sheet current. While at small island width the current density at
the current sheet proper between the Y-points is larger than along the
separatrix around the island, during the later evolution the current density
becomes weaker on the inner separatrix and stronger on the outer, until
the former shrinks to zero. It can easily be seen that $q_\infty(r) \geq 1$, i.e., the
unstable twist has, in fact, been eliminated by the resistive kink mode
reconnection. From the definitions of q and ψ_* we have

$$q^{-1} = \frac{R}{rB_z}\left(\psi'_* + \frac{r}{R}B_z\right) = \frac{R}{rB_z}\psi'_* + 1. \qquad (4.106)$$

Since $\psi'_\infty(r) \leq 0$, one finds $q_\infty(r) \geq 1$. At the origin, $\psi_\infty(r)$ is rather flat.
For the usual case of a quadratic dependence of $\psi_0(r)$ at the resonant
radius r_1, $\psi_0 \propto (r - r_1)^2$, area conservation, (4.103), around this radius
gives

$$rdr = 2r_1 dr_e = 2r_1\sqrt{2d\psi/\psi''_0},$$

hence $d\psi \propto r^2 dr^2$, i.e.,

$$\psi_\infty \propto r^4, \qquad (4.107)$$

and $q_\infty(0) = 1$ from (4.106). The change of the current density and the
safety factor profiles is illustrated in fig. 4.17(b).

4.4 Coalescence instability

The theory of the tearing mode discussed in section 4.1, in particular
the saturation island width, is only relevant if the mode corresponds
to the longest wavelength permitted in the magnetic configuration. This
condition is satisfied for certain helical modes in a cylindrical plasma
column of axial periodicity length $L_z = 2\pi R$ (a topological torus of major
radius R), namely the modes with mode numbers (m, n) with no common
divisor, such as (2,1), (3,1), (3,2). In other cases the nonlinear tearing mode
will be dominated by a different, more rapid process, the coalescence of
neighboring magnetic islands. This happens, for example, to a (4,2) mode,
which is reduced by coalescence to a (2,1) mode. In a long sheet pinch,
coalescence is the governing dynamical process.

Both the tearing and the coalescence instability result from the same
physical mechanism, the attractive force between parallel currents. While
the tearing mode is a slow process of current-density condensation from a
uniform (in y) distribution in the original sheet pinch, coalescence results
from the interaction of these current-density blobs, which is the stronger

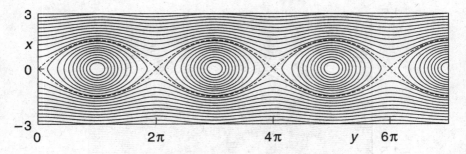

Fig. 4.18.   Islated sheet pinch solution (4.109) with $\epsilon = 0.71$.

the larger the blobs, i.e., the larger the island width. Contrary to the tearing mode coalescence is an ideal instability, though its later evolution is determined by current-sheet reconnection, similar to the nonlinear behavior in the double tearing mode or the resistive kink mode.

Island coalescence in a resistive plasma has been studied in a corrugated neutral-sheet equilibrium, illustrated in fig. 4.18:

$$\psi(x, y) = \frac{B_\infty}{k} \ln[\cosh(kx) + \epsilon \cos(ky)]. \qquad (4.108)$$

B_∞ is the field intensity for $|x| \to \infty$ and ϵ is a measure of the island width w_I given by $\cosh(w_I k/2) = 1 + 2\epsilon$, $w_I k \simeq 4\sqrt{\epsilon}$ for $w_I k \ll 1$. This configuration belongs to a general class of solutions of the equilibrium equation $\nabla^2 \psi = k^2 e^{-\psi}$, which can be written in the form $\psi = -\ln \{8|f'|^2/(1+|f|^2)^2\}$, where $f(z)$ is an arbitrary complex function of $z = x + iy$ (Fadeev et al., 1965). Solution (4.108) results from the choice $f = \sqrt{2k/B_\infty}(\epsilon + e^{kz})$, where the corresponding current density is

$$j = \nabla^2 \psi = B_\infty k \frac{1 - \epsilon^2}{[\cosh(kx) + \epsilon \cos(ky)]^2}. \qquad (4.109)$$

The current maxima are located at the minima of ψ, i.e., at the O-points of the islands along the y-axis, $y = (2n + 1)\pi$. Hence the configuration corresponds to a chain of current-density blobs which, if perturbed from their equilibrium positions, attract each other. The equilibrium is ideally unstable with respect to pairwise attraction for any island width (Pritchett & Wu, 1979).

Numerical simulations of island coalescence (Biskamp & Welter, 1980) show that the process can be divided into an ideal MHD phase, where the islands are freely accelerated toward each other leading to flux compression and formation of a current sheet, and a phase of quasi-stationary current-sheet reconnection. The length Δ of the sheet is determined by the geometry

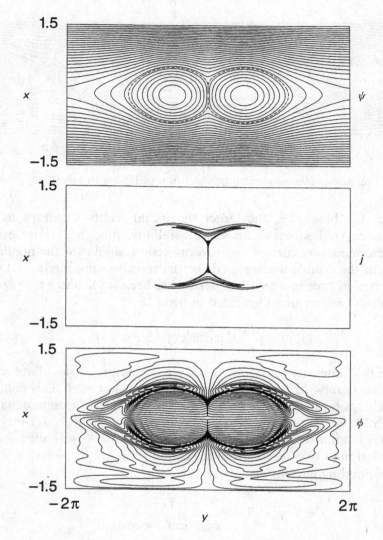

Fig. 4.19.   Contours of ψ, j, ϕ during island coalescence.

of the system, $\Delta \simeq w_l$. For intermediate values of the resistivity, $\eta = 10^{-2} - 10^{-4}$, an interesting self-similar behavior is observed. Reducing η, the increase of the field B_0 in front of the sheet and the slowing down of the upstream velocity u_0 into the sheet,

$$B_0 \sim \eta^{-1/3}, \quad u_0 \sim \eta^{1/3}, \tag{4.110}$$

are such that the reconnection rate $E = u_0 B_0$ is independent of η. This does not, however, imply that the reconnection process has any resemblance to

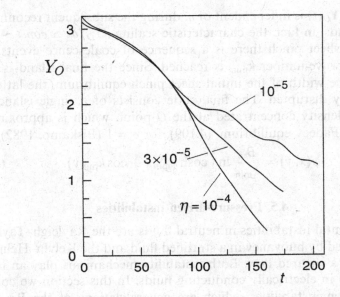

Fig. 4.20. Position of the O-point $Y_O(t)$ during island coalescence for three values of η. Ideal phase $t \lesssim 75$, reconnection phase $t > 75$.

the Petschek mechanism (section 3.3.4), since reconnection takes place at a macroscopic current sheet.

Obviously the scaling (4.110) can be valid only over a certain η-range, since the value of B_0 will not exceed the maximum field B_{max}, which would be reached in the ideal case $\eta = 0$ for $u_0 = 0$, when the attractive motion is reversed because of the repelling force exerted by the compressed field. Hence, the above scaling breaks down when B_0 approaches B_{max}. For smaller values of η, typically $\eta < 10^{-4}$, one finds the Sweet–Parker scaling expected for current-sheet reconnection (3.157),

$$\overline{E} \sim \overline{u_0} \sim \eta^{1/2}, \quad \overline{B_0} \simeq B_{max} \sim \eta^0. \tag{4.111}$$

The island configuration during coalescence is illustrated in fig. 4.19. It is found that the current sheet exhibits the same features as in the simulations of driven reconnection (see fig. 3.12), in particular the complex structure at the sheet edge.

In (4.111) the overlines indicate the time average, since for small η the motion of the island plasma is modulated by an oscillation resulting from the kinetic energy gained during the initial free acceleration phase. The behavior is indicated in fig. 4.20 for three values of the resistivity, showing the change Y_O of the position of the O-point during coalescence, $Y_O = \pi$ corresponding to the initial unperturbed island positions, $Y_O = 0$ to the final coalesced state. While during the first phase $t \lesssim 75$ of free acceleration, $Y_0(t)$ is independent of η, during the subsequent reconnection phase one finds, in fact, the characteristic scaling $dY_0/dt \simeq const \sim \eta^{1/2}$.

In a long sheet pinch there is a sequence of coalescence events, until the smallest wavenumber k_{min} is reached. Since the final island is much wider than the width of the initial sheet pinch equilibrium, the latter will be completely disrupted. The final state consists of a single island with the current density concentrated at the O-point, which is approximately given by the Fadeev equilibrium, (4.109), for $\epsilon = 1$ (Biskamp, 1982)

$$\psi(x, y) = \frac{B_\infty}{k_{min}} \ln \left(\cosh k_{min} x + \cos k_{min} y \right). \tag{4.112}$$

4.5 Pressure-driven instabilities

The fundamental instabilities in neutral fluids are the Rayleigh–Taylor instability caused by buoyancy in a stratified fluid and the Kelvin–Helmholtz instability of a sheared flow. Both instability mechanisms play an important role also in electrically conducting fluids. In this section we consider pressure-driven instabilities, which are generalizations of the Rayleigh–Taylor process, in particular interchange and ballooning modes as well as the finite-β tearing mode. Section 4.6 is devoted to shear flow instabilities in plasmas.

4.5.1 Interchange and ballooning modes

In the presence of gravity $\mathbf{g}$ an inverse density stratification (heavy fluid on top of light fluid), $\mathbf{g} \cdot \nabla \rho_0 < 0$, is unstable. The growth rate is largest for small-scale modes $kL_\rho \gg 1$, where $L_\rho = \rho_0/\rho_0'$ is the density scale-height,

$$\gamma = \left(\frac{g\rho_0'}{\rho_0} \frac{k_\perp^2}{k^2} \right)^{1/2} \simeq \left(\frac{g}{L_\rho} \right)^{1/2}, \tag{4.113}$$

where $k_\perp$ is the wave vector component perpendicular to $\mathbf{g}$. This is called the Rayleigh–Taylor instability.

In strongly magnetized plasmas, gravity is usually negligible. There is, however, an effect which acts in a similar way, the centrifugal force in a curved magnetic field. Contrary to the situation in the Taylor–Couette system, where the centrifugal force depends on the flow velocity, in a plasma it is caused by the thermal motion of the particles. The pseudo-gravitational force is

$$\rho \mathbf{g}_{eff} = -2p\boldsymbol{\kappa},^* \tag{4.114}$$

---

[*] The effective force perpendicular to the magnetic field results not only from the centrifugal force $-mv_\parallel^2\boldsymbol{\kappa}$ due to the parallel particle motion but also from the force on the magnetic moment $\mu_B = mv_\perp^2/2B$ of the quasi-particle or guiding center in an inhomogeneous field, $-\mu_B \nabla_\perp B = -\mu_B B\boldsymbol{\kappa}$, using (3.41), hence $\rho \mathbf{g}_{eff} = -\rho(\overline{v_\parallel^2} + \frac{1}{2}\overline{v_\perp^2})\boldsymbol{\kappa} = -2p\boldsymbol{\kappa}$.

where $\kappa = \mathbf{b} \cdot \nabla \mathbf{b}$ is the field-line curvature. In a finite-β plasma the appropriate equations are the high-β reduced MHD equations (3.59)–(3.61). Linearization gives, with $\partial_t = \gamma$,

$$\gamma \rho_0 \nabla_\perp^2 \phi = \mathbf{B}_0 \cdot \nabla j_1 + 2(\mathbf{b} \times \kappa) \cdot \nabla p_1, \qquad (4.115)$$

$$\gamma \psi_1 = \mathbf{B}_0 \cdot \nabla \phi + \eta \nabla_\perp^2 \psi_1, \qquad (4.116)$$

$$\gamma p_1 = -(\mathbf{b} \times \nabla \phi) \cdot \nabla p_0.^* \qquad (4.117)$$

Since in this section we restrict consideration to small-scale modes, $k_\perp L_p \gg 1$, $L_p = p_0/p_0'$ = pressure scale-height, one has $\mathbf{B}_1 \cdot \nabla j_0 \ll \mathbf{B}_0 \cdot \nabla j_1$, hence the kink-mode term $\mathbf{B}_1 \cdot \nabla j_0$ has been omitted in (4.115). In the local approximation one neglects the variation of the coefficients, which corresponds to the Boussinesq approximation in fluid dynamics (see, e.g., Drazin & Reid, 1981). The perturbation can be Fourier transformed, $\nabla f \to i k f$, such that (4.115)–(4.117) become algebraic. Choosing $\kappa = \kappa \mathbf{e}_x$, $\nabla p_0 = p_0' \mathbf{e}_x$, $\mathbf{b} = \mathbf{e}_z$, and neglecting η for the time being, we obtain the dispersion relation

$$\gamma^2 = \frac{2\kappa p_0'}{\rho_0} \frac{k_y^2}{k_\perp^2} - k_\parallel^2 v_A^2, \qquad (4.118)$$

where $v_A^2 = B_0^2/\rho_0$ and $k_\perp, k_\parallel$ refer to the direction of the magnetic field. For κ pointing along the pressure gradient, $\kappa p_0' > 0$, convective instability arises, if the line-bending stabilizing term $k_\parallel^2 v_A^2$ is sufficiently small,

$$\kappa \beta' > k_\parallel^2 \qquad (4.119)$$

with $\beta' = 2p_0'/B_0^2$. (If $\kappa p_0' > 0$, the curvature is called *unfavorable* or *bad*, since the resulting convective instability is obviously detrimental to confinement in a plasma device. The opposite case $\kappa p_0' < 0$ is called *favorable* curvature.) In real magnetic configurations, however, $k_\parallel$ cannot, in general, become arbitrarily small. A lower limit of $k_\parallel$ arises because of two different effects: (a) the extent along a field line, where κ is destabilizing, is finite; (b) neighboring field lines are not strictly parallel, since there is, in general, magnetic shear.

For a mode localized in the region of destabilizing curvature we have $k_\parallel \sim L_\kappa^{-1}$, where L_κ is the extent of this region along the field line. In a tokamak plasma, field-line curvature is mainly due to toroidal curvature, hence $\kappa \simeq -\mathbf{e}_R/R$, where $\mathbf{e}_R$ is in the direction of the major torus radius R. A field line winding around the plasma column with a twist $T = q^{-1}$ thus passes alternately through regions of unfavorable curvature on the torus outside and favorable curvature on the torus inside, the distance

---

* We use the same normalization as before, but since in a stratified medium the mean density may depend on position, ρ_0 is written explicitly.

between both being $L_\kappa = \pi R q$. A perturbation will grow, if it is confined to the outside. Hence, from (4.119), a necessary condition for this so-called *ballooning instability* is, qualitatively,

$$-\beta' > \frac{1}{Rq^2}. \tag{4.120}$$

Modes not localized in regions of destabilizing curvature feel only the average curvature of the field line $\bar\kappa$. These modes are called *interchange modes*, since they correspond simply to an interchange of the positions of neighboring flux tubes together with their plasma content. Interchange modes may be unstable in a cylindrical pinch, where the curvature is constant along the field lines, $\kappa = \kappa^c \mathbf{e}_r$, $\kappa^c \simeq -B_\theta^2/rB_z^2 = -r/(Rq)^2$. In a tokamak plasma $\bar\kappa$ consists of two parts, the average toroidal curvature, which is stabilizing, since the favorable contribution on the torus inside is slightly larger than the unfavorable on the outside, $\bar\kappa^t \simeq r/R^2$, and the cylindrical curvature, κ^c. For a tokamak of small inverse-aspect-ratio $\epsilon = r/R \ll 1$ and circular cross-section one obtains (Shafranov & Yurchenko, 1968)

$$\bar\kappa = \kappa^c(1 - q^2) = -\frac{r}{R^2 q^2}(1 - q^2). \tag{4.121}$$

Hence interchange modes are not excited in a tokamak plasma if $q^2 > 1$, which is the case for the major part of the plasma cross-section. In a reversed-field pinch, however, where the toroidal field is much weaker and hence the field-line twist much larger than in a tokamak, $q^2 \ll 1$, the cylindrical curvature dominates, such that interchange modes may be unstable.

The second effect limiting $k_\parallel$ from below, the magnetic shear, is rather universally present. If the perturbation is aligned with a particular field line $\mathbf{k} \cdot \mathbf{B} = 0$, it is in general no longer so on the neighboring field lines. Let us first discuss this effect qualitatively for a cylindrical plasma. In the vicinity $x = r - r_s$ of the resonant radius r_s we have

$$\mathbf{k} \cdot \mathbf{B} = \frac{m}{r}B_\theta \left(1 - \frac{n}{m}q(r)\right) \simeq -\frac{m}{r_s^2}B_\theta(r_s)\hat{s}x, \tag{4.122}$$

where $\hat{s} = d\ln q/d\ln r$ is the shear parameter or simply the shear, see (3.18). Since the radial mode width is typically of the order of the poloidal wavelength, $(m/r_s)x \sim 1$, we find from (4.122)

$$k_\parallel \sim \hat{s}B_\theta/B_z r_s = \hat{s}/Rq, \tag{4.123}$$

which on substitution in (4.119) gives the condition for interchange instability in a cylindrical plasma

$$\kappa^c\beta' > \alpha\frac{\hat{s}^2}{R^2 q^2}, \quad \text{i.e., } -\beta' > \alpha\frac{\hat{s}^2}{r}, \tag{4.124}$$

where α is a numerical factor of order unity. If $\hat{s} = O(1)$, as in a tokamak, we see that interchange modes require much higher β, $\beta \sim 1$, than ballooning modes, (4.120).

For a rigorous treatment of localized modes in a general equilibrium configuration it is convenient to use flux coordinates chosen in such a way that one of the two coordinates spanning a flux surface is in the direction of the magnetic field. The problem then reduces to the solution of an ordinary differential equation along the field line. Consider a flux tube of small but finite diameter encompassing the field line, about which the mode is localized. In a local Cartesian coordinate system the sheared magnetic field has the form

$$\mathbf{B}_0 = B_0[\mathbf{e}_z + (x/L_s)\mathbf{e}_y], \tag{4.125}$$

where L_s is the shear length, $L_s^{-1} = \hat{s}/Rq$ in a cylindrical plasma. We now transform to a twisted coordinate system (Beer *et al.*, 1995):

$$\xi = x, \quad \eta = y + (x/L_s)z, \quad \zeta = z, \tag{4.126}$$

such that ζ is the coordinate along the field line, which gives

$$\mathbf{B} \cdot \nabla = B_0 \partial_\zeta, \quad \nabla_\perp^2 = \partial_x^2 + \partial_y^2 = [\partial_\xi + (\zeta/L_s)\partial_\eta]^2 + \partial_\eta^2. \tag{4.127}$$

Neglecting η, we can substitute ψ_1 and p_1 from (4.116) and (4.117),

$$\gamma^2 \rho_0 \nabla_\perp^2 \phi = \mathbf{B}_0 \cdot \nabla \nabla_\perp^2 (\mathbf{B}_0 \cdot \nabla \phi) + 2(\mathbf{b} \times \boldsymbol{\kappa}) \cdot \nabla[(\mathbf{b} \times \nabla p_0) \cdot \nabla \phi]. \tag{4.128}$$

In the twisted coordinates, (4.126), the coefficients depend only on the parallel coordinate ζ, the dependence on x in the operator $\mathbf{B}_0 \cdot \nabla$ being eliminated. Hence we can Fourier transform in the perpendicular directions $\partial_\xi \to ik_\xi$, $\partial_\eta \to ik_\eta$, such that (4.128) becomes an ordinary differential equation. k_ξ, k_η are the covariant components of $\mathbf{k}_\perp$,

$$\mathbf{k}_\perp = k_\xi \nabla \xi + k_\eta \nabla \eta = [k_\xi + (\zeta/L_s)k_\eta]\mathbf{e}_x + k_\eta \mathbf{e}_y, \tag{4.129}$$

$$\nabla_\perp^2 = -k_\perp^2 = -[k_\xi + (\zeta/L_s)k_\eta]^2 - k_\eta^2. \tag{4.130}$$

Introducing normal and geodesic curvature components,

$$\boldsymbol{\kappa} = \kappa_n \mathbf{e}_x + \kappa_g \mathbf{e}_y,$$

the operator in the last term in (4.128) becomes

$$(\mathbf{b} \times \boldsymbol{\kappa}) \cdot \nabla_\perp = i(\mathbf{b} \times \boldsymbol{\kappa}) \cdot \mathbf{k}_\perp = i\{\kappa_n k_\eta - \kappa_g[k_\xi + (\zeta/L_s)k_\eta]\}.$$

Hence (4.128) assumes the form,

$$\frac{\gamma^2}{v_A^2} F^2 \phi = \frac{d}{d\zeta} F^2 \frac{d}{d\zeta} \phi + \beta'\left(\kappa_n - \frac{\zeta - \zeta_0}{L_s}\kappa_g\right)\phi, \tag{4.131}$$

where we have used the notation

$$F^2 = \frac{k_\perp^2}{k_\eta^2} = 1 + \frac{(\zeta - \zeta_0)^2}{L_s^2}, \quad \zeta_0 = -\frac{k_\xi}{k_\eta} L_s.$$

Equation (4.131) is called the ballooning equation. The curvature components κ_n, κ_g and the shear length L_s are functions of ζ. Equation (4.131) must, in general, be solved numerically, but some special cases can be discussed analytically. The condition for instability of interchange modes, Mercier's criterion (Mercier, 1960), is obtained by considering the asymptotic solution for $\zeta \to \infty$ (see, e.g., Freidberg, 1987). In the cylindrical case this reduces to Suydam's instability criterion (Suydam, 1958),

$$\kappa^c \beta' > \frac{1}{4L_s^2} = \frac{1}{4} \frac{\hat{s}^2}{R^2 q^2}, \tag{4.132}$$

which specifies the numerical factor in the qualitative relation (4.124), or

$$\beta' \kappa^c L_s^2 \equiv D_R > \frac{1}{4}. \tag{4.133}$$

In the case of a large-aspect-ratio tokamak of circular cross-section, Mercier's criterion becomes (Shafranov & Yurchenko, 1968)

$$\overline{\kappa} \beta' \equiv \kappa^c (1 - q^2) \beta' > \frac{1}{4} \frac{\hat{s}^2}{R^2 q^2}. \tag{4.134}$$

Effect of finite resistivity Let us now include finite resistivity, which permits the magnetic field to slip across the plasma and thus reduce the stabilizing line-bending effect. The dispersion relation in the local approximation, (4.118), now reads

$$\gamma^2 = \frac{2\kappa p_0'}{\rho_0} - \frac{k_\parallel^2 v_A^2}{1 + \eta k_\perp^2 / \gamma}, \tag{4.135}$$

hence for $\eta k_\perp^2 / \gamma > 1$ the stabilizing term is reduced. We now show that in an inhomogeneous system, where $k_\parallel$ arises because of magnetic shear, the instability threshold is, in fact, independent of the shear effect, only the growth rate depends on η. From (4.116) we see, by use of (4.130), that the resistive term does not introduce additional differential operators,

$$\psi_1 = \frac{\mathbf{B}_0 \cdot \nabla \phi}{\gamma + \eta k_\perp^2}, \tag{4.136}$$

such that the ballooning equation (4.131) becomes

$$\frac{\gamma^2}{v_A^2} F^2 \phi = \frac{d}{d\zeta} \frac{F^2}{1 + (\eta k_\perp^2 F^2 / \gamma)} \frac{d}{d\zeta} \phi + \beta' \left(\kappa_n - \frac{\zeta - \zeta_0}{L_s} \kappa_g \right) \phi = 0. \tag{4.137}$$

Equation (4.137) can be solved in a way methodically similar to that used in tearing mode theory, section 4.1.1. For $\eta k_\perp^2 / \gamma \ll 1$ the problem exhibits two disparate scale-lengths along the field line, the equilibrium scale L_s, and the resistive scale defined by $\eta k_\perp^2 F^2 / \gamma \sim 1$, $L_\eta = L_s (\gamma / \eta k_\perp^2)^{1/2} \gg L_s$. In analogy to the tearing mode boundary layer problem, one solves (4.137) separately on both scales using appropriate approximations. Since the general theory, which also includes finite compressibility (Correa-Restrepo, 1982), would lead beyond the scope of this presentation, we only outline the method and sketch the results. The theory applies to toroidal configurations, where field lines wind around the plasma indefinitely, making the coefficients in the ballooning equation (4.137) periodic functions of ζ, apart from the explicit secular ζ dependence. In the resistive range one formally introduces two scales ζ and $z = \epsilon_\eta^{-1} \zeta$, $\epsilon_\eta = (\eta k_\perp^2 / \gamma)^{1/2}$, expanding the solution in the form

$$\phi(\zeta) = \phi_0(z) + \epsilon_\eta \phi_1(\zeta, z) + \cdots, \tag{4.138}$$

where the dependence of the ϕ_j's on ζ is assumes to be periodic. $\phi_0(z)$ is the envelope of the eigenfunction. Consider first a cylindrical plasma column, where $\kappa_g = 0$ and $\kappa_n = \kappa^c = const.$ In this case we obtain the equation for the envelope $\phi_0(z)$ directly from (4.137), since there is no explicit dependence on the equilibrium scale ζ. Let us write (4.137) in the following non-dimensional form

$$\hat{\gamma}^2 \hat{z}^2 \phi_0 = \frac{d}{d\hat{z}} \frac{\hat{z}^2}{1 + (\hat{\eta}/\hat{\gamma})\hat{z}^2} \frac{d\phi_0}{d\hat{z}} + D_R \phi_0, \tag{4.139}$$

where

$$\hat{z} = \frac{z}{L_s}, \quad \hat{\gamma} = \gamma \frac{L_s}{v_A}, \quad \hat{\eta} = \eta k_\perp^2 \frac{L_s}{v_A}, \quad D_R = \beta' \kappa^c L_s^2, \tag{4.140}$$

and $L_s = Rq/\hat{s}$ in cylindrical geometry. Moreover, we restrict ourselves to conditions, where γ is sufficiently small such that $(\hat{\eta}/\hat{\gamma})\hat{z}^2 \gg 1$. In this case (4.139) reduces to the Weber equation

$$\frac{d^2 \phi_0}{du^2} + (\Lambda - u^2) \phi_0 = 0, \tag{4.141}$$

with $u = \hat{z}(\hat{\eta}\hat{\gamma})^{1/4}$ and

$$\Lambda = \frac{\hat{\eta}^{1/2}}{\hat{\gamma}^{3/2}} D_R. \tag{4.142}$$

The harmonic oscillator solutions lead to the eigenvalues

$$\Lambda = \Lambda_n = 1 + 2n, \quad n = 0, 1, \ldots.$$

The largest growth rate corresponds to $n = 0$, where $\phi_0 = \exp(-u^2/2)$,

$$\hat{\gamma} = \hat{\eta}^{1/3} D_R^{2/3} \tag{4.143}$$

or

$$\gamma = \left(\eta k_\perp^2\right)^{1/3} \left(\frac{v_A}{L_s}\right)^{2/3} (D_R)^{2/3}.$$

Since the eigenvalues Λ_n are positive, it follows from (4.142) that the condition for instability is

$$D_R > 0. \tag{4.144}$$

Compared with Suydam's criterion, (4.132) or (4.133), the stabilizing line-bending effect induced by the magnetic shear is eliminated due to reconnection. Equation (4.141) is only valid for sufficiently small γ, $\hat{\eta}F^2/\hat{\gamma} \gg 1$, or $D_R \ll 1$ using the width of ϕ along the field line and the growth rate, (4.143). For larger D_R the unity in the denominator of $F^2/(1 + \hat{\eta}F^2/\hat{\gamma})$ cannot be neglected. In this case D_R in the growth rate, (4.143), has to be replaced,

$$D_R \to \tfrac{1}{2} - \sqrt{\tfrac{1}{4} - D_R}, \tag{4.145}$$

which also indicates that the present analysis of the resistive interchange mode is valid only for $D_R < \tfrac{1}{4}$, i.e., for ideally stable conditions, see (4.133). Ideally unstable modes are practically not affected by a small resistivity.

In a general magnetic configuration, where the coefficients κ_n, κ_g, L_s in (4.137) depend on the equilibrium scale ζ, the envelope $\phi_0(z)$ is determined by the solubility condition for ϕ_2 in the expansion (4.138). The resulting equation, containing only averages over the equilibrium scale, can be solved analytically, but the algebra is rather involved (Correa-Restrepo, 1982).

On the equilibrium scale ζ, inertia and resistivity can be neglected, since we are only interested in ideally stable configurations where the resistive growth time is much longer than the Alfvén time, so that the solution for marginal ideal stability suffices, $\phi(\zeta) = \phi_{\text{id}}(\zeta)$. Since, however, the coefficients depend on the details of the equilibrium, a numerical treatment is, in general, required. To obtain a smooth overall solution, the separate solutions must be matched asymptotically by identifying the logarithmic derivatives of $\phi_0(z)$ for $z \to 0$ with that of $\phi_{\text{id}}(\zeta)$ for $|\zeta| \to \infty$, which gives the dispersion relation. Because of finite compressibility the growth rate may become complex (see also section 4.5.2 below). The main results are:

(a) If the average curvature is destabilizing $\beta'\overline{\kappa}_n > 0$ or $D_R > 0$, there are always unstable modes, the resistive interchange modes with a growth rate scaling $\gamma \sim \eta^{1/3}$ as in the case of cylindrical symmetry. The instability threshold is $D_R = 0$, hence shear stabilization is eliminated by resistivity. Contrary to the long-wavelength tearing modes, small-scale pressure-driven modes are dominantly electrostatic, $\mathbf{B}_0 \cdot \nabla\phi + \eta j_1 \simeq 0$ in (4.116), which means that the magnetic perturbation ψ_1 produced by the current density j_1 can be neglected. In a tokamak, the average curvature $\bar{\kappa}_n = \kappa^c(1 - q^2)$ is destabilizing only for $q < 1$, i.e., close to the magnetic axis, while in the reversed field pinch with $|q| \ll 1$, $\bar{\kappa}_n = \kappa^c$ is always destabilizing.

(b) Also in the case of stabilizing average curvature $\beta' \bar{\kappa}_n < 0$ or $D_R < 0$, unstable modes exist. These are no longer driven by the basic electrostatic Rayleigh–Taylor mechanism, but by magnetic reconnection. Growth rates are slower than resistive interchange modes, scaling with larger exponents of η.

4.5.2 Nonlinear evolution of pressure-driven modes

The nonlinear dynamics of pressure-driven modes is quite different from that of current-driven kink modes. While the latter are dominated by reconnection, which slows down the nonlinear evolution compared with the linear growth rate as discussed in section 4.3.2,[*] pressure-driven instabilities tend to accelerate nonlinearly by a self-focusing process, which avoids reconnection. This behavior has been analyzed by Cowley & Artun (1997) for the simple model of a plane stratified plasma with a Rayleigh–Taylor-unstable density profile $\mathbf{g} \cdot \nabla\rho < 0$. Gravity is chosen in the negative x-direction $\mathbf{g} = -g\mathbf{e}_x$. The system is stabilized by a horizontal magnetic field $\mathbf{B} = B_0(x)\mathbf{e}_z$, which is line-tied at two boundary plates $z = 0, L_z$, hence $k_\parallel \sim L_z^{-1}$ in (4.118). (A linear stability analysis has been performed by Zweibel & Bruhwiler, 1992.) Cowley & Artun use the Lagrangian form of ideal MHD to derive an approximate theory for the early nonlinear evolution. After some rather tedious algebra the authors arrive at an equation for the amplitude $\xi(x, y, t)$ of the vertical plasma displacement $\xi_x(x, y, z, t) = \xi(x, y, t)f(z)$:

$$\partial_t^2 \partial_y^2 \xi = \gamma^2(x)\partial_y^2 \xi + c_1\partial_x^2\xi + c_2\partial_x^2\overline{\xi^2}\,\partial_y^2\xi + c_3\partial_y^2\xi^2.^{\dagger} \tag{4.146}$$

The function $f(z)$ describes the sinusoidal variation along the field accounting for the line-tying boundary condition $f(0) = f(L_z) = 0$. The parallel mode structure is not expected to change strongly during the early nonlinear phase, since the nonlinear dynamics affects primarily the

---

[*] This is, however, only true for resistivity-governed reconnection considered in this chapter. Noncollisional kink modes may exhibit an explosive behavior, see chapter 6.4.3.
[†] Here we give Cowley's equation in a slightly simplified form.

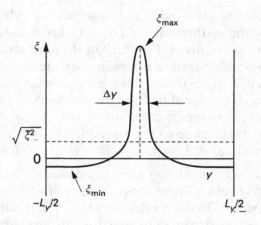

Fig. 4.21. Schematic plot of the finger-like structure of the asymptotic solution of (4.147).

perpendicular scales. Hence $f(z)$ is assumed constant in time, which allows us to average over z thus reducing the problem to the two-dimensional equation (4.146).

The first term on the r.h.s. of this equation is the linear Rayleigh–Taylor drive (4.118), $\gamma^2 > 0$ corresponding to instability, $\gamma^2 < 0$ to stable oscillation. The second term comprises the quasi-linear effect, which is stabilizing, while the last term acts as a nonlinear drive leading to explosive growth. Ignoring the quasi-linear term, the x-dependence becomes parametric. Integration over y and further simplification leads us to the following basic equation

$$\partial_t^2 \xi = \gamma^2 \xi + \xi^2 - \overline{\xi^2}, \tag{4.147}$$

where the overline indicates the y-average. In the nonlinear regime $\xi \gg \gamma^2$, (4.147) has the similarity solution

$$\xi = 6(t_0 - t)^{-2}. \tag{4.148}$$

The time $t_0 = t_0(y)$ depends on the initial conditions. If the maximum of ξ is at $y = 0$, then near the maximum $t_0(y) \simeq t_0 + \frac{1}{2} t_0'' y^2$, such that the width of ξ shrinks to zero as t approaches the finite-time singularity t_0, $\Delta y \sim \sqrt{(t_0 - t)/t_0''}$, as can easily be seen. Hence an initially smooth function $\xi(y, t = 0)$ develops a finger-like structure shown schematically in fig. 4.21. In the solution (4.148) the average $\overline{\xi^2}$ in (4.147) has been neglected, which is justified for $t \to t_0$, where $\overline{\xi^2}/\xi_{max}^2 \sim \Delta y / L_y \ll 1$. Cowley & Artun have also solved numerically the full nonlinear equation (4.146). While the scaling observed for the maximum amplitude is similar to (4.148), the

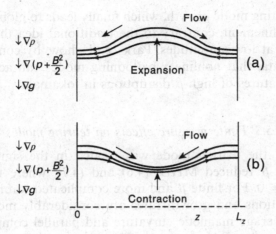

Fig. 4.22. Schematic drawings of the essential processes in (a) a rising flux tube, and (b) a falling flux tube.

width of the "finger" shrinks much more rapidly, $\Delta y \sim (t_0 - t)^2$, than predicted by the approximate equation (4.147), $\Delta y \sim (t_0 - t)^{1/2}$. Moreover, the vertical mode width increases $\Delta x \sim (t_0 - t)^{-0.4}$, hence the mode expands into the linearly stable layers of the system.

The physics leading to this explosive growth is illustrated in fig. 4.22. The stratification is such that $\nabla \rho$ points upward, and $\nabla(p + \frac{1}{2}B^2)$ and **g** point downward. The rising flux tube moves into a region of reduced ambient pressure, hence the cross-section broadens giving rise to reduced density at the tube apex, which makes the tube even more buoyant. In addition, the decreased magnetic field reduces the restoring field-line bending effect. In the adjacent (in y) falling flux tube the nonlinear effects are stabilizing, as the field is compressed. Hence the upward motion is strongly favored, leading to the finger-like structures just discussed.

The dynamics found in the Rayleigh–Taylor-unstable magnetized slab studied by Cowley & Artun is qualitatively very similar to the nonlinear evolution of ballooning modes in a high-β tokamak plasma, as demonstrated in three-dimensional MHD simulations of a toroidal plasma column by Park *et al.* (1995). Here, a slowly growing $n = 1$ kink mode leads to a helically distorted three-dimensional quasi-equilibrium. Where the helical shift is toward the torus outside, a steep pressure gradient is generated, on which high-n ballooning modes become unstable, exhibiting explosive nonlinear growth with a finger-like structure much in the same way as in Cowley's model. The presence of magnetic shear, not included in the simple slab model, also induces a self-focusing of the pressure bulge along field lines. Because of this localization, shear stabilization is not effective in saturating mode growth, which finally leads to global destruction of pressure confinement, contrary to the traditional idea that small-scale modes saturate at low amplitudes. Park *et al.* show, by comparison with experimental data, that nonlinear ballooning modes can account for the characteristic features of high-β disruptions in tokamaks.

4.5.3 Finite pressure effects on tearing modes

In section 4.1.1 the tearing mode was treated in the simplest possible framework, low-β reduced MHD, (4.6) and (4.7), where the instability threshold is $\Delta' = 0$. For finite β and more complicated toroidal geometry, threshold conditions and growth rates are considerably modified due to the effects of average magnetic curvature and parallel compressibility. A general theory has been developed by Glasser *et al.* (1975). Consideration is restricted to the solution in the resistive layer, which must be matched to the ideal external solution. The quantity Δ', which depends on the equilibrium configuration, must, in general, be obtained numerically and is considered as a given parameter in the analytic treatment. The situation is further complicated by the presence of several resonant surfaces, the usual case in a tokamak plasma because of toroidal coupling (Connor *et al.*, 1991). Here, we limit consideration to a low-β circular-cross-section, large-aspect-ratio tokamak, where the dispersion relation reads (Glasser *et al.*, 1975)

$$\widehat{\Delta}' = 2\pi \frac{\Gamma(\frac{3}{4})}{\Gamma(\frac{1}{4})} \frac{\widehat{\gamma}^{5/4}}{\widehat{\eta}^{3/4}} \left(1 - D_R \frac{\widehat{\eta}^{1/2}}{\widehat{\gamma}^{3/2}}\right) \tag{4.149}$$

with the same definition of $\widehat{\gamma}$ and $\widehat{\eta}$ as in (4.140), and $D_R = \beta' \overline{\kappa}_n L_s^2$, $L_s = Rq/\widehat{s}$, $\overline{\kappa}_n = \kappa^c(1 - q^2)$, $\widehat{\Delta}' = \Delta'/k$, $k = m/r$. In particular $k v_A/L_s = (nq'/q)v_A/R$. If $D_R > 0$, there is always one unstable root for any value of Δ', the resistive interchange mode. Since Δ' is independent of η, this root is given by the vanishing of the factor in brackets on the r.h.s., as the factor in front, $\widehat{\gamma}^{5/4}/\widehat{\eta}^{3/4} \propto \eta^{-1/3}$, becomes large for small η, hence

$$\widehat{\gamma} = \widehat{\eta}^{1/3} D_R^{2/3},$$

which is identical with the dispersion relation for resistive interchange modes (4.143).

For $D_R = 0$ we find the standard tearing mode in a cylindrical plasma

$$\widehat{\gamma} = 0.55 \widehat{\eta}^{3/5} (\widehat{\Delta}')^{2/5},$$

which is the usual expression (4.73) inserting the definitions of $\widehat{\gamma}, \widehat{\eta}, \widehat{\Delta}'$. The condition for instability is $\Delta' > 0$.

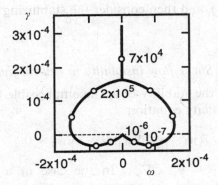

Fig. 4.23. Solution of the dispersion relation (4.149): γ and ω as functions of S. The numbers are the corresponding S values (from Hender et al., 1987).

The case $D_R < 0$ is the one usually encountered in a tokamak plasma. Here the instability requires Δ' to exceed a finite positive threshold $\Delta' > \Delta_c$. At the threshold γ is purely imaginary,

$$\gamma = \pm i\omega_c.$$

The quantities ω_c and Δ_c are easily obtained by separating (4.149) into real and imaginary parts

$$\widehat{\omega}_c = \widehat{\eta}^{1/3}\left[\tan(\pi/8)|D_R|\right]^{2/3}, \tag{4.150}$$

$$\widehat{\Delta}_c = \sqrt{2}\pi\mathrm{cosec}(\pi/8)\frac{\Gamma(\frac{3}{4})}{\Gamma(\frac{1}{4})}\frac{\widehat{\gamma}^{5/4}}{\widehat{\eta}^{3/4}} = 1.88\frac{|D_R|^{2/3}}{\widehat{\eta}^{1/3}}. \tag{4.151}$$

We see that $\Delta_c \to \infty$ for $\eta \to 0$. Since Δ' is independent of η, there are no unstable modes for negative D_R in this limit. However, the η-dependence of Δ_c is relatively weak, such that, in practice, for η-values relevant in tokamak research, Δ_c is still reasonably small. Figure 4.23 gives the complex growth rate as function of $S = \eta^{-1}$. The standard tearing mode scaling is approximately valid for $\widehat{\gamma} > \widehat{\eta}^{1/3}|D_R|^{2/3}$ or $|D_R| < (\widehat{\eta}\widehat{\Delta}'^3)^{2/5}$.

4.6 Shear flow instability

In conventional MHD stability theory, consideration is restricted to magnetostatic configurations. By contrast, magnetic reconnection involves strong plasma flows, in particular along current sheets and separatrices. An important question therefore concerns the stability of such systems. In this section we first give a brief introduction to shear flow instabilities in neutral fluids, which is a central theme in fluid mechanics (see, e.g.,

Drazin & Reid, 1981), and then consider the stabilizing effect of a parallel magnetic field.

4.6.1 Shear flow instability in neutral fluids

Neglecting viscosity, the stability of an incompressible flow is determined by the linearized vorticity equation

$$\partial_t w_1 + \mathbf{v}_0 \cdot \nabla w_1 + \mathbf{v}_1 \cdot \nabla w_0 = 0 \tag{4.152}$$

with $w_1 = \nabla^2 \phi_1$, $\mathbf{v}_1 = \mathbf{e}_z \times \nabla \phi_1$. In the case of a plane shear flow $\mathbf{v}_0 = \mathbf{e}_y V(x)$ we can make the ansatz $\phi_1 = \phi(x) \exp\{ik(y - ct)\}$, where we introduce the complex phase velocity $c = \omega/k = c_r + ic_i$,[*] since the advective term $\mathbf{v}_0 \cdot \nabla w_1$ gives rise, in general, to complex frequencies $\omega = \omega_r + i\gamma$, hence the growth rate is $\gamma = kc_i$. Choosing $k > 0$, instability occurs for $c_i > 0$. Equation (4.152) now becomes

$$(c - V)(\phi'' - k^2\phi) + V''\phi = 0, \tag{4.153}$$

which is called the Rayleigh equation. Contrary to small-scale pressure-driven modes discussed in section 4.5.1, (4.153) does not give rise to localized modes depending only on the local value of V'', but the instability involves the entire shear-flow profile $V(x)$. A necessary condition for instability is that $V(x)$ should have at least one inflexion point $V'' = 0$ (Rayleigh, 1880). This condition can be derived from the energy integral obtained by dividing (4.153) by $(c - V)$, multiplying by ϕ^* and integrating over the entire system. The criterion excludes, for instance, linear or parabolic velocity profiles from instability.

In magnetic reconnection processes, regions of high velocity shear or high vorticity are confined to narrow channels or sheets with no nearby boundaries. Here inflection points of the velocity, or extrema of the vorticity always arise, so that Rayleigh's criterion is satisfied. Consider first a simple vortex sheet corresponding to the velocity profile

$$V(x) = \begin{cases} V_1 & x < 0 \\ V_2 & x > 0. \end{cases} \tag{4.154}$$

The perturbation of the configuration is conveniently described by the perpendicular displacement $\xi_x = \xi$, which is related to the velocity perturbation,

$$v_x = -ik\phi = \partial_t \xi + ikV\xi, \tag{4.155}$$

---

[*] In this section c is the phase velocity of a fluid perturbation, not the velocity of light. This notation is chosen to conform with the literature.

hence

$$\phi = \begin{cases} (c - V_1)\xi & x < 0 \\ (c - V_2)\xi & x > 0. \end{cases} \tag{4.156}$$

Integration of (4.153) across the sheet gives

$$\int_{-\epsilon}^{\epsilon} [(c - V)(\phi'' - k^2\phi) + V''\phi]\,dx = c\phi'|_{-\epsilon}^{\epsilon} + \int_{-\epsilon}^{\epsilon} (V''\phi - V\phi'')\,dx$$

$$= (c - V)\phi'|_{-\epsilon}^{\epsilon} = 0 \tag{4.157}$$

Outside the sheet we have $V'' = 0$, and ϕ follows from the equation

$$\phi'' - k^2\phi = 0. \tag{4.158}$$

Since $\phi \to 0$ for $|x| \to \infty$, the solution is

$$\phi = \begin{cases} A_- e^{kx} & x < 0 \\ A_+ e^{-kx} & x > 0, \end{cases} \tag{4.159}$$

and, by use of (4.156),

$$\phi' = \begin{cases} k(c - V_1)\xi & x < 0 \\ -k(c - V_2)\xi & x > 0. \end{cases} \tag{4.160}$$

Insertion into (4.157) gives the dispersion relation

$$(c - V_1)^2 + (c - V_2)^2 = 0, \tag{4.161}$$

whence

$$c = \frac{V_1 + V_2}{2} \pm i\frac{|V_1 - V_2|}{2}. \tag{4.162}$$

A vortex sheet is always unstable, which is called the Kelvin–Helmholtz instability (Kelvin, 1871; Helmholtz, 1868). The term is, however, also used in a broader sense for general shear-flow instabilities. The increase of the growth rate $\gamma \propto k$ with k is limited by the finite width of the sheet δ, $k\delta \sim 1$, where the maximum of γ is reached, $\gamma_{\max} \sim |V_1 - V_2|/\delta$.

In reconnection theory the most intense flows are narrow jets, which correspond to vorticity double layers. Here even the long-wavelength characteristics $k\delta \ll 1$ depend on the jet profile. We limit the discussion to the symmetric case $V(-x) = V(x)$. For the standard current-sheet profile $j = \operatorname{sech}^2 x$ we have shown, section 3.2.3, that the flow along the sheet has the same profile. Let us therefore study the example

$$V(x) = \operatorname{sech}^2 x, \tag{4.163}$$

which is known in fluid mechanics as the Bickley jet (Bickley, 1937). For a symmetric jet the Rayleigh equation (4.153) has two classes of solutions,

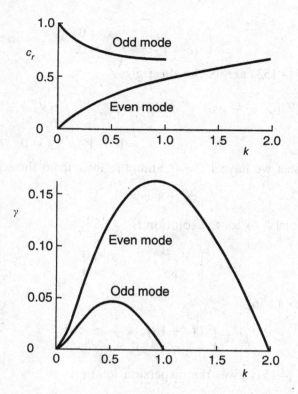

Fig. 4.24. Phase velocity c_r and growth rate γ for the inviscid Bickley jet (after Drazin & Reid, 1981).

even, with $\phi'(0) = 0$, and odd, with $\phi(0) = 0$. The even mode corresponds to a kinking of the jet and is the analog of the Kelvin–Helmholtz mode in a vortex sheet, (4.162). The odd mode implies a pinching of the jet, since the perpendicular velocity has opposite directions for $x < 0$ and $x > 0$. It is intuitively clear that the even or kink mode is more easily excited.

The critical or marginally stable solutions can be obtained analytically for the Bickley jet; the even mode is

$$\phi^{\text{even}} = \text{sech}^2 x, \quad k = 2, \; c = 2/3, \tag{4.164}$$

and the odd mode

$$\phi^{\text{odd}} = \sinh x \, \text{sech}^2 x, \quad k = 1, \; c = 2/3, \tag{4.165}$$

as can easily be verified by insertion into the Rayleigh equation. It is interesting to note that the phase velocities of both marginal modes equal the jet velocity at the inflexion point x_s, $V(x_s) = 2/3$, $V''(x_s) = 0$.

Instability occurs for any wave vector below the marginal values, but the eigenvalues c have to be computed numerically. The result is shown in fig. 4.24, where the phase velocities and the growth rates are plotted for both modes. The maximum growth rate is of the order

$$\gamma_{max} \simeq 0.1V',\qquad(4.166)$$

while for small k the growth rate is linear in k, $\gamma \sim k\Delta V$, where ΔV is a characteristic internal velocity difference. This behavior is qualitatively similar to the Kelvin–Helmholtz instability, (4.162).

Up to now we have neglected viscosity μ, which is only justified for sufficiently high Reynolds number $Re = V_0\delta/\mu$. In a resistive current sheet, however, the width δ is small such that Re may only be moderately high. Including viscosity in Rayleigh's equation leads to the Orr–Sommerfeld equation (Orr, 1907; Sommerfeld, 1908):

$$(V - c)\nabla^2\phi - V''\phi + i(\mu/k)\nabla^2\nabla^2\phi = 0,\qquad(4.167)$$

with $\nabla^2\phi = \phi'' - k^2\phi$ and normalization such that $\mu = Re^{-1}$. The critical or marginally stable Reynolds numbers $Re_c(k)$ have been computed numerically (Silcock, 1975). As can be expected, Re_c is minimal for $k \sim 1$ close to the maximum growth rate, the even mode is nearly independent of the viscosity, $Re_c^{even} \sim 5$, while the odd mode, which is more localized with steeper gradients, is rather strongly affected by viscosity, hence the required Reynolds number is higher, $Re_c^{odd} \sim 10^2$.

4.6.2 Effect of magnetic field on the Kelvin–Helmholtz instability

Plasmas are usually embedded in a magnetic field, which has to be taken into account when considering the instability of sheared plasma flows. A field perpendicular to the flow direction has only a weak effect on the instability, in particular in the incompressible limit a perpendicular field does not appear in the dynamics at all. By contrast, a parallel field will be bent by the motions associated with the instability, which requires energy and hence has a stabilizing influence. As we shall see, the instability will be completely stabilized, if the Alfvén velocity in the embedding parallel field exceeds the flow velocity. Consider the incompressible MHD equations (4.6), (4.7), linearized about the stationary state $\mathbf{v}_0 = V(x)\mathbf{e}_y$, $\mathbf{B}_0 = B(x)\mathbf{e}_y$ and write the perturbation in terms of the stream function ϕ, $\mathbf{v}_1 = \mathbf{e}_z \times \nabla\phi$, and the flux function ψ, $\mathbf{B}_1 = \mathbf{e}_z \times \nabla\psi$,

$$(c - V)\nabla^2\phi + V''\phi + B\nabla^2\psi - B''\psi + i(\mu/k)\nabla^2\nabla^2\phi = 0,\qquad(4.168)$$

$$(c - V)\psi + B\phi + i(\eta/k)\nabla^2\psi = 0.\qquad(4.169)$$

In the ideal limit $\eta = \mu = 0$, these equations can be combined into one for $f = \psi/kB$ (Lau & Liu, 1980):

$$\frac{d}{dx}\Big[(c - V)^2 - v_A^2\Big]\frac{df}{dx} - k^2\Big[(c - V)^2 - v_A^2\Big]f = 0.\qquad(4.170)$$

First consider a vortex sheet, (4.154), embedded in a homogeneous field $B(x) = B_0$, which can be solved exactly. Integration of (4.168) across the sheet gives

$$(c - V)\phi'|_{-\epsilon}^{\epsilon} + B_0\psi'|_{-\epsilon}^{\epsilon} = 0,$$

or, by use of $\psi' = -B_0\phi'/(c - V)$ from (4.169),

$$\left(c - V - \frac{B_0^2}{c - V}\right)\phi'\Big|_{-\epsilon}^{\epsilon} = 0.$$

Inserting (4.160) gives the dispersion relation

$$(c - V_1)^2 + (c - V_2)^2 = 2v_A^2, \tag{4.171}$$

which has the solution

$$c = \frac{V_1 + V_2}{2} \pm \sqrt{v_A^2 - \frac{1}{4}(V_1 - V_2)^2}. \tag{4.172}$$

Here we have used that $B_0 = v_A$ in our normalization. Hence applying a homogeneous parallel magnetic field with $v_A > \frac{1}{2}|V_1 - V_2|$ stabilizes the Kelvin–Helmholtz instability of a vortex sheet. The physical interpretation of this stabilization is that in this case any perturbation of the plasma flow is simply carried away as a propagating Alfvén wave, before it can grow.

These results have been generalized by Miura & Pritchett (1982), who consider a vortex sheet of finite width, choosing the velocity profile $V(x) = \tanh x$, including finite plasma compressibility. The latter is found to have always a stabilizing effect, which is, however, important only in high-β plasmas for Mach numbers $M \gtrsim 1$.[*]

In several more recent papers the nonlinear evolution of a magnetized vortex sheet has been studied (Malagoli *et al.*, 1996; Frank *et al.*, 1996; Jones *et al.*, 1997). The main result is that in the nonlinear regime the stabilizing effect of a parallel magnetic field is still stronger. Even for field intensities significantly below the limit for linear instability the nonlinear dynamics is rather weak. For the restricted length L_y considered in these

---

[*] The Kelvin–Helmholtz instability is different from the parallel-velocity shear instability (D'Angelo, 1965), which has recently regained considerable attention as a mechanism for anomalous transport in tokamak plasmas (see, McCarthy *et al.*, 1997). This is not an MHD instability, but an electrostatic mode in a low-β plasma driven at small scales, where the electrons have approximately a Boltzmann distribution $n_e \propto e^{e\phi/T_e}$ or, in linearized form, $\tilde{n}_e/n_0 = e\phi/T_e$, which follows from the parallel component of the electron equilibrium equation $e\nabla_\parallel\phi - \nabla_\parallel p_e/n = 0$ assuming isothermal behavior. In the local approximation the growth rate γ is given by $\gamma^2 = k_\parallel k_\perp(cT_e/eB)v'_\parallel - k_\parallel^2 c_s^2$, where the sound wave exerts a similar damping influence as does the Alfvén wave in the MHD case, (4.172).

studies the system in fact settles in a quasi-laminar state with a broadened velocity profile, such that $kv_A > \Delta V/\delta$ for all modes k allowed, $kL_y \geq 1$, hence the system is now linearly stable. This quasi-linear process involves magnetic reconnection to straighten out the bent field, which will affect the nonlinear time-scale. Only for very low intensity is the magnetic field simply carried along passively, which automatically generates the small-scale magnetic structures necessary for reconnection, while the overall process is similar to the nonmagnetic Kelvin–Helmholtz instability.

4.6.3 Instability of a magnetized jet

Let us now return to the Bickley jet, (4.163), embedded either in a homogeneous parallel field B_0 or, in view of the application to a resistive current sheet, in the sheared field

$$B_{0y}(x) = B_0 \tanh x. \tag{4.173}$$

As in the case of a vortex sheet, the parallel field has a strongly stabilizing effect on the Kelvin–Helmholtz instability (Wang *et al.*, 1988; Lee *et al.*, 1988; Dahlburg *et al.*, 1998; Biskamp *et al.*, 1998a). The linearized equations (4.168), (4.169) have been solved numerically. There are again two classes of eigenmodes:

(i) ϕ even (kink mode), ψ odd,

(ii) ϕ odd (pinch mode), ψ even.

Since, for finite B_0, the phase velocity of the unstable modes is found to be only slightly larger than in the nonmagnetic jet, we consider only the growth rate. Let us first discuss the ideal case $\eta = \mu = 0$. In fig. 4.25, $\gamma(k)$ is plotted for different values of B_0 for (a) the kink mode and (b) the pinch mode. While the kink mode is progressively stabilized with increasing field B_0, both at small and high k, such that the marginally stable wave vector is finite, $k_c^{kink} \simeq 0.8$, the pinch mode is stabilized only from the high-k side, such that $k_c^{pinch} = 0$. It is interesting to note that the pinch mode, which has the lower growth rate at small B_0, remains unstable at a significantly higher field than the kink mode: $B_c^{pinch} = 0.96$, whereas $B_c^{kink} = 0.46$. Hence there is a broad range $0.46 \leq B_0 \leq 0.96$, where only the pinch mode is unstable. In this range the instability is driven primarily by the free magnetic energy of the configuration (4.173), as can be seen by comparison with the case of a homogeneous magnetic field, where this free energy source is absent and where the marginal B_0 values for both modes are quite close: $B_c^{kink} = 0.35$, $B_c^{pinch} = 0.42$.

The release of magnetic energy is strongly affected by the ideal constraint of flux conservation. We therefore expect that the pinch mode, especially

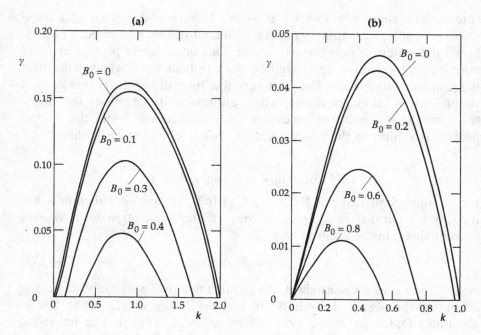

Fig. 4.25. Ideal growth rate of the Bickley jet embedded in the magnetic field of a current sheet, (4.173), for different values of B_0. (a) Kelvin–Helmholtz kink mode, (b) pinch mode.

in the high-field range $B_0 > 0.5$, will be modified significantly by allowing resistive diffusion. In fact one finds that, while the kink mode is practically unchanged by finite resistivity, the pinch mode, i.e., the even-ψ mode, assumes the structure of the tearing mode. Figure 4.26 contains plots of the growth rate for $B_0 = 0.8$ and $\eta = 3 \times 10^{-4}$ (full line) together with the ideal growth rate (dashed line) and that of the pure tearing mode in the absence of the jet (dash-dotted line). The surprising feature is the second hump extending up to $k = 2.1$, far beyond the stability limits of both the Kelvin–Helmholtz mode and the tearing mode. Hence the combined action of the sheared flow and the sheared magnetic field destabilizes a high-k tearing mode. In this regime the growth rate scales as $\gamma \propto \eta^{3/5}$, as expected for an ideally stable mode.

The nonlinear evolution of the Kelvin–Helmholtz instability of a magnetized jet agrees qualitatively with that of a vortex sheet. Even for values of B_0 significantly below the marginal one $B_c^{kink} = 0.46$, where the linear growth rate is only slightly reduced, the nonlinear stabilizing effect of the magnetic field is dramatic. The jet is just somewhat broadened, thus reducing the flow speed, since the total plasma flux is conserved, until $B_0/V_{max} > 0.46$, where the kink mode becomes stable. The perturbations

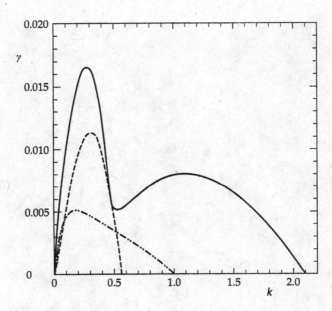

Fig. 4.26. Growth rate γ of the pinch-tearing mode for $B_0 = 0.8$, $\eta = 3 \times 10^{-4}$ (full), the ideal pinch mode (dashed), and the tearing mode in a static sheet pinch (dash-dotted) (from Biskamp *et al.*, 1998a).

subsequently decay due to viscous and resistive dissipation. By contrast, the nonmagnetic jet is completely disrupted. Figure 4.27 illustrates this difference. It shows gray-scale plots of the vorticity: (a) the (slightly perturbed) initial jet, (b) the nonlinear disruption of the jet for $B_0 = 0$, and (c) the nonlinearly broadened jet for finite field $B_0 = 0.3$.

While, in a homogeneous or weakly inhomogeneous field, the jet is completely stable for $B_0/V_{max} \gtrsim 0.4$, in the strongly sheared field, (4.173), characteristic of a current sheet, only the Kelvin–Helmholtz mode is stable. Here, the pinch-tearing mode remains unstable leading to magnetic islands which, due to their advective motion, assume a drop-like shape with a blunt leading edge. Since small islands moving with the central jet velocity are faster than bigger ones, they catch up and coalesce with the latter, until only the largest islands remain. Such islands are called plasmoids, to which we return in section 4.7.2.

4.7 Instability of a resistive current sheet

4.7.1 Threshold condition for tearing instability

The results of the preceding section indicate that a resistive current sheet is stable against Kelvin–Helmholtz instability of the sheared flow along

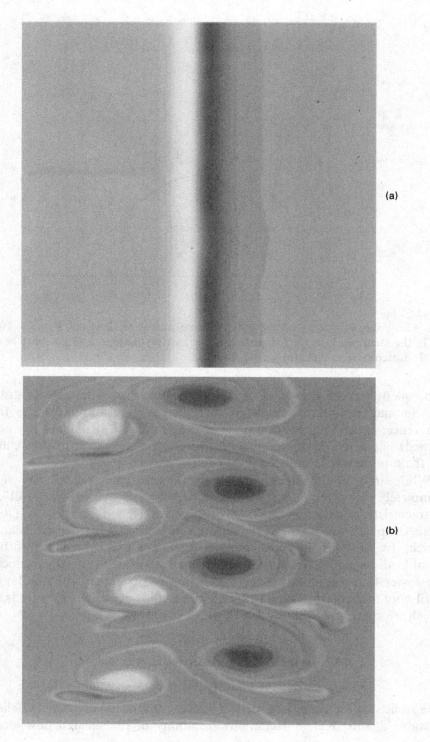

Fig. 4.27.   For caption see facing page.

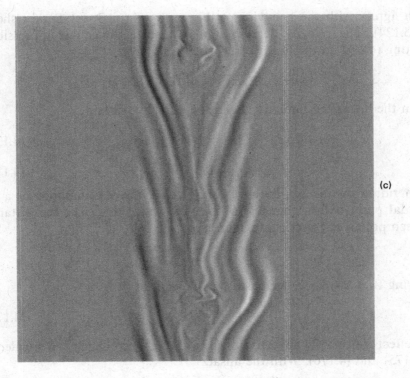

(c)

Fig. 4.27. Gray-scale plots of the vorticity illustrating the nonlinear evolution of the Bickley jet. (a) Initial state; (b) nonlinear state of the non-magnetized jet; and (c) nonlinear broadening of a magnetized jet for $B_0 = 0.3$ (from Biskamp *et al.*, 1998a).

the sheet, since $v \leq v_A$, as explained in section 3.2.2. However, the most important candidate for instability in a sheet current, the tearing mode, is not stabilized by the velocity shear. In a static sheet pinch, for which the tearing mode is usually considered (section 4.1), the instability condition $ka < 1$ implies that the configuration becomes unstable if the sheet length Δ exceeds one or two wavelengths of the marginally stable tearing mode, corresponding to an aspect ratio $\Delta/a \gtrsim 10$. To account for the apparent stability of current sheets of much larger aspect ratio, as observed in numerical simulations, a mechanism other than velocity shear must be active. The problem has been studied by Bulanov *et al.* (1979), who realized that the only stabilizing effect is due to the *non-uniformity* of the flow *along* the sheet. The authors assume a parallel velocity increasing linearly along the sheet

$$\mathbf{v}_0(y) = \Gamma y \mathbf{e}_y, \tag{4.174}$$

which agrees with the behavior of the outflow in a Sweet–Parker sheet, see (3.124). The velocity perturbation is again assumed incompressible. Inserting (4.174) in the linearized vorticity equation

$$\partial_t w_1 + \mathbf{v}_0 \cdot \nabla w_1 + w_1 \nabla \cdot \mathbf{v}_0 = \mathbf{B}_0 \cdot \nabla j_1 + \mathbf{B}_1 \cdot \nabla j_0{}^*$$

and in the linearized induction equation, (4.6), we obtain

$$(\partial_t + \Gamma y \partial_y + \Gamma) w_1 = \mathbf{B}_0 \cdot \nabla j_1 + \mathbf{B}_1 \cdot \nabla j_0, \qquad (4.175)$$

$$(\partial_t + \Gamma y \partial_y) \psi_1 = \mathbf{B}_0 \cdot \nabla \phi_1 + \eta \nabla^2 \psi_1. \qquad (4.176)$$

When riding on the equilibrium flow, (4.174), the wavenumber of a sinusoidal perturbation changes in time, the wavelength, i.e., the distance between points of equal phase, increases

$$\dot{\lambda} = \Gamma \lambda$$

implying that $k = 2\pi/\lambda$ decreases

$$\dot{k} = -\Gamma k. \qquad (4.177)$$

This effect cancels the advective term $\mathbf{v}_0 \cdot \nabla$, i.e., the explicit y-dependence in (4.175) and (4.176). With the ansatz

$$\psi_1 = \psi\, e^{\gamma t + ik(t)y}, \quad w_1 = \nabla^2 \phi_1 = (\phi'' - k^2 \phi) e^{\gamma t + ik(t)y}, \qquad (4.178)$$

we obtain $\partial_t \psi_1 = (\gamma + i\dot{k}y)\psi_1$, hence $(\partial_t + \mathbf{v}_0 \cdot \nabla)\psi_1 = \gamma \psi_1$, using relation (4.177). We also have

$$(\partial_t + \Gamma y \partial_y + \Gamma) w_1 = (\gamma + \Gamma)\phi_1'' - (\gamma - \Gamma)k^2 \phi_1.$$

Thus (4.175) and (4.176) become

$$(\gamma + \Gamma)\phi'' - (\gamma - \Gamma)k^2 \phi = ikB_0(\psi'' - k^2 \psi) - ikB_0'' \psi, \qquad (4.179)$$

$$\gamma \psi - ikB_0 \phi = \eta(\psi'' - k^2 \psi). \qquad (4.180)$$

Bulanov *et al.* write the equations in terms of the perturbed magnetic field $B_x = -ik\psi$ and the perpendicular plasma displacement $\xi = \int v_x dt$, where $v_x = -ik\phi$. While, in a homogeneous system, this would only mean multiplying the equations by ik, in the presence of inhomogeneous flow, where $k = k_0 e^{-\Gamma t}$, the growth rate of B_x and v_x is reduced by Γ, $\gamma' = \gamma - \Gamma$, and $\xi = (\gamma' + \Gamma)v_x$. In these variables, (4.179) and (4.180) assume

---

* Because of the compressibility of the equilibrium flow, (4.174), the $\nabla \cdot \mathbf{v}_0$ term has to be included in the vorticity equation.

the form

$$(\gamma' + \Gamma)(\gamma' + 2\Gamma)\xi - \gamma'(\gamma' + \Gamma)\xi = ik(B_x'' - k^2 B_x) - ikB_0'' B_x, \quad (4.181)$$

$$(\gamma' + \Gamma)(B_x - ikB_0\xi) = \eta(B_x'' - k^2 B_x). \quad (4.182)$$

Applying the same procedure as in section 4.1.1 or 4.2.1, we find the dispersion relation

$$(\gamma' + \Gamma)^4(\gamma' + 2\Gamma) = \gamma_0^5, \quad (4.183)$$

where γ_0 is the growth rate in the static sheet pinch. Solving (4.183) one finds that the tearing mode becomes stable for $\Gamma > 0.87\gamma_0$. By contrast, the dispersion relation following from the ψ, ϕ system, (4.179) and (4.180), $\gamma^4(\gamma + \Gamma) = \gamma_0$, predicts a significantly higher threshold value of Γ. Hence we are faced with an ambiguity in the stability problem, which requires us to specify whether we wish to consider stabilization of B_x, i.e., the magnetic energy associated with the tearing mode, or of ψ, i.e., the magnetic flux. We prefer the former definition, since MHD stability is usually discussed in the energy picture.

The stabilizing effect of the inhomogeneous outflow can easily be understood. The tearing instability corresponds to a local condensation or bunching of the current density, which is counteracted by the stretching due to the flow. One may assume that the mode is effectively stabilized if the change of the wavelength during one growth time exceeds a substantial fraction of the wavelength, say one-half, $\Gamma/\gamma_0 > 0.5$. Hence the sheet is tearing-mode stable for $\Gamma > 0.5\gamma_{max}$. In the asymptotic limit of large S one has $\gamma_{max} \simeq 0.5(\tau_A\tau_\eta)^{-1/2}$ for $k_0\delta \simeq S^{-1/4}$, (4.24). As discussed in sections 3.2.2 and 3.2.3, the following relations are valid in a resistive current sheet:

$$\Gamma = v_A/\Delta \simeq u/\delta \simeq \eta/\delta^2 = \tau_\eta^{-1}$$

and the local Lundquist number is

$$S_0 = \tau_\eta/\tau_A = v_A\delta/\eta = \Delta/\delta.$$

Hence stability requires

$$\tau_\eta^{-1} \gtrsim 0.25(\tau_\eta\tau_A)^{-1/2}$$

or

$$\delta/\Delta \gtrsim 0.05.$$

Since, however, for such low values of S_0 the maximum growth rate is substantially smaller than predicted by the asymptotic expression, we may estimate that a resistive current sheet is susceptible to tearing instability only for aspect ratio

$$S_0 = \Delta/\delta \gtrsim 10^2. \quad (4.184)$$

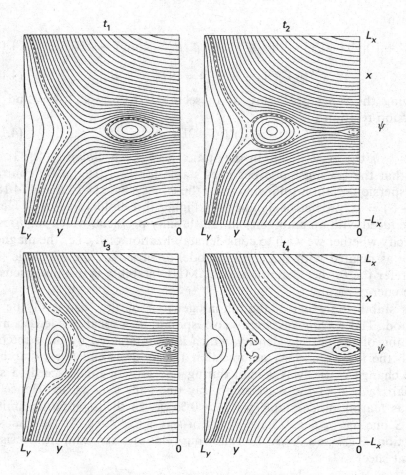

Fig. 4.28. Plasmoid generation in a tearing-unstable Sweet–Parker sheet. The figure shows ψ-contours at times $t_1 < t_2 < t_3 < t_4$ (from Biskamp, 1986).

The tearing instability problem in a current sheet has also been investigated by Phan & Sonnerup (1991), who assumed that the flow is diverted in the sheet along the z-direction instead of the y-direction. Since there is no stabilizing flow in this case, the stability limit obtained, $S_0 \simeq 12.5$, is close to that of a static sheet.

4.7.2 Plasmoids

From the scaling laws of driven reconnection, (3.151) and (3.154), $\Delta/\delta \sim \eta^{-2}$ (for variable sheet length Δ), or, (3.150), $\Delta/\delta \sim \eta^{-2/3}$ (for fixed Δ), it is clear that the tearing mode will arise in a Sweet–Parker sheet for sufficiently small η. Let us therefore consider its nonlinear behavior. The

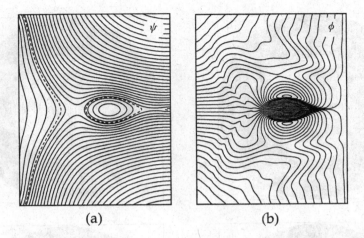

Fig. 4.29. (a) Magnetic structure and (b) flow pattern of a plasmoid (from Biskamp, 1986).

usual appearance of the tearing mode in a static sheet pinch is that of a chain of magnetic islands. In a Sweet–Parker sheet the more typical case is that of one or a few isolated islands, which are convected along the sheet, growing in a self-similar way both in width and in length being stretched by the inhomogeneous flow. Such isolated islands are called plasmoids. Figure 4.28 illustrates the repetitive generation and convection of plasmoids. Shown is a configuration of driven reconnection such as in fig. 3.16 (for clarity the lower symmetric quadrant is added in the plot). Changing the boundary conditions at $y = 0$ slightly, from $\partial_y \psi = 0$ to $\psi(x, t) = \psi(x) + Et$, where E is the externally imposed average reconnection rate and $\psi(x)$ is the ψ-profile taken from case (c) of fig. 3.16, is sufficient to destroy stationarity and give rise to repetitive generation of plasmoids. The dynamics of the plasmoid becomes clear from fig. 4.29(b), where the density of streamlines demonstrates the high speed of the plasmoid along the original sheet compared with the velocity in the ambient medium.

Plasmoid formation and ejection is a general phenomenon, which occurs if the sheet aspect ratio Δ/δ becomes large enough. Figure 4.30 shows such an event during the nonlinear evolution of the resistive kink mode. It also occurs in the extended noncollisional current sheets dominated by electron inertia. Fully developed plasmoids assume an asymmetric drop-like shape with a blunt front and a long tail corresponding to a current sheet, which may be thinner than the original stationary one and which is hence particularly sensitive to further plasmoid generation. Since such secondary slim plasmoids propagate faster than the primary bulky one, the former will catch up and finally coalesce with the latter. Thus the tearing

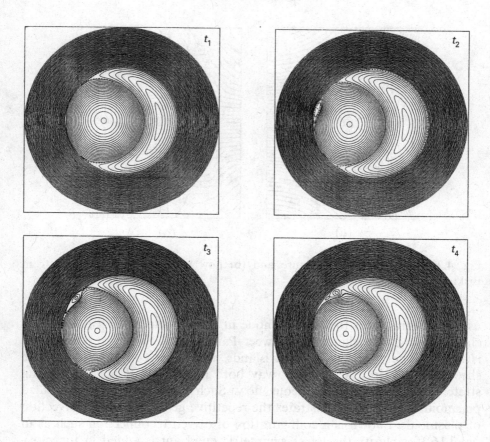

Fig. 4.30.   Plasmoid dynamics in the current sheet of the nonlinear resistive kink mode (from Biskamp, 1986).

instability can strongly increase the reconnection efficiency compared with that in a stationary current sheet.

Generation of plasmoids is a widespread phenomenon. Plasmoids are an essential feature in many rapid MHD processes, such as an erupting magnetic arcade in the solar corona, the standard model of a large flare (see, e.g., Biskamp, 1993a, chapter 10), or the dynamics in the geomagnetic tail connected with the substorm phenomon, which we discuss in chapter 8. Tearing instability (and the resulting plasmoid dynamics) are also an important mechanisms for the generation of small-scale fluctuations in MHD turbulence (see, for instance, Biskamp, 1993a, chapter 7).

5

Dynamo theory

Dynamo theory deals with the generation or, more precisely, the amplification and sustaining of magnetic fields by motions in an electrically conducting fluid. The topic is of eminent practical importance, since, as is generally believed, the magnetic fields in astrophysical objects as diverse as planets, stars, accretion discs and galaxies are generated in this way.

The property of the conducting medium to be a fluid is important, since dynamo action by the motion of solid ducts is commonplace as the general means of electric power generation. Here the simplest model is the disc dynamo proposed by Bullard (1955) sketched in fig. 5.1, which shows the essential features of dynamo action. The basic equations are

$$L\dot{I} + RI = c_1\Omega I, \tag{5.1}$$

$$\dot{\Omega} = F - c_2 I^2, \tag{5.2}$$

where L and R are the inductance and the resistance of the system, F is the external torque on the disc, I is the current generating the magnetic

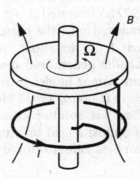

Fig. 5.1. Bullard's disc dynamo.

145

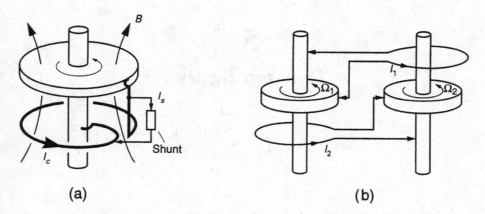

Fig. 5.2. Disc dynamo models exhibiting spontaneous field reversals: (a) shunted disc dynamo; (b) coupled disc dynamos.

field penetrating the disc, and c_1, c_2 are geometrical factors. For small I the angular frequency Ω of the disc is constant, such that

$$I = I_0 \exp(c_1\Omega - R)t.$$

I is growing exponentially, if Ω is sufficiently large to overcome the resistive damping. In the nonlinear regime there is a steady-state solution

$$I = \sqrt{F/c_2}, \quad \Omega = R/c_1.$$

It is easy to see from fig. 5.1 that a seed field in the opposite direction will also be amplified. If, however, the rotation is reversed $\Omega \to -\Omega$, field amplification does not occur, hence dynamo action depends on the relative sense of the current loop winding and the disc rotation. The general nonlinear solution is oscillatory, but does not reverse the sign of I or B. Field reversal does, however, occur in a shunted disc dynamo (Malkus, 1972; Robbins, 1977), illustrated in fig. 5.2(a). Here, the current I flowing in the disc is branched into the current I_c in the coil around the the axis and the current I_s in the shunt. Written in terms of I, I_c, and the angular velocity Ω, the generalized disc dynamo equations are equivalent to the well-known Lorenz model, a truncation of the equations of two-dimensional convection (Lorenz, 1963). Hence for certain values of the external parameters, there are solutions oscillating in a nonperiodic form between two states of opposite field orientation. Field reversals also occur in a system of two coupled Bullard disc dynamos (Rikitake, 1958), illustrated in fig. 5.2(b).

In a disc dynamo the current is flowing in a multiply connected region consisting of the wire and the disc. Dynamo action in a conducting fluid occupying a singly connected region is more difficult to achieve, even the very existence of a dynamo effect in such a fluid has been questioned for a considerable period. We start this chapter with an outline of kinematic theory, section 5.1, discussing the topological properties of the fluid motions necessary for dynamo action. The essential requirement is finite (kinetic) helicity – the flow must, loosely speaking, be sufficiently twisting – since plane or axisymmetric flows cannot lead to field amplification, which is the essence of the anti-dynamo theorems. A major step from the proof-of-principle models to a theoretical tool for practical applications has been the introduction of mean-field electrodynamics (MFE). Here the fluid motions are assumed small-scale and turbulent, as they are expected to be in most systems of interest. The details of these motions have no effect on the behavior of the large-scale magnetic fields, only two average quantities being important, the turbulent kinetic energy leading to an anomalous magnetic diffusion and dissipation, and the helicity which through the α-effect may give rise to field amplification. MFE is the framework for most theoretical studies of stellar or galactic dynamos.

While the basic mechanisms of small-scale field amplification, namely flux tube stretching and twisting, do not require reconnection, the generation of large-scale observable fields from small-scale ones do, hence reconnection is a basic ingredient of dynamo theory. Though not introduced explicitly, MFE of course involves reconnection, as will easily be seen.

In its best-known and commonly used form MFE is a kinematic theory, where the fluid turbulence is determined by nonmagnetic mechanisms, for instance thermal convection in stellar dynamos. Since the magnetic field is primarily amplified at small scales, the small-scale fluid motion will be modified by the magnetic force even at still low mean magnetic energy, which reduces very efficiently their contribution to the α-effect, such that progressively larger turbulent scales become involved. This process requires a self-consistent MHD description.

Section 5.2 introduces self-consistency in the framework of mean-field theory by modeling the reaction on the fluid motion in a phenomenological way. Then an overview is given of the present theory of the solar dynamo, which is essentially restricted to MFE. Parker's original idea of an $\alpha\Omega$-dynamo working in the solar convection zone (Parker, 1955a), which was considered as a solid theoretical basis for many years, is again put into question by recent helioseismological evidence, such that solar dynamo theory remains a challenging problem.

In section 5.3, direct numerical simulations of convection in a spherical shell are presented. Though the properties of the fluid are highly idealized and the Rayleigh numbers are still rather low, these simulations give a reliable picture of the basic dynamo mechanism which is not affected by ad hoc assumptions. Of all the astrophysical dynamo systems, the properties of the Earth's liquid core come closest to the conditions simulated in these computations. I therefore briefly review the present understanding of the geodynamo.

High-Rayleigh number convection typical, for instance, for stellar convection zones, involve fully developed MHD turbulence, which is discussed in section 5.4. This is essentially a sequel of chapter 7 of my previous book (Biskamp, 1993a). After a brief introduction to the basic properties of MHD turbulence, the section is concentrated on recent developments in 3D turbulence.

Dynamo theory is a vast field of its own, which cannot fully be covered in this brief presentation. Many aspects are touched only cursorily or not at all, in particular the more mathematical aspects of dynamo theory, which have developed into a topic in their own right rather independent of the more practically oriented fluid mechanical approach (see, e.g., Bayly & Childress, 1987, 1988; Finn & Ott, 1988). For a more comprehensive treatment of dynamo theory I refer the reader to several special treatises, e.g., Moffatt (1978), Parker (1980), Zeldovich *et al.*, (1983), Roberts & Soward (1992).

5.1 Kinematic dynamo theory

In kinematic theory the fluid velocity $\mathbf{v}(\mathbf{x}, t)$ is given, and the problem consists in determining the development of the magnetic field, i.e., we have to consider only the induction equation

$$\partial_t \mathbf{B} = \nabla \times (\mathbf{v} \times \mathbf{B}) + \eta \nabla^2 \mathbf{B}, \tag{5.3}$$

which is a linear homogeneous equation for $\mathbf{B}$. Kinematic dynamo theory answers the question about the stability of a nonmagnetic flow to infinitesimal magnetic disturbances, which is formally similar to the stability theory discussed in chapter 4. We will learn about the conditions that must be satisfied for magnetic field amplification and to a certain extent also about the spatial structure of this field. Nonlinear saturation by reaction of the magnetic field on the fluid motion is considered in section 5.2.

The basic physics of the terms in (5.3) is similar to that in (5.1) for the simple disc dynamo, which suggests that for sufficiently large $\mathbf{v}$ dynamo action should be possible. Since, on the other hand, the topology of the setup in the disc dynamo plays an important role in its functioning, we expect that in the fluid problem the structure of the flow must satisfy certain requirements.

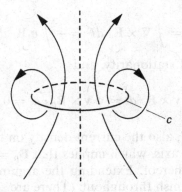

Fig. 5.3. Magnetic configuration illustrating Cowling's theorem.

5.1.1 Nonexistence theorems and special dynamo solutions

In fluid dynamics, in particular in MHD, problems are often simplified by assuming two-dimensional geometry. In dynamo theory, however, this proves to be too restrictive. Consider the simplest case of plane geometry $\partial_z = 0$, where **B** has the form

$$\mathbf{B}(x, y) = \mathbf{B}_p + \mathbf{e}_z B_z, \quad \mathbf{B}_p = \mathbf{e}_z \times \nabla \psi,$$

such that (5.3) reduces to two equations for ψ and B_z

$$\partial_t \psi + \mathbf{v} \cdot \nabla \psi = \eta \nabla^2 \psi, \tag{5.4}$$

$$\partial_t B_z + \mathbf{v} \cdot \nabla B_z = \mathbf{B}_p \cdot \nabla v_z + \eta \nabla^2 B_z, \tag{5.5}$$

assuming, as commonly done, incompressibility $\nabla \cdot \mathbf{v} = 0$. While the axial field B_z can be amplified by the poloidal field $\mathbf{B}_p$, via the first term on the r.h.s. of (5.5), there is no reaction of B_z on $\mathbf{B}_p$. In fact the poloidal magnetic energy decays, as can be shown directly by multiplying (5.4) by ψ and integrating over space and time

$$\int_0^\infty dt \int B_p^2 dV \le \frac{1}{\eta} \int \psi^2 dV \Big|_{t=0} = const, \tag{5.6}$$

hence the driving term in (5.5) tends to zero and so does the axial field energy $\int B_z^2 dV$. A similar argument can be used in the axisymmetric case. It is, however, instructive to recapitulate briefly the proof presented by Cowling (1934) to demonstrate his famous theorem "A steady-state axisymmetric field cannot be maintained". An axisymmetric field has at least one magnetic axis C, where the poloidal field vanishes, see fig. 5.3. Integration of Ohm's law along this closed line gives

$$\oint \mathbf{E} \cdot d\mathbf{l} + \oint \mathbf{v} \times \mathbf{B} \cdot d\mathbf{l} = \oint \eta \mathbf{j} \cdot d\mathbf{l}. \tag{5.7}$$

Since

$$\oint \mathbf{E} \cdot d\mathbf{l} = \int \nabla \times \mathbf{E} \cdot d\mathbf{F} = - \int \partial_t \mathbf{B} \cdot d\mathbf{F} = 0$$

because of the assumed stationarity, and

$$\oint \mathbf{v} \times \mathbf{B} \cdot d\mathbf{l} = \oint \mathbf{v} \times \mathbf{B}_p \cdot d\mathbf{l} = 0,$$

because of $\mathbf{B}_p = 0$ on C, also the current density on the r.h.s. of (5.7) must vanish on the magnetic axis, which implies that $\mathbf{B}_p = 0$ not only on C but in a full surrounding thereof. Extending the argument to this region, it follows that $\mathbf{B}_p$ must vanish throughout. (There are some subtleties, which must be considered in a rigorous proof, but these do not change the basic result.)

The preceding discussion might suggest that the anti-dynamo theorem applies to all symmetric field configurations, i.e., those exhibiting an ignorable coordinate. This is, however, not true for the most general case, that of helical symmetry $r, u = l\phi + kz$, where $\mathbf{B}(r, u)$ has the form

$$\mathbf{B} = \mathbf{h} \times \nabla \psi_H + \mathbf{h}f,$$

$$\mathbf{h} = \frac{r}{l^2 + k^2 r^2} \nabla r \times \nabla u, \quad \mathbf{h} \cdot \nabla f = 0,$$

$\psi_H = krA_\phi - lA_z$ being the helical flux function, (3.37), which follows the equation

$$\partial_t \psi_H + \mathbf{v} \cdot \nabla \psi_H = \eta \left(L \psi_H - \frac{2kl}{l^2 + k^2 r^2} f \right). \tag{5.8}$$

Here L is the generalized Laplacian

$$L \equiv (l^2 + k^2 r^2) \left(\frac{1}{r} \partial_r \frac{r}{l^2 + k^2 r^2} \partial_r + \frac{1}{r^2} \partial_u^2 \right),$$

while f, the field component in the helical direction, obeys the equation

$$\partial_t f + \mathbf{v} \cdot \nabla f = F\{f, \psi_H\}. \tag{5.9}$$

F is a rather complicated linear functional of f and ψ_H generalizing the r.h.s. of (5.5), which we do not need to specify. The important point to note is that the ψ_H equation (5.8) now contains a reaction of f on ψ_H absent both in plane geometry ($k = 0$) and axisymmetry ($l = 0$), which allows dynamo action. An explicit steady-state helical dynamo solution has been given by Lortz (1968). Such helical fields have a certain relevance for the dynamo effect in the reversed-field pinch (RFP) (for a review of RFP theory see Biskamp, 1993a; Ortolani & Schnack, 1993).

Steady spatially periodic flows have received particular attention as efficient dynamos, that most commonly studied being the so-called *ABC* flow

$$\mathbf{v} = (A \sin z + C \cos y, \ B \sin x + A \cos z, \ C \sin y + B \cos x),$$

which has the property of maximal helicity since $\mathbf{v} = \mathbf{w}$, $\mathbf{w} = \nabla \times \mathbf{v}$. (A flow with $\mathbf{w} = \lambda \mathbf{v}$ is called Beltrami flow, which is the hydrodynamic analog of a force-free magnetic field $\mathbf{j} = \lambda \mathbf{B}$). A special case of a two-dimensional *ABC* dynamo was first analyzed by Roberts (1972). Helicity is the essential requirement for dynamo action, as is obvious from the α-effect introduced in section 5.1.3. It should, however, be noted that the presence of helicity does not guarantee field amplification or sustaining, since an axisymmetric flow, which does not give rise to dynamo action, also has finite helicity for nonzero azimuthal velocity v_ϕ, $H^K = 2 \int dV v_\phi (\nabla \times \mathbf{v}_p)_\phi$, as can easily be checked.

The first dynamo solution in a conducting sphere with a nonvanishing dipole field outside the sphere has been developed by Herzenberg (1958). Here the velocity consists of two spherical rotors of constant angular velocity embedded in the sphere, with the fluid otherwise at rest. Assuming the rotor radii to be small compared with the distance between their centers makes an analytic calculation of the field possible. (The analysis can be generalized to n rotors, see for instance Moffatt (1978), where references to the original papers are found.) While the actual calculations have been restricted to special rotor configurations and orientations, dynamo action seems to occur for rather general conditions, though the magnetic field may not be stationary but oscillating.

Models like Herzenberg's rotor configuration are of course very artificial, serving only as a proof-of-principle of the existence of motions that give rise to dynamo action. Before moving on to consider systems of more practical relevance, it is useful to illustrate the basic process of field amplification in a simple mechanistic model.

5.1.2 *The rope dynamo*

Vainshtein & Zeldovich (1972) proposed the following elementary process called a rope dynamo, which demonstrates the possibility of exponential growth of the magnetic field, see fig. 5.4. Consider a closed flux tube of cross-section F carrying the flux ϕ. Stretching the tube to twice its original length reduces the cross-section by one-half because of mass conservation (assuming incompressibility), while the flux remains unchanged. Twisting and folding the tube as indicated in the figure leads to a tube with the original cross-section but twice the flux. Repeating the process results in

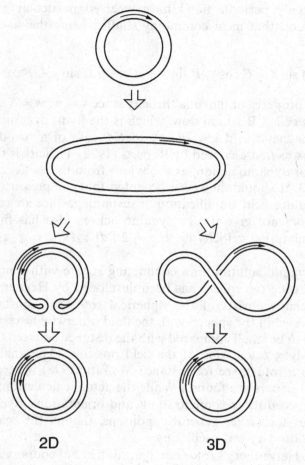

Fig. 5.4.   Basic mechanism of the rope dynamo in 2D and in 3D.

exponential growth of the flux

$$\phi_n = \phi_0 e^{n \ln 2}.$$

It is interesting to note that this stretch–twist–fold process does not involve reconnection, it occurs on the fastest possible time-scale – the advective time – and appears to be typical for the initial amplification of a seed field in real systems. Though of course under realistic conditions the process of folding back cannot be perfect, which will make the field configuration more and more filamentary, the Vainshtein–Zeldovich process is, at least in principle, possible. In two-dimensional geometry, by contrast, only stretching can occur. Whereas a flux tube of the original cross-section can be generated, as indicated in fig. 5.4, not only does this require reconnection but, more importantly, the field inside the combined tube is now anti-parallel, hence only the magnetic energy not the flux has increased. By repeating the process, the cross-sections of the unidirectional-field filaments shrink, until resistive diffusion leads to field annihilation, prohibiting further growth of the energy, which ultimately decays, consistent with the anti-dynamo theorem.

The rope dynamo is the prototype of a so-called fast dynamo,[*] as contrasted with a slow dynamo wherein the growth rate γ depends on the resistivity, or more precisely on the magnetic Reynolds number $Re_M = vL/\eta$, $\gamma \sim Re_M^{-\nu}$, $0 < \nu \leq 1$. This point has attracted considerable attention in the astrophysical community since, because of the typically huge value of Re_M slow dynamos may need more time than is available to produce, for instance, the galactic magnetic field. However, the strict distinction between fast and slow dynamos results from the use of strongly simplified models and should not be overemphasized. It will become clear subsequently that the generation of observable large-scale fields, such as the dipole field of the Earth, requires reconnection and hence cannot be fast in the strict sense, but that, on the other hand, turbulence may give rise to reconnection rates independent of the Reynolds number.

5.1.3 Mean-field electrodynamics

In the dynamo models considered so far the velocity field $\mathbf{v}(\mathbf{x}, t)$ was treated as a known function, which has been chosen in a rather ad hoc way. We now take a first step toward self-consistency. In most systems of practical relevance, both in astrophysical and laboratory plasmas, the fluid motions responsible for dynamo action are caused by some instability – buoyancy-driven in stars and planets or current-driven in the reversed-field pinch – and are therefore expected to become turbulent for the typical high Reynolds numbers. Moreover, these fluid motions have spatial scales which are small compared with the observed large-scale magnetic fields, these motions generate. This suggests a two-scale statistical approach called mean-field electrodynamics (MFE), which was developed in a pioneering paper by Steenbeck *et al.*, (1966) (for a more recent review see Krause & Rädler, 1980).

One assumes that $\mathbf{v}$ and $\mathbf{B}$ can be decomposed into an average part

---

[*] The Vainshtein–Zeldovich picture of the rope dynamo is oversimplified, as noticed by Moffatt & Proctor (1985), since by the twisting (= kinking in the language of section 2.4) of the rope the field inside the rope becomes more and more twisted because of magnetic helicity conservation. At some point this internal twist will relax, which involves reconnection. Hence the rope dynamo can only be fast, i.e., independent of η, for a finite time.

varying only on the large scale and a small-scale fluctuating part:

$$\mathbf{v} = \mathbf{v}_0 + \tilde{\mathbf{v}}, \quad \mathbf{B} = \mathbf{B}_0 + \tilde{\mathbf{B}}, \tag{5.10}$$

$$\langle \tilde{\mathbf{v}} \rangle = \langle \tilde{\mathbf{B}} \rangle = 0.$$

Insertion into the induction equation yields equations for the mean and the fluctuating parts $\mathbf{B}_0$ and $\tilde{\mathbf{B}}$:

$$\partial_t \mathbf{B}_0 = \nabla \times (\mathbf{v}_0 \times \mathbf{B}_0) + \nabla \times \boldsymbol{\epsilon} + \eta \nabla^2 \mathbf{B}_0, \tag{5.11}$$

$$\partial_t \tilde{\mathbf{B}} = \nabla \times (\mathbf{v}_0 \times \tilde{\mathbf{B}} + \tilde{\mathbf{v}} \times \mathbf{B}_0) + \nabla \times [(\tilde{\mathbf{v}} \times \tilde{\mathbf{B}}) - \langle \tilde{\mathbf{v}} \times \tilde{\mathbf{B}} \rangle] + \eta \nabla^2 \tilde{\mathbf{B}}, \tag{5.12}$$

where

$$\boldsymbol{\epsilon} = \langle \tilde{\mathbf{v}} \times \tilde{\mathbf{B}} \rangle \tag{5.13}$$

is the mean electromotive force (the negative electric field, note the sign convention) induced by the turbulent fluctuations. By solving (5.12), $\boldsymbol{\epsilon}$ can be expressed in terms of $\tilde{\mathbf{v}}$ and $\mathbf{B}_0$. An explicit solution is obtained in the quasi-linear approximation, where the nonlinear term in the fluctuation equation (5.12) is neglected, which is formally justified when the fluctuation amplitudes are sufficiently small. Moreover, one usually assumes, for convenience, that the dominant scales of the turbulent fluctuations are still large compared to the dissipative scales, so that the diffusion terms in (5.12) can be omitted. We thus find

$$\tilde{\mathbf{B}}(\mathbf{x}, t) = \int_{-\infty}^{t} dt' \nabla \times [\tilde{\mathbf{v}}(\mathbf{x}', t') \times \mathbf{B}_0], \tag{5.14}$$

where the time-integral is along the Lagrangian orbit $\mathbf{x}' = \mathbf{x}(t')$ and the weak space–time dependence of the mean field is neglected. Insertion into (5.13) gives

$$\boldsymbol{\epsilon} = \int_{-\infty}^{t} dt' \langle \tilde{\mathbf{v}}(\mathbf{x}, t) \times \{ \nabla \times [\tilde{\mathbf{v}}(\mathbf{x}', t') \times \mathbf{B}_0] \} \rangle. \tag{5.15}$$

On the assumption that $\tilde{\mathbf{v}}$ is statistically independent of $\mathbf{B}_0$, $\boldsymbol{\epsilon}$ is a linear functional of $\mathbf{B}_0$. Making use of the separation of scales we can conclude that $\boldsymbol{\epsilon}$ depends only on the local properties of $\mathbf{B}_0$, i.e., on $\mathbf{B}_0$ and its first-order spatial derivatives,

$$\epsilon_i = \alpha_{ij} B_{0j} + \beta_{ijk} \partial_j B_{0k}. \tag{5.16}$$

A further simplifying assumption usually made for convenience is that the turbulent field $\tilde{\mathbf{v}}$ is isotropic, whence $\alpha_{ij} = \alpha \delta_{ij}$ and $\beta_{ijk} = -\beta \epsilon_{ijk}$ (the choice of the sign will soon become clear), which reduces (5.16) to the form

$$\boldsymbol{\epsilon} = \alpha \mathbf{B}_0 - \beta \nabla \times \mathbf{B}_0, \tag{5.17}$$

where α is a pseudo-scalar ($\alpha \to -\alpha$ for $\mathbf{x} \to -\mathbf{x}$) and β is a scalar, since $\mathbf{B}$ is an axial vector. The r.h.s. of (5.15) is most conveniently evaluated by using explicit vector components $\epsilon = \{\epsilon_i\}$. Because of (5.17), ϵ_1 contains only B_{01} and the derivatives $\partial_2 B_{03}$, $\partial_3 B_{02}$,

$$\epsilon_1 = \int_{-\infty}^{t} dt' \left(\langle \tilde{v}_2 \partial_1 \tilde{v}_3' \rangle - \langle \tilde{v}_3 \partial_1 \tilde{v}_2' \rangle \right) B_{01} - \int_{-\infty}^{t} dt' \left(\langle \tilde{v}_2 \tilde{v}_2' \rangle \partial_2 B_{03} - \langle \tilde{v}_3 \tilde{v}_3' \rangle \partial_3 B_{02} \right).$$
(5.18)

Isotropy implies invariance under cyclic permutations, hence

$$\langle \tilde{v}_2 \partial_1 \tilde{v}_3' \rangle - \langle \tilde{v}_3 \partial_1 \tilde{v}_2' \rangle = \langle \tilde{v}_2 (\partial_1 \tilde{v}_3' - \partial_3 \tilde{v}_1') \rangle = -\tfrac{1}{3} \langle \tilde{\mathbf{v}} \cdot \nabla \times \tilde{\mathbf{v}}' \rangle,$$

and

$$\langle \tilde{v}_1 \tilde{v}_1' \rangle = \langle \tilde{v}_2 \tilde{v}_2' \rangle = \langle \tilde{v}_3 \tilde{v}_3' \rangle = \tfrac{1}{3} \langle \tilde{\mathbf{v}} \cdot \tilde{\mathbf{v}}' \rangle.$$

Comparison with (5.17) gives the expressions

$$\alpha = -\tfrac{1}{3} \int_{-\infty}^{t} dt' \langle \tilde{\mathbf{v}} \cdot \nabla \times \tilde{\mathbf{v}}' \rangle = -\tfrac{1}{3} \tau_\alpha \langle \tilde{\mathbf{v}} \cdot \nabla \times \tilde{\mathbf{v}} \rangle = -\tfrac{1}{3} \tau_\alpha H^K,$$
(5.19)

$$\beta = \tfrac{1}{3} \int_{-\infty}^{t} dt' \langle \tilde{\mathbf{v}} \cdot \tilde{\mathbf{v}}' \rangle = \tfrac{1}{3} \tau_\beta \langle \tilde{v}^2 \rangle = \tfrac{1}{3} \tau_\beta E^K,$$
(5.20)

where $\tau_\alpha \simeq \tau_\beta$ are the velocity correlation times, H^K is the kinetic helicity, (3.35), and E^K the kinetic energy of the turbulence. While β is positive, α may assume either sign depending on the sign of the helicity. The mean-field equation now becomes

$$\partial_t \mathbf{B}_0 = \nabla \times (\mathbf{v}_0 \times \mathbf{B}_0) + \nabla \times \alpha \mathbf{B}_0 + (\eta + \beta) \nabla^2 \mathbf{B}_0,$$
(5.21)

assuming $\beta = const.$ Since for the large-scale field $\mathbf{B}_0$ the collisional resistivity η has a negligible effect, it is usually omitted in the mean-field equation.

It has been mentioned before that the build-up of large-scale magnetic fields requires reconnection, since the basic advective dynamo mechanism idealized by the stretch–twist–fold process, can only amplify small-scale fields. Though in the derivation of the coefficients α and β in the mean-field equation (5.21) resistivity is formally neglected, violation of magnetic flux conservation is implicitly introduced by discarding the nonlinear contribution, since only the full equation (5.12) satisfies flux conservation (for $\eta = 0$).

5.1.4 α^2- and $\alpha\Omega$-dynamos

To date, the theoretical treatment of the dynamo systems of main interest in astrophysics, especially the origin and properties of stellar and

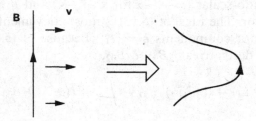

Fig. 5.5.   Generation of toroidal field by differential rotation $\mathbf{B}_p \cdot \nabla\Omega$.

galactic magnetic fields, is essentially limited to kinematic theory. Here the main tool is the mean-field equation (5.21), which for axisymmetric mean field and flow is usually written in cylindrical coordinates (r, ϕ, z) in the following form:

$$\partial_t A + \frac{1}{r}\mathbf{v}_p \cdot \nabla r A = \alpha B_\phi + \eta_T \left(\nabla^2 - \frac{1}{r^2}\right) A, \qquad (5.22)$$

$$\partial_t B_\phi + r\mathbf{v}_p \cdot \nabla \frac{1}{r} B_\phi = r\mathbf{B}_p \cdot \nabla\Omega + \mathbf{e}_\phi \cdot (\nabla \times \alpha\mathbf{B}_p) + \eta_T \left(\nabla^2 - \frac{1}{r^2}\right) B_\phi \quad (5.23)$$

with

$$\mathbf{B} = \mathbf{B}_p + \mathbf{e}_\phi B_\phi, \quad \mathbf{B}_p = \nabla \times A\mathbf{e}_\phi,$$
$$\mathbf{v} = \mathbf{v}_p + r\Omega\mathbf{e}_\phi,$$

where the $1/r^2$ term in the diffusion operator $(\nabla^2 - r^{-2})A = \partial_r r^{-1}\partial_r rA + \partial_z^2 A$ results from the derivative of the unit vector $\partial_\phi \mathbf{e}_\phi = -\mathbf{e}_r$. It is understood that $\mathbf{B}, \mathbf{v}$ refer to the large-scale mean quantities, the subscript 0 being omitted. The notation $v_\phi = r\Omega$ is chosen following convention to emphasize the role of the differential rotation $\nabla\Omega$, and η_T $(= \beta)$ is the turbulent resistivity.

We see immediately that for $\alpha = 0$ there is no feedback from the toroidal field B_ϕ to the poloidal field B_p, such that the latter must decay and eventually also the former, in accordance with Cowling's theorem. Hence $\alpha \neq 0$ is required for dynamo action. In the B_ϕ equation (5.23) two effects produce a coupling to the poloidal field. The action of the term $\mathbf{B}_p \cdot \nabla\Omega$, which is called the Ω-effect, is easily understood – it is the generation of a toroidal field from a poloidal field by differential rotation, fig. 5.5. The α-effect $\nabla \times \alpha\mathbf{B} \simeq \alpha\mathbf{j}$ is illustrated in fig. 5.6. Here α corresponds to a twisting motion with positive helicity (a), which leads to a reduction of $\mathbf{B}$, or negative helicity (b), which leads to growth of $\mathbf{B}$. In particular, the α-term in (5.22) gives the reaction of the toroidal on the poloidal field missing for purely axisymmetric flows, which closes the circular process of overall field amplification.

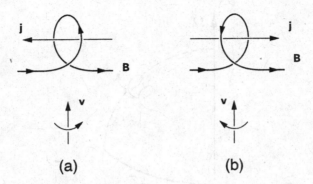

(a)   (b)

Fig. 5.6.   Illustration of the α-effect. Weakening or strengthening of the magnetic field, $\partial_t \mathbf{B} \simeq \alpha \mathbf{j}$, by a helical flow with (a) $H^K = -\alpha > 0$; (b) $H^K < 0$.

A natural cause for nonvanishing α is thermal convection in a rotating sphere. One can estimate the magnitude and sign of α using the mixing-length approximation for the turbulence. Helicity is proportional to the two vectors characterizing the convective turbulence, the angular frequency $\mathbf{\Omega}$ and the stratification $\nabla \ln \rho$. Hence by simple dimensional arguments we find $H^K \sim \tilde{v} l \mathbf{\Omega} \cdot \nabla \ln \rho$, where $\tilde{v}$ is a typical fluctuation velocity and l is the correlation length or the mixing length, which is of the order of the density scale-height, $l \sim L_\rho = |\nabla \ln \rho|^{-1}$. The sign of H^K can be obtained by the following reasoning (Parker, 1955a). Since a buoyantly rising fluid element expands, the action of the Coriolis force leads to a left-hand helical motion on the northern hemisphere (cf. the anti-cyclonic motion, i.e., in the sense opposite to the rotation, around a high-pressure system in the Earth's atmosphere), which corresponds to negative helicity. In the downdrafts the fluid is compressed, hence the motion induced by the Coriolis force is cyclonic, but since also the radial velocity is reversed, the helicity is again negative. By the same argument, the helicity is positive on the southern hemisphere (fig. 5.7). Since by definition the correlation time is $\tau = l/\tilde{v}$, we obtain, from (5.19),

$$\alpha \simeq -\tfrac{1}{3} l^2 \mathbf{\Omega} \cdot \nabla \ln \rho \simeq \tfrac{1}{3} l \Omega \cos \theta, \qquad (5.24)$$

where θ is the colatitude, the usual poloidal angle in spherical coordinates. Hence α is positive in the northern hemisphere and negative in the southern. These characteristics of the convective turbulence in a rotating sphere have important consequences for the solar dynamo mechanism, as we will discuss in section 5.2.2.

The turbulent resistivity is usually estimated by the simple expression

$$\eta_T \sim \tfrac{1}{3} \tilde{v} l. \qquad (5.25)$$

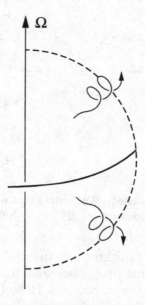

Fig. 5.7.   Helical motions of buoyantly rising fluid elements in the northern and southern hemisphere.

The relative importance of the two dynamo terms in (5.23) is character-ized by two Reynolds-number-type dimensionless quantities denoted by C_α and C_Ω,

$$C_\alpha = \alpha_0 R/\eta_T, \quad C_\Omega = \Delta\Omega R^2/\eta_T, \qquad (5.26)$$

where α_0 and $\Delta\Omega$ are the maximum values of α and of the radial variation of Ω, respectively, and R is the radius of the sphere or the disc. For $C_\alpha \gg C_\Omega$ the α-effect dominates and the differential rotation can be ignored in (5.23). Such a system is called an α^2-dynamo. In steady state, (5.22) gives the estimate $B_p \sim C_\alpha B_\phi$, and (5.23) gives $B_\phi \sim C_\alpha B_p$, hence the value of C_α for field sustainment is $C_\alpha \sim 1$, and the poloidal and toroidal magnetic fields are of the same order of magnitude,

$$B_\phi \sim B_p. \qquad (5.27)$$

In the opposite limit $C_\Omega \gg C_\alpha$ one can neglect the α-term in the B_ϕ-equation, which now gives the steady-state condition $B_\phi \sim C_\Omega B_p$. Together with $B_p \sim C_\alpha B_\phi$ from (5.22) we thus have $D = C_\alpha C_\Omega \sim 1$, where D is called the dynamo number, and

$$B_\phi \sim (C_\Omega/C_\alpha)^{1/2}D^{1/2}B_p \sim (C_\Omega/C_\alpha)^{1/2}B_p \gg B_p. \qquad (5.28)$$

Such a system is called an $\alpha\Omega$-dynamo, which carries a toroidal field much larger than the poloidal field. An $\alpha\Omega$-dynamo is therefore also characterized as "strong-field dynamo", compared with an α^2-dynamo, where both fields have similar magnitude, which is called "weak-field dynamo". In many astrophysical objects, such as the Sun or the Earth, the toroidal field is unobservable, concealed in the interior, since only the poloidal field extends outside the conductive region. In this case, the toroidal field can only be calculated or estimated theoretically using the observable properties of the object. From these it is concluded that both the geodynamo and the solar dynamo are of $\alpha\Omega$-type.

5.1.5 Free dynamo modes

The kinematic dynamo equations (5.22), (5.23) are linear in the magnetic field and thus for stationary mean velocity fields v and dynamo parameters α, η_T and given boundary conditions have eigenmodes proportional to $e^{\gamma t}$. In general, v and α are functions of space, which would require to solve the equations numerically. Some important qualitative properties can, however, already be disclosed in the local or WKB approximation, where $v_p, \nabla\Omega, \alpha, \eta_T$ and the geometrical factors r can be considered constant. The concept of dynamo waves has been introduced by Parker (1955a) and Yoshimura (1975) to explain the period and migration direction of the solar magnetic field. With the usual Fourier ansatz $A, B_\phi \propto \exp\{i\mathbf{k}\cdot\mathbf{x}\}$ one obtains

$$(\gamma + \eta_T k^2 + i\mathbf{k}\cdot\mathbf{v}_p)A - \alpha B_\phi = 0, \tag{5.29}$$

$$(\gamma + \eta_T k^2 + i\mathbf{k}\cdot\mathbf{v}_p)B_\phi + [(i\mathbf{k}\times r\nabla\Omega)_\phi - \alpha k^2]A = 0, \tag{5.30}$$

using $\nabla \times \mathbf{B}_p = -k^2 A$. The simplest case is that of the α^2-dynamo in the absence of a mean flow, where the solvability condition of (5.29) and (5.30) gives the dispersion relation

$$(\gamma + \eta_T k^2)^2 = \alpha^2 k^2. \tag{5.31}$$

Hence the mode is purely growing or decaying, $\text{Im}\{\gamma\} = 0$. Dynamo action $\gamma \geq 0$ occurs, if $|\alpha| > \eta_T k$, which is independent of the sign of α.

In the opposite case of an $\alpha\Omega$-dynamo, where the α-term in (5.30) can be neglected, we find the dispersion relation

$$(\gamma + \eta_T k^2 + i\mathbf{k}\cdot\mathbf{v}_p)^2 = -2i\Gamma, \tag{5.32}$$

where

$$\Gamma = \tfrac{1}{2}\alpha(\mathbf{k}\times r\nabla\Omega)_\phi. \tag{5.33}$$

The character of the solution depends on the sign of Γ, i.e., the direction of $\mathbf{k}$ for a given sign of α. First take $\Gamma > 0$. With $\sqrt{-2i} = \pm(1-i)$ the solution of (5.32) is

$$\gamma = -\eta_T k^2 \pm \Gamma^{1/2} - i(\mathbf{k}\cdot\mathbf{v}_p \pm \Gamma^{1/2}), \tag{5.34}$$

i.e., $\text{Re}\{\gamma\} \geq 0$ for the upper sign and $\Gamma^{1/2} \geq \eta_T k^2$. Contrary to the α^2-dynamo the magnetic field is in general oscillating with frequency $\omega = -\text{Im}\{\gamma\} = \mathbf{k} \cdot \mathbf{v}_p + \Gamma^{1/2}$ corresponding to so-called dynamo waves. For sufficiently small poloidal velocity $\mathbf{v}_p$, the wave propagates in the $+\mathbf{k}$-direction, but may reverse its direction if $\mathbf{v}_p$ is large, $\mathbf{k} \cdot \mathbf{v}_p = -\Gamma^{1/2}$ leading to a stationary wave pattern.

In the case $\Gamma < 0$ the solution of the dispersion relation reads, using $\sqrt{2i} = \pm(1 + i)$,

$$\gamma = -\eta_T k^2 \pm |\Gamma|^{1/2} - i(\mathbf{k} \cdot \mathbf{v}_p \mp |\Gamma|^{1/2}), \tag{5.35}$$

such that the frequency of the dynamo wave is $\omega = \mathbf{k} \cdot \mathbf{v}_p - |\Gamma|^{1/2}$. For $\mathbf{v}_p = 0$ the dynamo wave (upper sign in (5.35)) now travels in the $-\mathbf{k}$-direction.

Besides the frequency and the propagation direction of the dynamo wave, the eigenmode equations (5.29), (5.30) also predict the phase relation between the poloidal and the toroidal fields. For the most relevant case of marginal stability and $\mathbf{v}_p = 0$, (5.29) gives

$$A = i(\alpha/\omega)B_\phi$$

and hence for $B_r = -ik_z A$

$$B_r = \frac{\alpha}{\omega/k_z} B_\phi. \tag{5.36}$$

We shall see in section 5.2.2 that both the prediction for the phase velocity of the dynamo wave and for the phase between B_r and B_ϕ have stringent consequences for the solar dynamo mechanism.

Let us finally discuss the symmetry properties of the eigenmodes of (5.22) and (5.23). With the usual symmetry assumptions with respect to the equator $z = 0$ or colatitude $\theta = 90°$, α and v_θ odd, v_r and Ω even, the eigenmode parities are either

$$A \text{ even}, \quad B_\phi \text{ odd}, \quad \text{(dipole)}$$

or

$$A \text{ odd}, \quad B_\phi \text{ even}, \quad \text{(quadrupole)}.$$

As will be seen later, the dipole mode is the relevant one for most astrophysical dynamos.

Since the pioneering work of Steenbeck *et al.* (1966), the kinematic mean-field theory has become the standard tool for modeling dynamo action in various astrophysical objects. Choosing the appropriate geometry and profiles $\alpha(\mathbf{x})$, $\Omega(\mathbf{x})$ based on the properties of the underlying convective turbulence and consistent with observations, one considers the dynamo mode with the lowest dynamo number $D = C_\alpha C_\Omega$. This procedure gives an approximate picture of the spatial structure and the time variability of the magnetic field of the object. Though the conventional assumption of isotropy of the turbulence, which leads to the simplification $\alpha_{ik} = \alpha\delta_{ik}$, is hardly ever really justified, anisotropy does not seem to change the dynamo behavior qualitatively.

5.2 Mean-field MHD dynamo theory

If dynamo action occurs kinematic theory tells us that the magnetic field with a certain spatial structure corresponding to a particular dynamo mode grows exponentially in time. To describe the saturation of this process the reaction on the fluid through the Lorentz force must be included, such that the fluid motion adjusts self-consistently.

In this section we will first discuss some phenomenological models which have been introduced in the framework of mean-field theory. We then give an overview of the present understanding of the solar dynamo, the prototypical application of mean-field theory, pointing out the difficulties of the theory which have emerged in the light of the recent helioseismological observations.

5.2.1 Nonlinear quenching processes and magnetic buoyancy

Nonlinear effects enter the dynamo process in two different ways: through a finite-amplitude modification of the fluctuating quantities invalidating the quasi-linear approximation, which is called micro-feedback, and through the Lorentz force acting on the mean flow, called macro-feedback. The most important process in the first category is the nonlinear reduction of α, called α-quenching, which acts directly on the development of the mean magnetic field $\mathbf{B}_0$ in the induction equation (5.21). The physical origin of this process is the Alfvén-wave effect, which is considered in more detail in section 5.4. Basically it means that, in the presence of a mean field, small-scale velocity and magnetic field fluctuations tend to be coupled, forming Alfvén waves, (3.22), $\tilde{\mathbf{B}} \simeq \pm\tilde{\mathbf{v}}$ in the present normalization. This correlation, which competes with the local (in k-space) decorrelating interactions of the small-scale fluctuations, is the stronger the larger the mean field. Since the electromotive force $\epsilon = \langle \tilde{\mathbf{v}} \times \tilde{\mathbf{B}} \rangle$ vanishes for pure Alfvén waves, ϵ and in particular α, should decrease as the mean field grows due to the α-term, which ultimately leads to a saturated dynamo state.

A rigorous theory of α-quenching has not yet been developed. Instead the process is modeled in a simple phenomenological way. Since the Alfvén-wave effect depends only on the magnitude of the mean field, α

must be a decreasing function of B_0. The following ansatz is used in many applications (e.g., Rüdiger, 1973; Ivanova & Ruzmaikin, 1977)

$$\alpha = \alpha_0/(1 + c_M B_0^n), \tag{5.37}$$

where c_M is a constant and small n, $n = O(1)$, implies a smooth reduction and large n an abrupt quenching.[*]

To discuss the reaction on the mean flow, we consider the momentum equation, again in the incompressible limit,

$$\rho(\partial_t \mathbf{v}_0 + \mathbf{v}_0 \cdot \nabla \mathbf{v}_0) = \mathbf{B}_0 \cdot \nabla \mathbf{B}_0 - \nabla P_0 + \mathbf{g}\rho - \rho \nabla \cdot \mathscr{Q} + \rho \nu \nabla^2 \mathbf{v}_0. \tag{5.38}$$

Here $\mathscr{Q}$ is the total stress tensor due to small-scale fields,

$$Q_{ij} = \langle \tilde{v}_i \tilde{v}_j \rangle - \langle \tilde{B}_i \tilde{B}_j \rangle, \tag{5.39}$$

where the kinetic part $R_{ij} = \langle \tilde{v}_i \tilde{v}_j \rangle$ is called the Reynolds stress tensor and the magnetic part $M_{ij} = \langle \tilde{B}_i \tilde{B}_j \rangle$ the Maxwell tensor. In the kinematic regime $Q_{ij} = R_{ij}$, which acts as an inhomogeneous term in the $\mathbf{v}_0$-equation much in the same way as ϵ does in the induction equation (5.11). It is important to note that due to rotation, i.e., to the Coriolis force, and to stratification the turbulence is not isotropic, hence, in general, the off-diagonal terms do not vanish, $\langle \tilde{v}_i \tilde{v}_j \rangle \neq 0$ for $i \neq j$. To lowest order in the the two-scale approximation R_{ij} can be written in the form

$$R_{ij} = \Lambda_{ijk}\Omega_k - N_{ijkl}\partial_k v_{0_l}, \tag{5.40}$$

where Ω_k is the angular velocity vector. Relation (5.40) is analogous to the expression (5.16) for ϵ_i. While N_{ijkl} corresponding to β_{ijk} can be reduced to a turbulent viscosity ν_T,[†] the tensor $R_{ij}^\Lambda = \Lambda_{ijk}\Omega_k$ is the analog of the α-effect and is the main source of differential rotation sustained against viscous decay. In spherical coordinates only two components are non-zero

$$\begin{aligned} R_{r\phi}^\Lambda &= \Lambda^V \Omega \sin\theta \\ R_{\theta\phi}^\Lambda &= \Lambda^H \Omega \cos\theta. \end{aligned} \tag{5.41}$$

The coefficients Λ^V, Λ^H are functionals of the vertical and horizontal velocity fluctuations $\langle \tilde{v}_\phi^2 \rangle$, $\langle \tilde{v}_\theta^2 \rangle$, $\langle \tilde{v}_r^2 \rangle$. For horizontally symmetric turbulence one finds (Rüdiger, 1989)

$$\Lambda^V \simeq 2\tau(\langle \tilde{v}_\phi^2 \rangle - \langle \tilde{v}_r^2 \rangle), \quad \Lambda^H \simeq 0. \tag{5.42}$$

---

[*] It should be noted that a reduction of ϵ implies also a reduction of the turbulent resistivity η_T, a process often ignored in the discussion of nonlinear dynamo effects.

[†] Because of the anisotropy of the turbulence one has, strictly speaking, a viscosity tensor, where vertical and horizontal viscosities are different.

(For more general nonsymmetric conditions $\Lambda^H \simeq 2\tau(\langle \tilde{v}_\phi^2 \rangle - \langle \tilde{v}_\theta^2 \rangle)$.) If Λ^V is independent of θ, $\Lambda^V = \Lambda^V(r)$, only radial differential rotation can be sustained. To describe latitudinal differential rotation as observed for instance on the solar surface, Λ^V must also depend on θ.

When the magnetic field becomes sufficiently strong, the Maxwell tensor M_{ij} must be taken into account. In fact, due to the Alfvén-wave effect the small-scale magnetic fluctuations will reduce the total stress tensor Q_{ij} (5.39), which vanishes for pure Alfvén waves $\widetilde{\mathbf{B}} = \pm \widetilde{\mathbf{v}}$. Though not generally acknowledged, this micro-feedback, which we call Λ-quenching, appears to be an efficient saturation of dynamo action, since it tends to eliminate the source of differential rotation in the $\alpha\Omega$-process.

The strongest macro-feedback is due to the Lorentz force associated with the mean field $\mathbf{B}_0$ and affects directly the mean flow $\mathbf{v}_0$ in the induction equation (5.21). This process, which was first discussed by Malkus & Proctor (1975) for the α^2-dynamo, seems to be even more important for rapidly rotating systems, where the $\alpha\Omega$-process dominates. Here the reaction on the differential rotation Ω is most significant, whence the name Ω-quenching. Numerical mean-field dynamo studies including this effect have been performed by Brandenburg et al. (1990, 1991).

A nonlinear saturation process of a different kind arises through magnetic buoyancy, which tends to remove magnetic flux from the region of dynamo action. This concept has been introduced by Parker (1955b) to explain the emergence of magnetic flux tubes above the solar surface, forming sunspots (for details, see Parker, 1980). Since the gas pressure in a flux tube has to be lower than the external pressure because of overall cross-field pressure balance

$$\tfrac{1}{2}B^2 = p_{\text{ex}} - p_{\text{in}}, \qquad (5.43)$$

the density in the tube is lower than the ambient density, $\rho_{\text{in}}/\rho_{\text{ex}} = (p_{\text{in}}/p_{\text{ex}})^{1/\gamma} < 1$ ($\gamma = 1$ for isothermal conditions in the presence of rapid heat conduction along the tube). Hence there is a buoyancy force per unit volume

$$\mathbf{F} = \mathbf{g}(\rho_{\text{ex}} - \rho_{\text{in}}),$$

which will drive these flux tubes upward (and finally across the surface). Buoyant motion will be counteracted by magnetic tension as well as by friction due to small-scale turbulent eddies. The former is inversely proportional to the length L of the tube, $\mathbf{B} \cdot \nabla \mathbf{B} \sim B^2/L$, while the latter is proportional to the surface area of the tube. Hence longer and thicker flux tubes rise faster. We will come back to this point in section 5.2.2.

While the amplitude of the small-scale magnetic fluctuations is probably limited by the Alfvén-wave effect $\widetilde{B} \lesssim \widetilde{v}$, which can be called small-scale or dynamic equipartition, the final magnitude of the large-scale magnetic field is difficult to estimate. The field often greatly exceeds dynamic equipartition, since, due to density draining, the field can reach the static equilibrium intensity, $B_0^2 \sim p \gg v_0^2$. Moreover, the toroidal field may continue to build up, becoming more and more force-free, $\mathbf{j_0} \times \mathbf{B_0} \ll |j_0||B_0|$, hence even $B_0^2 \gg p$ is possible. If this process is dynamically accessible, only convective transport of B out of the production region can limit the field intensity, either by magnetic buoyancy or by poleward convection (as a strained rubber band slides on a smooth sphere).

5.2.2 *The solar dynamo*

Let us briefly summarize the main observations concerning the magnetic field of the Sun (for more details see, e.g., Priest, 1984):

(a) The most obvious manifestation of this field are sunspots, which usually occur in pairs in a latitude belt of $\simeq \pm 30°$ about the equator. The two partners of a pair have opposite magnetic polarity; they appear to be connected by field lines, which emerge through the solar surface in one spot and submerge again in the partner spot. The orientation of the connecting field is essentially in azimuthal direction ϕ, with one orientation in the northern hemisphere and the reversed one in the southern (Hale's polarity law).

(b) Apart from the azimuthal field associated with sunspots, there is a weaker poloidal component which is distributed on a grainy network all over the solar surface. Its average radial component also reverses sign from northern to southern hemisphere, hence has an approximate dipole character.

(c) The most remarkable feature of the solar magnetic field is its regular time variation. Both azimuthal and poloidal field reverse sign every 11 years. This gives rise to the sunspot cycle, starting from a minimum number of spots (close to zero), proceeding to sunspot maximum and back to minimum. The full period including both polarities is 22 years.

(d) The B_ϕ and B_r components are in anti-phase, i.e., $B_\phi B_r < 0$ for most of the time. While the field inside the sunspot latitude belt migrates toward the equator (which gives rise to the famous butterfly diagram, fig. 5.8), the more diffusely distributed field outside this belt migrates toward the poles.

(e) Because of the very intermittent spatial distribution, the field intensity varies strongly. In sunspots the field reaches several kilogauss, which is about the limit given by cross-field equilibrium $B^2/8\pi \sim p$

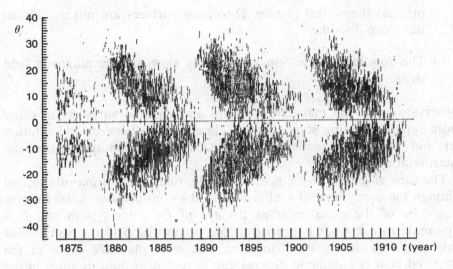

Fig. 5.8. Butterfly diagram (from Maunder, 1913). It shows the incidence of sunspots as a function of colatitude and time. A vertical bar is plotted for each degree and each 27-day period if, during this period, at least one sunspot is observed in this angular range. The figure demonstrates the migration of sunspots toward the equator during the 11-year cycle.

in the photosphere. Also in the thin flux tubes of the network called magnetic knots, where most of the remaining field is concentrated, the field seems to reach this magnitude. Since sunspots and network knots cover less than one percent of the solar surface, the spatially averaged field is much weaker, of the order of a few gauss.

(f) The Sun possesses an extended convection zone. While the hot central part of the solar interior is in static equilibrium dominated by radiative transport, in the cooler outer shell opacity is increased by ionization and recombination processes, which gives rise to steeper temperature gradients and the onset of thermal convection (see section 5.3.1). The lithium depletion observed on the Sun suggests that the convection zone reaches down to the lithium-burning region, i.e., to about 0.7 of the solar radius. The visible manifestation of the turbulent convection are the granular structure of the photosphere and the associated motions.

(g) The Sun is rotating with a mean period of 27 days; the rotation is not rigid, the surface value of the angular velocity being about 10 percent larger at the equator than at the pole. The variation of Ω inside the convection zone, in particular the radial dependence, has recently been determined from helioseismology. Contrary to the original theoretical picture, Ω-contour surfaces are not cylindrical but more disc-like.

(h) The rotation of the Sun is relatively slow and its magnetic field weak.

Observations of the magnetic intensity and its time variation in other more rapidly rotating stars shows a close relation between the rotation rate and the mean magnetic field, $\langle B \rangle \propto \Omega$, which corroborates the dynamo interpretation.

The time variation of the field essentially rules out a primordial origin (though the decay time of such a field in the non-turbulent solar interior would be of the same order as the age of the solar system and it is legitimate to ask what happened to the primordial field the Sun should have captured during its formation). The most plausible cause of the observed field is a dynamo process due to turbulent fluid motions in the convection zone. From the characteristics of sunspots and the poloidal field this should be an $\alpha\Omega$-dynamo: differential rotation transforms a poloidal into a toroidal field (fig. 5.5), of which the sunspot field is the direct manifestation, while the helicity of the convective turbulence regenerates the poloidal field. A theory of the solar dynamo should account for the solar cycle time (at least give the correct order of magnitude), the equatorward sunspot drift and the phase relation between B_ϕ and B_r. Because of the extremely high Rayleigh number (for a definition of the Rayleigh number and other quantities see section 5.3.1), direct numerical simulations of the entire convection zone are out of the question and one has to resort to a mean-field approach characterized primarily by the dynamo coefficients α, η_T (for recent reviews of mean-field solar dynamo theory see Rädler, 1990; Stix, 1991; Gilman, 1992; Brandenburg, 1994). Typical values of the velocity fluctuation amplitude and the correlation length in the photosphere are $\tilde{v} \sim 1\,\text{km/s}$ and $l \sim 10^3\,\text{km}$. In the convection zone l increases with depth, while $\tilde{v}$ decreases such that $\tilde{v}l$ is roughly constant, hence by (5.25)

$$\eta_T \sim \tfrac{1}{3}\tilde{v}l \sim 3 \times 10^2\,\text{km}^2/\text{s}. \tag{5.44}$$

By its very nature α is a more complicated effect, which is usually estimated by $\alpha \sim \epsilon_\alpha\Omega l$, (5.24). Here ϵ_α is a numerical factor, which will be discussed further below. Since l must be taken for the region of maximum dynamo action which, as we will see, is close to the bottom of (or even below) the convection layer, we assume $l \sim 3 \times 10^4\,\text{km}$ and, with $\Omega \sim 3 \times 10^{-6}\text{s}^{-1}$, one obtains

$$\Omega l \sim 0.1\,\text{km/s}. \tag{5.45}$$

Since the original paper by Parker (1955a), the fundamental concept in solar dynamo theory is a dynamo wave of stationary amplitude with appropriate frequency and propagation direction. The condition $\mathrm{Re}\{\gamma\} = 0$ in (5.34) and (5.35) gives the frequency (for negligible poloidal flow $\mathbf{v}_p$)

$$\omega = \Gamma^{1/2} = \eta_T k^2, \tag{5.46}$$

where $k = 2\pi/L$ and L is the global scale of the dynamo process, which is of the order of the solar radius R. Hence, using (5.44), the cycle time would be

$$\tau \simeq R^2/2\pi\eta_T \simeq 3 \times 10^8\,\mathrm{s} \simeq 10\,\mathrm{yr} \tag{5.47}$$

which, compared with the observed value $\tau = 22$ yr, is of the correct order of magnitude.

The imaginary part of (5.34) or (5.35) gives a relation between α and the radial scale-length l_Ω of the differential rotation $\partial_r \ln \Omega = l_\Omega^{-1}$, $\omega = \Gamma^{1/2} = (\alpha k R \Omega/l_\Omega)^{1/2}$, (5.33). Writing $kR \simeq 2\pi$ and $\alpha = \epsilon_\alpha l\Omega$, we find the relation

$$\sqrt{2\pi}\,\frac{\Omega}{\omega} \simeq 300 = \left(\frac{l_\Omega}{\epsilon_\alpha l}\right)^{1/2}. \tag{5.48}$$

Hence for large helicity $\epsilon_\alpha \sim 1$ the differential rotation needed for dynamo action is very low, $l_\Omega \sim 3 \times 10^9$ km corresponding to $R\partial_r \ln \Omega \sim 10^{-3}$–$10^{-4}$, which is not plausible. To avoid this conclusion the effective α should be strongly reduced compared with the estimate (5.24), i.e., $\epsilon_\alpha \sim 10^{-2}$ or $\alpha \sim 1\,\mathrm{m/s}$.

The more stringent condition is posed by the observed equatorward migration, which corresponds to $\Gamma < 0$. (This can be seen directly from (5.35): since $\mathrm{Re}\{\gamma\} = 0$ requires the upper sign, a positive (poleward) meridional flow would reduce the frequency, hence the inherent migration of the dynamo wave is equatorward.) This implies $\alpha\partial_r\Omega < 0$. As the arguments for $\alpha > 0$ are quite strong (see section 5.1.4), this seems to require $\partial_r\Omega < 0$, i.e., Ω increasing with depth (sub-rotation). (The alternative possibility $\alpha < 0$, $\partial_r\Omega > 0$ (super-rotation) is ruled out also because in this case $B_r B_\phi > 0$ from (5.36) contrary to the observation.) Observational results from helioseismology (fig. 5.9) show sub-rotation only below the poles, while below the equator there is a (weak) super-rotation. These properties of the differential rotation have also be obtained theoretically by solving the mean-field equation for Ω (see, e.g., Küker *et al.*, 1993). There seems now to be rather general agreement that the only way out of this dilemma is to locate the main dynamo action at the bottom of the convection zone or even in the convectively stable overshoot region below the convection zone. Fully three-dimensional numerical simulations of magnetoconvection, which will be discussed in section 5.3, reveal that in fact the magnetic field, mainly toroidal, is strongest in this region.

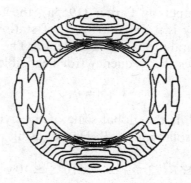

Fig. 5.9.   Contours of the solar differential rotation Ω (from Libbrecht, 1988).

The field intensity depends on the mechanism by which the field is transported across the convection zone to the surface of the Sun. The process must be sufficiently rapid so that the Coriolis force does not tilt the field substantially out of the azimuthal direction. Basically two different mechanisms are conceivable. The field could be swept along with rapid updrafts. In this case the field intensity at the bottom would have to be a few kilogauss. If this transport process is not effective, magnetic buoyancy has to be the dominant mechanism, in which case the field has to be higher by at least one order of magnitude to provide the necessary buoyancy force. The question about the dominant dynamo process on the Sun is far from being decided. It might be that the mean-field theory itself is not the appropriate framework and that a more direct approach is required.

5.3 MHD theory of thermal convection

5.3.1 Thermal convection in a rotating sphere

The engine for dynamo action in stars and planets is (very probably) convection of an electrically conducting fluid under the influence of the Coriolis force. Convection is usually driven thermally, but even if it has a different physical origin, as in the Earth's interior, the theoretical treatment is formally very similar. Let us first discuss the properties of thermal convection in a rotating system ignoring the magnetic field. Here a particular goal is to understand the generation of differential rotation which, as we have seen before, is of crucial importance for the dynamo process.

We write the basic equations in a coordinate frame rotating with angular velocity $\mathbf{\Omega}$:

$$\partial_t \rho + \nabla \cdot \rho \mathbf{v} = 0, \qquad (5.49)$$

$$\partial_t \mathbf{v} + \mathbf{v} \cdot \nabla \mathbf{v} = -\frac{1}{\rho} \nabla p + \mathbf{g} + 2\mathbf{v} \times \mathbf{\Omega} + \tfrac{1}{2} \nabla(\mathbf{\Omega} \times \mathbf{r})^2 - \nabla \cdot \mathbf{\pi}, \tag{5.50}$$

$$\partial_t T + \mathbf{v} \cdot \nabla T = -(\gamma - 1) T \nabla \cdot \mathbf{v} + \kappa \nabla^2 T + \Phi, \tag{5.51}$$

$$p = 2(\rho/m_i) k_B T, \tag{5.52}$$

where the third and the fourth term on the r.h.s. of (5.50) are the Coriolis force and the centrifugal force, π is the viscous stress tensor, and Φ is the viscous dissipation rate. As before, the fluid is assumed to be sufficiently dilute to obey the ideal gas law.

A fluid heated from below becomes convectively unstable, if the temperature profile is too steep. Neglecting rotation, viscosity and heat conduction, instability arises for a super-adiabatic temperature gradient,

$$-\frac{dT}{dz} > -(\gamma - 1) \frac{T}{\rho} \frac{d\rho}{dz}. \tag{5.53}$$

Relation (5.53) can easily be understood. Consider a fluid element of volume δV which, originally located at height z, is slightly displaced upward to $z + \delta z$. It is decompressed to the ambient pressure at the new height $p(z + \delta z) = p(z) + p'\delta z$, $p' \equiv dp/dz < 0$. Since heat conduction is neglected, the change of state δp, $\delta \rho$ is adiabatic

$$\frac{\delta \rho}{\rho} = \frac{1}{\gamma} \frac{\delta p}{p} = \frac{1}{\gamma} \frac{p'}{p} \delta z. \tag{5.54}$$

If the density of the displaced fluid element is smaller than the surrounding density

$$\rho + \delta \rho < \rho(z + \delta z) = \rho + \rho'\delta z, \tag{5.55}$$

there is a buoyancy force pushing the fluid element still further up, i.e., the fluid is convectively unstable. Insertion of (5.54) into (5.55) and use of the relation $p'/p = (\rho'/\rho) + T'/T$ gives the instability condition, (5.53).

Since fluid velocities are usually much smaller than the sound speed, the fluid motions can be taken as incompressible. It is therefore useful to simplify (5.49)–(5.52) by eliminating the sound wave. In the Boussinesq approximation one assumes, in addition, that the density perturbation is small compared to the mean density and that the wavelength is short compared to the density scale-height $kL_\rho \gg 1$, $L_\rho^{-1} = |\nabla \ln \rho_0|$, such that ρ_0 can be considered constant. We introduce the smallness parameter ϵ, writing $\rho v^2 = \epsilon p$, i.e., $v = \epsilon^{1/2} c_s$, where c_s is the sound speed. The quantities ρ, T vary on the convective time-scale, hence $\partial_t \sim \epsilon^{1/2}$, and we write $\rho_0 + \epsilon \tilde{\rho}$, $T = T_0 + \epsilon \tilde{T}$. By contrast, p relaxes on the sound time-scale and hence is in instantaneous equilibrium, such that

$$\tilde{\rho} = (\partial \rho_0 / \partial T)_p \tilde{T} = -\alpha \rho_0 \tilde{T}, \tag{5.56}$$

where in an ideal gas the thermal expansion coefficient is $\alpha = T^{-1}$. To zeroth order, (5.50) gives the barometric equation (the centrifugal force, which is small in most cases of interest, can formally be incorporated in the total pressure)

$$\nabla p_0 = \mathbf{g}\rho_0, \tag{5.57}$$

relating the global profiles of p_0 and ρ_0. In order ϵ, (5.49)–(5.52) reduce to

$$\nabla \cdot \mathbf{v} = 0, \tag{5.58}$$

$$\partial_t \mathbf{v} + \mathbf{v} \cdot \nabla \mathbf{v} = -\frac{1}{\rho_0} \nabla p - \alpha \mathbf{g} T + 2\mathbf{v} \times \boldsymbol{\Omega} + \nu \nabla^2 \mathbf{v}, \tag{5.59}$$

$$\partial_t T + \mathbf{v} \cdot \nabla T = \kappa \nabla^2 T + \Phi, \tag{5.60}$$

where we have used the equilibrium relation (5.57) in (5.59) and omitted the tildes. Equations (5.58)–(5.60) are called the Boussinesq equations.

A somewhat more general approach is provided by the anelastic approximation (see, e.g., Gilman & Glatzmaier, 1981), which still eliminates the sound wave but relaxes the scale requirement by allowing convective wavelengths to be of the order of the equilibrium scales. The essential change consists of replacing the incompressibility condition, (5.58), by

$$\rho_0^{-1} \nabla \cdot (\rho_0 \mathbf{v}) = \nabla \cdot \mathbf{v} + \mathbf{v} \cdot \nabla \ln \rho_0 = 0, \tag{5.61}$$

which is similar to the quasi-incompressibility condition, (3.48), of the perpendicular velocity in a strong curved magnetic field.

The physics contained in the Boussinesq equations becomes clearer when we write the equations in dimensionless form by introducing the normalizations

$$\mathbf{x}/d \to \mathbf{x}, \quad t\kappa/d^2 \to t, \quad T/\Delta T \to T, \tag{5.62}$$

where d is the extent of the convection zone, i.e., the thickness of the convectively unstable shell, and ΔT is the change of the temperature across the shell:[*]

$$\partial_t \mathbf{v} + \mathbf{v} \cdot \nabla \mathbf{v} = -\nabla p - Pr\,Ra\,T\mathbf{e}_g + Pr\,Ta^{1/2}\mathbf{v} \times \mathbf{e}_\Omega + Pr\,\nabla^2 \mathbf{v}, \tag{5.63}$$

$$\partial_t T + \mathbf{v} \cdot \nabla T = \nabla^2 T, \tag{5.64}$$

neglecting the heat source. The dimensionless parameters in (5.63) are the Rayleigh number

$$Ra = \frac{\alpha g \Delta T d^3}{\kappa \nu}, \tag{5.65}$$

---

[*] For a thick shell, the radial variation of $g \propto r$ has to be included.

the Prandtl number

$$Pr = \frac{\nu}{\kappa},$$ (5.66)

and the Taylor number

$$Ta = \left(\frac{2\Omega d^2}{\nu}\right)^2.$$ (5.67)

The Rayleigh number is the ratio of the convective and the diffusive heat flux, while the Taylor number is the square of the ratio of the viscous diffusion time and the rotation period. (Instead of the Taylor number, one often uses the Ekman number $E = Ta^{-1/2}$.) Equation (5.63) tells us that for high Taylor number the convective flux $\propto \mathbf{v}$ decreases with increasing Ω and hence the critical Rayleigh number for onset of convection must also increase. Balancing gravity with the Coriolis force in (5.63) and inserting $\mathbf{v}$ into (5.64) we obtain the scaling of the critical Rayleigh number

$$Ra_c \sim Ta^{1/2} kd,$$ (5.68)

where k is the wavenumber of a convective mode. In a rotating slab with $\boldsymbol{\Omega} \parallel \mathbf{g}$, where k is the horizontal wavenumber, the critical wavenumber is (Chandrasekhar, 1961)

$$k_c d \propto Ta^{1/6},$$ (5.69)

i.e., the wavelength of the unstable modes decreases with increasing Ω, as the Coriolis force tends to disrupt larger eddies. Insertion into (5.68) gives the relation

$$Ra_c \sim Ta^{2/3}.$$ (5.70)

Convection is a very efficient process of heat transport. Increasing Ra above Ra_c only increases the vigor of the convective motion, while the temperature profile remains close to the adiabate. For mildly supercritical conditions $(Ra - Ra_c)/Ra_c \lesssim 1$ steady convection cells are set up. In this regime the velocity is still small. For high Taylor number the Coriolis force dominates[*] and the velocity is determined by the geostrophic equation

$$2\rho_0 \mathbf{v} \times \boldsymbol{\Omega} = \nabla p,$$ (5.71)

or, taking the curl,

$$\nabla \times (\mathbf{v} \times \boldsymbol{\Omega}) = \boldsymbol{\Omega} \cdot \nabla \mathbf{v} = 0,$$ (5.72)

i.e., the velocity is constant along $\boldsymbol{\Omega}$, which is the famous Taylor–Proudman theorem (see, e.g., Chandrasekhar, 1961). Hence the convection cells are cylindrical columns parallel to $\boldsymbol{\Omega}$ (Taylor–Proudman columns), where the fluid spirals up or down. The convection pattern is illustrated in fig. 5.10. It

---

[*] This regime is also characterized by small Rossby number $Ro = v/\Omega L$, the ratio of inertia and Coriolis forces.

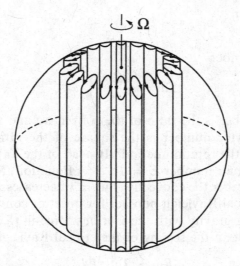

Fig. 5.10. Schematic drawing of Taylor–Proudman convection columns in a spherical shell (from Busse, 1970).

consists of a certain number m of pairs of columns, of which one partner has cyclonic (in the sense of the rotation Ω) flow and the other anti-cyclonic (opposite to Ω) flow. The mode number m increases with increasing Taylor number. The spherical geometry plays a role mainly through the boundaries, where the convection columns intersect the surface of the sphere. Theoretical treatments of the linear stability problem (Roberts, 1968; Busse, 1970; Busse, 1986) confirm the Taylor number scalings of Ra, (5.70), and of $k_c \propto m$, (5.69), while also specifying the dependence of the proportionality factors on the fluid properties, i.e., on Pr. These studies show that the convection columns or rolls are not circular in cross-section but elongated in the radial direction. An interesting feature is caused by the fact that the top and bottom boundaries of a column are not parallel but oppositely inclined with respect to the equatorial plane. This introduces a Rossby-wave-like behavior of the eigenmodes, which have a finite frequency at marginal stability,

$$\omega/\Omega \propto Ta^{-1/6} \tag{5.73}$$

i.e., the system is overstable. The columns drift in the azimuthal direction with a phase velocity roughly proportional to the slope of the boundaries. For radially extended columns, which occur especially at small Pr, the slope varies substantially over the cross-section, increasing outward, hence the outer parts of the columns are turned away from the radial and more into the azimuthal direction (Zhang, 1992). The prograde (= faster than the average rotation) phase velocity is, however, essentially restricted to linear modes. Nonlinearly saturated steady convection rolls, even for only moderately supercritical Rayleigh number, exhibit a clear retrograde drift (see, e.g., Kageyama *et al.*, 1993; Kitauchi *et al.*, 1997).

The convection rolls give rise to mean (= longitudinally averaged) azimuthal motions or zonal flows, i.e., differential rotation, through the Reynolds stress tensor. For conditions close to marginal stability, this mean flow is essentially constant on cylindrical surfaces. There is, however, substantial variation of the mean flow pattern depending on the Prandtl number. For small to moderately large $Pr \lesssim 10$ differential rotation is cylindrical with $d\Omega/dr > 0$, i.e., it is prograde ("eastward", also called super-rotation), on the outside, and retrograde (= "westward" or sub-rotation) on the inside. For large Pr a zonal flow pattern appears on the surface, which is reminiscent of the zonal flows observed in the atmospheres of the large planets Jupiter and Saturn.

In most astrophysical dynamo systems, stability conditions for convective motions are far above marginal, $Ra/Ra_c \gg 1$, where no steady or at least approximately regular convection pattern is to be expected. In fact, numerical simulations in the regime $Ra/Ra_c \sim 10^2$ and at relatively high Taylor numbers $Ta \lesssim 10^9$ (Sun & Schubert, 1995) exhibit a strongly time-dependent velocity field with a few major but rather irregular vertical columns concentrated near the inner shell boundary and embedded in a bath of small-scale turbulent motions. At fixed Ta the characteristics of the mean azimuthal flow depends strongly on Ra, as illustrated in fig. 5.11, which gives contours of the azimuthal velocity $\bar{v}_\phi$. While for $Ra/Ra_c = 25$ the pattern is essentially that found in the mildly supercritical regime with super-rotation at the equator and sub-rotation in the polar region, the $Ra/Ra_c = 50$ case shows a more complicated pattern with alternating regions of super- and sub-rotation on the surface, similar to the mildly supercritical but high-Prandtl number regime (Zhang, 1992). Because of this rather sensitive dependence on Ta, Ra/Ra_c and Pr, predictions for the differential rotation in the Sun or the planets are not yet possible, since Ra and Ta values achievable in present-day direct simulations are far below those prevailing in these astrophysical objects. In the present simulations no tendency toward a disc-like differential rotation pattern, as expected for the Sun (fig. 5.9), is recognizable.

5.3.2 The self-consistent MHD dynamo

The problem of dynamo action by convection in a rotating sphere has attracted much interest, mostly in the geophysical community, because of its significance for the geodynamo. Before discussing the latter more specifically in section 5.3.3, we want to understand the basic mechanisms of magnetic field amplification by thermal convection and the reaction on the

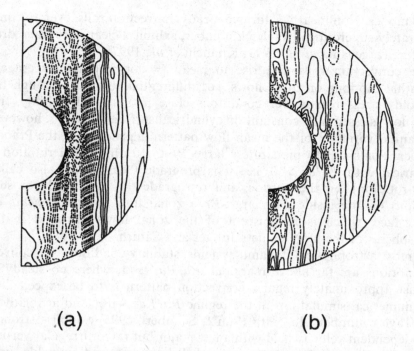

(a)   (b)

Fig. 5.11.   Contours of longitudinally averaged azimuthal velocity $\bar{v}_\phi$; solid contours represent super-rotation, dashed contours sub-rotation, for $Ta = 10^9$ and
$Pr = 1$. (a) $Ra/Ra_c = 25$, (b) $Ra/Ra_c = 50$ (from Sun & Schubert, 1995).

fluid motion. For large Taylor number, where inertia and viscous effects
can be neglected, the equation of motion reduces to the magnetostrophic
approximation

$$2\rho_0 \mathbf{v} \times \mathbf{\Omega} = \nabla p - \frac{1}{c} \mathbf{j} \times \mathbf{B}, \qquad (5.74)$$

generalizing the geostrophic balance (5.71). From this equation, Taylor
(1963) has derived an important constraint on the Lorentz force satisfied
in a magnetostrophic state. The conducting fluid is assumed to be confined
in a spherical vessel. We introduce cylindrical coordinates r, ϕ, z with the
z-axis being the axis of rotation, and take the integral of the ϕ-component
of (5.74) over a cylindrical surface $r = r_0$ up the the points $z = \pm z_0$, where
the cylinder pierces the spherical boundary. It can easily be shown that
the left side vanishes,

$$2\rho_0 \int_{r=r_0} (\mathbf{v} \times \mathbf{\Omega})_\phi r d\phi dz = 0,$$

hence we find the condition

$$\int_{r=r_0} (\mathbf{j} \times \mathbf{B})_\phi d\phi dz = 0. \qquad (5.75)$$

(The buoyancy force, omitted in (5.74), does not change this condition, since it has no ϕ-component.) A state satisfying (5.75) is called a Taylor state in dynamo theory.* The physical meaning of this constraint can be understood by returning to the full equation of motion. Although for small velocities the inertia effect is in general much weaker than the Coriolis force, there are special motions $v_\phi(r)$, where $(\mathbf{v} \times \boldsymbol{\Omega})_\phi$ vanishes. If (5.75) is not satisfied, large azimuthal accelerations occur. This will, in turn, lead to strong magnetic stresses through a radial field B_r, until the system settles in a state where (5.75) holds approximately with some torsional oscillations about this state. Hence it can be expected that a fully developed dynamo is close to a Taylor state.

To study the actual dynamics, even the simplest case of steady convection can only be studied numerically by three-dimensional direct simulation. The first global computations of this kind were performed by Gilman & Miller (1981), Gilman (1983) and Glatzmaier (1985a,b). Detailed analysis of the dynamo action near critical conditions is due to Zhang & Busse (1988, 1989, 1990; see also Wicht & Busse, 1997). These authors determine the critical Rayleigh number for onset of dynamo action, which depends on the Taylor number and the magnetic Prandtl number Pr_M, or the magnetic Reynolds number Re_M,

$$Pr_M = v/\eta, \quad Re_M = Lv/\eta. \tag{5.76}$$

It is pointed out that the Lorentz force of a finite dynamo field counteracts the effect of the Coriolis force and hence tends to soften the constraint imposed by the Taylor–Proudman theorem, (5.72). The dynamo field may thus increase the convection velocity and even allow subcritical dynamo action, i.e., a finite magnetic field below the critical Rayleigh number for linear dynamo action. However, because of the symmetry constraints imposed in the computations by Zhang & Busse it is difficult to draw conclusions for more strongly supercritical dynamo behavior.

Only recently have fully three-dimensional numerical simulations relevant to this regime been performed by St. Pierre (1993), Kageyama *et al.* (1995), Kagayama & Sato (1997), Glatzmaier & Roberts (1995, 1996, 1997), Kuang & Bloxham (1997). While a discussion of the results by the latter two groups is postponed to section 5.3.3, here we want to follow more closely the work by Kageyama *et al.*, which appears to illustrate the basic dynamo physics particularly clearly. (The fact that these authors solve the fully compressible MHD equations instead of making the usual

---

* This is a generalization of the static force-free field (2.40), to which a plasma tends to relax under the constraint of constant magnetic helicity, see section 2.4. Such a state $\nabla \times \mathbf{B} = \alpha \mathbf{B}$ with $\alpha = const$ is also called a Taylor state.

Boussinesq or anelastic approximations is insignificant, since the resulting Mach numbers remain small and hence compressibility effects are weak. The same goes for the unusual boundary condition on the magnetic field, the vanishing of the tangential component, applied in these studies.) The Rayleigh number is only mildly supercritical, $Ra = 2 \times 10^4$, for the Taylor number chosen, $Ta \simeq 6 \times 10^6$, giving rise to a coherent convection pattern, which consists of a certain number of pairs of Taylor–Proudman columns and convective heat transport smaller than conductive transport. The Prandtl number is $Pr = 1$.

Dynamo action depends on the magnetic Prandtl number, which varies in the interval $7 \le Pr_M \le 280$. Increasing Pr_M above the threshold value $Pr_{Mc} \simeq 13$, a magnetic seed field is amplified or, more specifically, a magnetic eigenmode grows exponentially with the growth rate

$$\gamma \propto \ln(Pr_M/Pr_{Mc}),   \tag{5.77}$$

i.e., the η-dependence is rather weak. The magnetic energy E^M grows up to saturation and subsequently exhibits irregular oscillations about a mean value $\overline{E^M}$. For the largest value of Pr_M used the saturation magnetic energy is much larger than the kinetic energy, $\overline{E^M}/E^K \sim 30$. While the kinetic energy remains almost constant being rather insensitive to the dynamo field, the convection *pattern* is strongly affected, as soon as E^M exceeds E^K. (Though a reaction by the Lorentz force could be expected, if E^M and E^K are comparable, the rather sharp transition at E^M/E^K precisely unity seems to be fortuitous in view of the complex structures of the magnetic and velocity fields.)

The transition is illustrated in fig. 5.12, which shows the time evolution of the magnetic energy for different magnetic Prandtl numbers, and in fig. 5.13, displaying the azimuthal drift of the (pairs of) convection columns. As long as $E^M < E^K$, the reaction is too weak to be visible, the drifts are steady and retrograde (westward) (note also the short period at the bottom of the figures, $t < 50$ (in units of the sound-crossing time), corresponding to the linear phase of the convective instability, where the drift is prograde, see the discussion in section 5.3.1). If E^M exceeds E^K, the drift reverses direction and the columns behave more irregularly, (d), becoming seemingly turbulent for $E^M \gg E^K$, (e). Case (c) is particularly interesting. After reaching E^K briefly at $t \simeq 1100$, E^M falls back temporarily allowing a regular convection state to re-establish itself, if only with a higher mode number, until, after a further transitional phase, the system settles finally in the nonlinear MHD regime with $E^M > E^K$.

Let us discuss the dynamo mechanism illustrated in fig. 5.14. The toroidal field is pulled into the gap between cyclonic and anticyclonic columns, where it is convected up close to the latter and down close to the former, thus generating a poloidal field component (a). In the

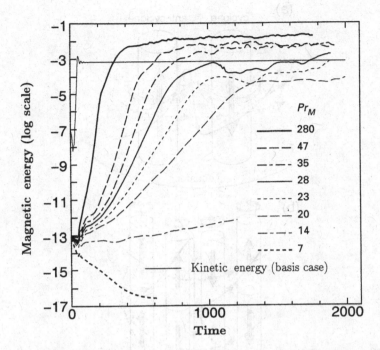

Fig. 5.12. Time evolution of the magnetic energy for eight different values of the magnetic Prandtl number $Pr_M = \nu/\eta$ (the viscosity remains fixed, $\nu = 2.8 \times 10^{-3}$). The kinetic energy is nearly identical in all cases with that of the nonmagnetic case, the horizontal line (from Kageyama *et al.*, 1995).

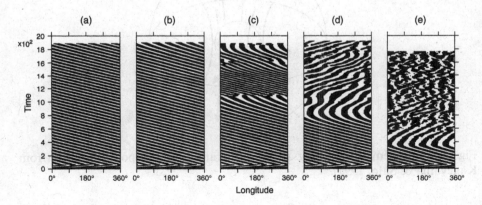

Fig. 5.13. Azimuthal hodographs of the convection columns in the equatorial plane for (a) $Pr_M = 20$, (b) 23, (c) 28, (d) 35, (e) 280 (from Kageyama *et al.*, 1995).

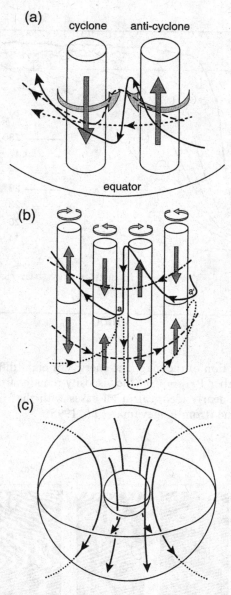

Fig. 5.14. Schematic drawing of the generation of the poloidal field (from Kageyama & Sato, 1997).

southern hemisphere the toroidal field and the flow directions are reversed, generating a poloidal field of the same direction as in the northern hemisphere (b). Reconnection finally gives an overall poloidal field (c), which may subsequently be deformed again into the toroidal direction

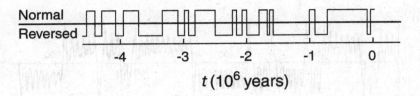

Fig. 5.15. Record of geomagnetic reversals over the last 4.5×10^6 yr (from Cox, 1969).

by differential rotation. We see that this coherent process functions as an $\alpha\Omega$-dynamo, the process (a) representing the "cyclonic event" (Parker, 1955a), which produces the α-effect in mean-field theory (cf. fig. 5.7).

5.3.3 The geodynamo

The geomagnetic field is certainly the best-known example of a magnetic field in an astrophysical body. In fact, it is more than just an object of scientific curiosity, because of its importance of protecting the Earth's surface from the bombardment by the solar wind and cosmic radiation. Of the inner planets, only the Earth carries a sizable magnetic field with a mean surface value of about 0.5 gauss. The field has a rather simple predominantly dipole structure with the magnetic moment oriented roughly along the Earth's rotation axis.

Though we probably know more about the interior of stars than about that of our own planet, there is now a fairly detailed and reliable picture of the Earth's internal structure resulting both from seismology and materials science. The main constituents are: the iron core, consisting of a solid inner core with a radius of 1250 km and a liquid outer core extending up to a radius of $R_c \simeq 3500$ km; and the solid mineral mantle covered by a thin crust. Only the core is electrically conducting with a magnetic diffusivity $\lambda \simeq 2\,\mathrm{m^2/s}$, which would give rise to a free decay time of any current flowing in the core $\tau \sim R_c^2/\lambda \sim 10^5$ years. Since paleomagnetic studies clearly indicate that the Earth carries a magnetic field since its early history, this field cannot be of primordial origin. Moreover the magnetization of minerals, in particular on the sea floor, shows that the geomagnetic field frequently reversed sign. The abrupt reversals occur, apparently randomly in time, after a period of almost steady field of average duration 2×10^5 yr (see, e.g., Bullard, 1968; Cox, 1969), fig. 5.15. A qualitatively similar behavior may result from disc dynamo models, as shown in fig. 5.16 obtained from Rikitake's system of two coupled disc dynamos, see fig. 5.2(b).

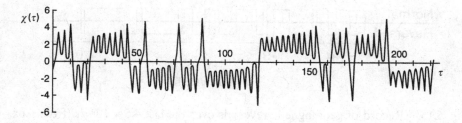

Fig. 5.16.   Time development of $X(\tau) \propto I_1$ in Rikitake's disc dynamo system (see fig. 5.2(b)), showing a nonperiodic sequence of reversals (from Cook & Roberts, 1970).

One concludes that there must be a dynamo process responsible for the geomagnetic field and its observed dynamics, which can only be located in the liquid outer core. The chemical composition of the outer-core material is that of an alloy consisting mainly of iron but mixed with some lighter elements such as Si, S and O. The engine driving the convection has its origin in the slow cooling of the Earth, which leads to (i) a freezing of the iron component onto the inner core with the corresponding release of latent heat at the bottom of the convection zone, and (ii) a melting of iron out of the mantle, hence a cooling due to negative latent heat at the top of the convection zone.* In addition to this thermal buoyancy source (Verhoogen, 1961) there is an even stronger gravity source caused by the buoyancy force on the light constituents of the alloy set free by the solidification of the iron (Braginskii, 1963).

The fact that the outer-core density is roughly uniform and the sound speed is by many orders higher than the convection velocity justifies the Boussinesq approximation in a numerical treatment of the geodynamo, in contrast to the highly stratified compressible solar convection zone (section 5.2.2). However, direct simulations using the true values of the dissipation coefficients are still far out of reach, because of the extreme values of the Taylor number $Ta \sim 10^{28}$ and the magnetic Prandtl number $Pr_M \sim 10^{-5}$ (the Prandtl number itself is $Pr \sim 1$), though these values should not be taken too literally, since the molecular kinematic viscosity is only very poorly known (the usual assumption is $v \sim 10^{-5}\,\mathrm{m}^2/\mathrm{s}$). In order to perform numerical simulations with the presently achievable spatial resolution, one uses an eddy viscosity and eddy thermal diffusivity resulting from small-scale l convective motions δv, $\bar{v} \sim \bar{\kappa} \sim (\delta v)^2/l \sim 10^3\,\mathrm{m}^2/\mathrm{s}$, which brings the Taylor number down to $\overline{Ta} \sim 10^{12}$.

---

* One thus finds the unfamiliar situation that the system is freezing at the hotter bottom and melting at the cooler top. The reason is that the melting temperature of the core material increases with pressure faster than the temperature increase along the adiabate, i.e., the temperature profile across the convection zone.

Using these assumptions, several numerical studies of the geodynamo have been performed by Glatzmaier & Roberts (1995, 1996, 1997). In these simulations the system is strongly supercritical $Ra/Ra_c \sim 10^3$, and the resulting convection is highly irregular. Field penetration into the inner core (assumed with the same magnetic diffusivity as the outer core) and the reaction on the dynamics in the outer core are included. The authors find generation of strong magnetic fields with energy $E^M \sim 10^3 E^K$ greatly exceeding the convective kinetic energy and a dominant dipole structure outside the convection zone with surface intensity of the correct order of magnitude. The inner core rotates faster (in an eastward direction in the rotating frame) than the mantle, in agreement with seismic observations, accelerated by the magnetic coupling to the high-latitude poleward convection along the bottom of the outer core (Glatzmaier & Roberts, 1996a).

Concerning the question of field reversal, however, no coherent picture has appeared as yet. While in their first paper Glatzmaier & Roberts found that the poloidal fields in the inner and the outer core are essentially opposite, their interaction giving rise to an inherent tendency toward reversal, which actually occurred once during the computed time of 4×10^4 yr, in their later studies they deduce that the field direction was very stable (and parallel to the core field).

A weak point in the numerical work of Glatzmaier & Roberts is the use of ad hoc hyperdiffusivities (viscous, magnetic, thermal) $\bar{v}(l) \propto \bar{v}l^3$, l = order of the spherical harmonic, which corresponds to adding a fifth-order derivative with respect to the colatitude θ.[*] The effect of such a choice is difficult to assess, in particular because of the strongly anisotropic character of this damping. At the least it leads to reduced effective values of the Taylor number.

Kuang & Bloxham (1997) have pointed out the effect of the velocity boundary conditions. Since the viscous sublayers are far too thin in the Earth's core to be resolvable by present computational means, these authors assume stress-free boundary conditions, the continuity of the tangential components $\partial_n v_t = 0$, instead of the physical no-slip conditions $v = 0$, thus effectively reducing the layer width to zero. They argue that this choice is more adequate than unrealistically broad viscous layers due to the no-slip conditions assumed in the Glatzmaier–Roberts computations, which overemphasize the effect of the viscosity. Kuang & Bloxham find in their simulations a stable dynamo behavior with $E^M \gg E^K$. The configuration is close to a Taylor state, hence the effects of viscosity and inertia are indeed weak. The field distribution is markedly different from that

---

[*] Hyperdiffusivities are often used in numerical turbulence studies (see section 5.4) in order to remove dissipation from the medium scales shifting it to the smallest possible scales.

found by Glatzmaier & Roberts, which they can essentially reproduce by applying no-slip boundary conditions. It thus appears that the difference in these models is primarily caused by the choice of boundary conditions and less by the different values of the nondimensional parameters such as the Taylor number.

5.4 MHD turbulence

In many convectively unstable systems Rayleigh numbers are far above critical, resulting in fully developed *magneto*hydrodynamic turbulence, since, because of the dynamo effect, magnetic fields are usually excited to finite intensity, reacting on the fluid motion. MHD turbulence occurs almost inevitably in plasmas in motion and is therefore a widespread phenomenon in astrophysics, for instance in the solar system, where we encounter turbulent magnetic fields in the convection zone and in the solar wind, and in accretion discs, where MHD turbulence is responsible for the angular momentum transport. It is interesting to note that the presence of a magnetic field may even enhance the tendency toward turbulence, for instance in accretion discs, where, in the absence of magnetic fields, no turbulence would be generated (Balbus & Hawley, 1991). In this section I restrict consideration mainly to some recent developments, in particular in 3D turbulence, referring for a broader introduction to chapter 7 of my previous book (Biskamp, 1993a).

5.4.1 Homogeneous MHD turbulence

The notion of homogeneous turbulence is a theoretical abstraction or idealization. It refers to the quasi-homogeneous properties of the turbulence in a small open subsystem L of the global inhomogeneous system away from boundary layers. The effect of turbulent structures of size exceeding L is treated as a homogeneous background in the subsystem. The convenient representation of the fields in homogeneous turbulence is the decomposition into Fourier modes

$$\mathbf{v}(\mathbf{x}) = \sum_{\mathbf{k}} \mathbf{v}_{\mathbf{k}} \, e^{i\mathbf{k}\cdot\mathbf{x}}, \quad \mathbf{B}(\mathbf{x}) = \sum_{\mathbf{k}} \mathbf{B}_{\mathbf{k}} \, e^{i\mathbf{k}\cdot\mathbf{x}}, \tag{5.78}$$

$$k_j = 2\pi n_j/L, \ n_j = 0, \pm 1, \pm 2, ...,$$

where the Fourier components have the physical meaning of field intensities at the scale $l = k^{-1}$. If the turbulence can also be assumed isotropic, the theoretical description and the interpretation of experimental and simulation results is further simplified considerably. However, whereas in hydrodynamic turbulence the assumption of isotropy is usually appropriate in a coordinate system moving with the mean flow, in an electrically conducting fluid a mean magnetic field $\mathbf{B}_0$ has a profound effect on the turbulent fluctuations. While the dynamics perpendicular to $\mathbf{B}_0$ may develop small-scale structures, spatial variations along $\mathbf{B}_0$ are smooth because of the stiffness of the magnetic field. Hence in the presence of a strong $\mathbf{B}_0$, *viz.*, in a low-β plasma, turbulence can be isotropic only in the perpendicular plane, i.e., the turbulence is quasi-two-dimensional, whereas fully isotropic turbulence can only be generated in a high-β plasma with no mean field.

The ideal invariants are characteristic quantities of a turbulence field. Here the quadratic invariants are most important, since these are particularly robust or "rugged" surviving finite truncations of the Fourier series, (5.78). In incompressible MHD there are three such invariants, which have been introduced in section 3.1.3: two mixed kinetic–magnetic quantities, the energy[*]

$$E = \tfrac{1}{2} \sum_{\mathbf{k}} \left(|\mathbf{v_k}|^2 + |\mathbf{B_k}|^2 \right) = \tfrac{1}{4} \sum_{\mathbf{k}} \left(|\mathbf{z_k^+}|^2 + |\mathbf{z_k^-}|^2 \right) = E^K + E^M \qquad (5.79)$$

and the cross-helicity

$$K = \sum_{\mathbf{k}} \mathbf{v_k} \cdot \mathbf{B_{-k}} = \tfrac{1}{4} \sum_{\mathbf{k}} \left(|\mathbf{z_k^+}|^2 - |\mathbf{z_k^-}|^2 \right) ; \qquad (5.80)$$

and a purely magnetic quantity, the magnetic helicity

$$H = \sum_{\mathbf{k}} \mathbf{A_k} \cdot \mathbf{B_{-k}}. \qquad (5.81)$$

Here $\mathbf{z^\pm} = \mathbf{v} \pm \mathbf{B}$ are the Elsässer fields, (3.11), the natural variables in incompressible MHD. In 2D, where the MHD equations are identical with the simple reduced equations (3.65) and (3.66), we also have three quadratic invariants, again energy and cross-helicity, but the magnetic helicity is now replaced by the mean-square magnetic potential

$$H^\psi = \sum_{\mathbf{k}} |\psi_{\mathbf{k}}|^2. \qquad (5.82)$$

The spectral densities E_k, K_k, H_k or H_k^ψ satisfy detailed balance relations, i.e., are conserved for each triad of modes $\mathbf{k}, \mathbf{p}, \mathbf{q}$ with $\mathbf{k} + \mathbf{p} + \mathbf{q} = 0$. Hence, when injected in a certain k-band, these quantities propagate or "cascade" in k-space, either to large k, which is called direct or normal cascade, or to small k, called inverse cascade. The latter gives rise to formation of large-scale structures and is an example of the general phenomenon

---

[*] Following conventions in turbulence theory we call the energy E in this section instead of W.

Table 5.1. *Ideal invariants and cascade directions. H^K is the kinetic helicity and Ω the enstrophy.*

Theory	3D		2D	
MHD	E_k	direct	E_k	direct
	K_k	direct	K_k	direct
	H_k	inverse	H_k^ψ	inverse
Navier–Stokes	E_k^K	direct	E_k^K	inverse
	H_k^K	direct	Ω_k	direct

of self-organization. The cascade directions can be determined from the absolute equilibrium distributions, i.e., the spectral equilibria of E_k etc. of a truncated ideal system of Fourier modes (see, e.g., Fyfe & Montgomery, 1976). While E_k, K_k have direct cascades, the magnetic quantities H_k, H_k^ψ are inversely cascading. Hence cascade properties are identical in 2D and 3D MHD, which reflects the basic similarity of 2D and 3D MHD turbulence, in contrast to hydrodynamic (Navier–Stokes) turbulence, where the inverse energy cascade in 2D results in the well-known formation of large eddies, a behavior which differs fundamentally from 3D turbulence, where no inverse cascade occurs. The cascade properties of MHD and Navier–Stokes theory are given in Table 5.1. It should, however, be noted that the formal identity of the cascades in 2D and 3D MHD suggests a somewhat oversimplified picture. Since H is an independent quantity, while H^ψ is strongly coupled to E, $E_k^M = -k^2 H_k^\psi$, MHD turbulence is expected to be more complex in 3D than in 2D, for instance the dynamo effect occurs only in 3D.

5.4.2 Selective decay and energy decay laws

When dissipation effects are switched on, the ideal invariants in MHD turbulence decay, but not all at the same rate. While the energy decay rate is fast, independent of the values of resistivity and viscosity, the decay of K and H or H^ψ is much slower. This behavior, called selective decay (see, e.g., Ting *et al.*, 1986), has a strong influence on the global turbulence dynamics. On a time-scale shorter than the decay time of K, for instance, K can be considered constant and the state to which the turbulence decays is determined by the variational principle

$$\delta[E - \alpha K] = \delta \left[\frac{1}{2} \int d^3x \left(v^2 + B^2 \right) - \alpha \int d^3x\, \mathbf{v} \cdot \mathbf{B} \right] = 0, \qquad (5.83)$$

where α is a Lagrange multiplier. Independent variation with respect to $\mathbf{v}$ and to $\mathbf{B}$ gives the relations $\mathbf{v} = \alpha\mathbf{B}$ and $\mathbf{B} = \alpha\mathbf{v}$, hence

$$\mathbf{v} = \pm\mathbf{B}. \tag{5.84}$$

Since in such a completely aligned or Alfvénic state the dynamics is turned off, as directly seen in the $z^\pm$ formulation, (3.12), this is also the final state apart from the very slow laminar diffusion. Strongly aligned turbulence is observed in the solar wind, and numerical simulations, both in 2D and 3D, demonstrate the general tendency toward an aligned state. Note that alignment is a local property not connected with large-scale structures, which is consistent with the direct cascade of the cross-helicity. The degree of alignment is measured by the velocity–magnetic field correlation $\rho = K/E$.

For weak initial correlation the tendency toward alignment is, however, superseded by a competing process of self-organization resulting from the effective constancy of H, or H^ψ in 2D. As shown in section 2.4, this leads to the variational principle $\delta[E - \alpha H] = 0$, the solution of which is a static force-free field, $\mathbf{j} = \alpha\mathbf{B}$. In 2D, the corresponding variational principle $\delta[E - \alpha^2 H^\psi] = 0$ has a similar linear solution $j = \alpha^2\psi$.

While smooth laminar fields decay exponentially, $E_k \sim e^{-2\mu k^2 t}$, the decay of turbulent fields shows a power-law behavior $E \sim t^{-n}$, independent of the Reynolds number. The exponent n is a characteristic parameter, which reflects the similarity properties of the turbulence. In hydrodynamic turbulence the exponent is difficult to predict, and no universal decay law seems to exist, as will be briefly discussed below. In MHD turbulence, however, the decay process is govered by self-organization due to selective decay, giving rise to definite values of n. The basic theory was developed by Batchelor (1969) for the enstrophy decay in 2D hydrodynamic turbulence and applied later to MHD turbulence by Hatori (1984). Let us first discuss the 2D case, where $H^\psi = const$, while the energy $E = E^K + E^M$ decreases. The macroscopic or integral scale l_0 of the turbulence can be defined by

$$H^\psi = E^M l_0^2. \tag{5.85}$$

In addition, numerical simulations show that in 2D MHD turbulence the ratio $\Gamma = E^K/E^M$ remains constant,[*] i.e., the turbulence decays in a self-similar way. The integral scale should be insensitive to the details of the

---

[*] There has been some discussion on this point in the literature. Certain simulation runs show a decrease of Γ, see, e.g., Kinney *et al.*, (1995), though the decay is much slower than that of the energy. Experience, however, suggests caution when drawing conclusions from individual runs, since one finds a significant scatter with some cases exhibiting quite anomalous features. Over the last decade we have performed many simulations of decaying 2D MHD turbulence with widely different parameters and initial states. The general impression is that Γ does remain nearly constant (an example is shown in fig. 5.17), quite differently from the 3D case.

turbulence depending only on the macroscopic quantities characteristic of the turbulent state, the energy E and the energy dissipation rate $\epsilon = -dE/dt^*$ (when restricting consideration to weak initial alignment, the cross-helicity does not enter). Hence, by simple dimensional arguments, l_0 satifies the scaling relation

$$l_0 \sim E^{3/2}/\epsilon. \tag{5.86}$$

Combining (5.85) and (5.86) yields

$$\frac{E}{1+\Gamma} \frac{E^3}{\epsilon^2} \sim H^\psi = const, \tag{5.87}$$

and thus, making use of $\Gamma = const$,

$$\epsilon \equiv -\frac{dE}{dt} \sim E^2, \tag{5.88}$$

which has the solution

$$E \sim (t - t_0)^{-1} \to t^{-1} \quad \text{for} \quad t \gg t_0. \tag{5.89}$$

Figure 5.17 presents results from a computer run of freely decaying 2D MHD turbulence with a resolution of 2048^2 modes, choosing an initial scale of the turbulence $k_0^{-1} = l_0$ much smaller than the system size, which allows the inverse cascade of H^ψ to evolve freely. Figure 5.17(a) shows ϵ/E^2, the ratio of the l.h.s. and the r.h.s. of (5.88), which is seen to be nearly constant, thereby confirming the decay law, (5.88). Also shown is the energy ratio Γ, fig. 5.17(b), which is indeed practically constant, and the ratio of viscous to resistive dissipation $\Delta = \epsilon^\mu/\epsilon^\eta = \mu \int w^2 d^2x/\eta \int j^2 d^2x$, fig. 5.17(c). (Variation of the magnetic Prandtl number $Pr_M = \mu/\eta$ suggests that Δ approaches unity for $Pr_M \to 0$.) The constancy of Δ over the entire time-span reflects the structure of the dissipative eddies, *viz.*, current and vorticity micro-sheets, which is consistent with the observation that, at the dissipative scales, kinetic and magnetic energy spectra are nearly equal, $E_k^K \simeq E_k^M$.

While, in 2D, finite E^M entails finite H^ψ, in 3D the magnetic helicity may take any value in the interval $0 \leq |H| \leq H_{max}$, where H_{max} is the maximum magnetic helicity for a given energy spectrum E_k^M. Selective decay is only effective for $H \neq 0$. Since MHD turbulence is usually generated in rotating systems, where Coriolis and buoyancy forces naturally lead to twisted field lines, finite H should be typical. In this case we can define the integral scale l_0 by

$$E^M l_0 = H = const. \tag{5.90}$$

---

*This is the conventional terminology. Strictly speaking, the energy decay *rate* is $-E^{-1} dE/dt$.

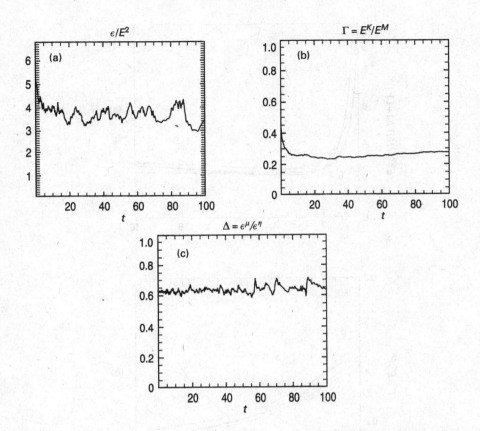

Fig. 5.17. (a) ϵ/E^2, (b) energy ratio Γ, and (c) ratio of dissipation rates Δ from a simulation run of 2D decaying MHD turbulence.

Assuming constancy of the energy ratio Γ, as in the 2D case, relations (5.86) and (5.90) lead to the decay law $-dE/dt \sim E^{5/2}$ with the asymptotic solution $E \sim t^{-2/3}$. However, extensive 3D numerical simulation studies indicate a slower decay, $E \sim t^{-n}$, $n \simeq 0.5$–0.6 (Biskamp & Müller, 1999). The discrepancy can be attributed to the fact that, contrary to the 2D case, the energy ratio Γ is not constant but decreases on the same time-scale as the energy itself. Let us therefore incorporate the variation of Γ in the theory. Assuming that the most important nonlinearities in the MHD equations arise from the advective terms,

$$-\frac{dE}{dt} \sim \mathbf{v} \cdot \nabla E \sim (E^K)^{1/2} \frac{E}{l_0},$$

substitution of l_0 from (5.90) gives

$$\frac{E^{5/2}}{H\epsilon} \frac{\Gamma^{1/2}}{(1+\Gamma)^{3/2}} = const. \tag{5.91}$$

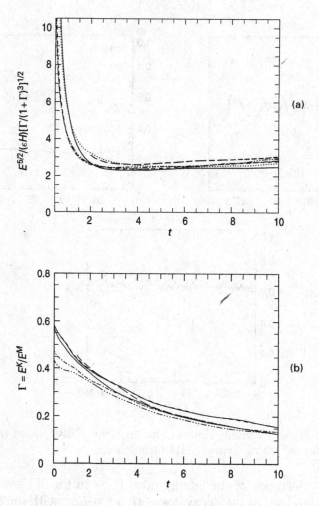

Fig. 5.18. Differential energy decay law (a) and energy ratio (b) for several simulations runs of 3D MHD turbulence (from Biskamp & Müller, 1999).

Figure 5.18(a) shows this expression for a number of simulation runs with different Reynolds numbers Re_M and different values of H. The near constancy of these curves for $t > 2$ (in units of the large-eddy turnover time), when the turbulence has become fully developed, and their small scatter shows that the relation (5.91) is well satisfied. From the same runs we also see the rapid decay of Γ, fig. 5.18(b), in particular we find $\Gamma(t) \propto E(t)$. Inserting this behavior in (5.91) gives the simple relation for the energy decay in 3D turbulence, valid asymptotically when $\Gamma \ll 1$,

$$-\frac{dE}{dt} \sim E^3, \tag{5.92}$$

whence the asymptotic decay laws

$$E \simeq E^M \sim t^{-1/2}, \quad E^K \sim t^{-1}. \tag{5.93}$$

Both decay laws are consistent with the numerical observations. The fact that the kinetic energy decays faster than the magnetic energy is consistent with the selective decay to a static force-free state.

The decay of nonhelical MHD turbulence $H \simeq 0$ is not governed by a selective decay process, which makes the situation more complicated, similar to that in (3D) hydrodynamic turbulence. There the classical approach starts from the Loitsianskii integral $\mathscr{L} = \int_0^\infty dl\, l^4 \langle v(x+l)v(x) \rangle$, which is assumed to be constant (see, e.g., Monin & Yaglom, 1975, Vol. II). Writing $\mathscr{L} = E l_0^5$ and assuming again $l_0 \sim E^{3/2}/\epsilon$, one immediately obtains $-dE/dt \sim E^{17/10}$ with the asymptotic solution $E \sim t^{-10/7}$. Since $\mathscr{L}$ can also be written in the form $\mathscr{L} = \int dk\, k^{-5} E_k$, where E_k is the energy spectrum, $\mathscr{L}$ is determined by the spectrum at the largest scales, which in general depends on the way of turbulence generation and hence cannot be universal. Moreover, the invariance of the Loitsianskii integral has been questioned (see, for instance, Frisch, 1995). For MHD turbulence a similar quantity has been considered, $\mathscr{L}_{\mathrm{MHD}} = \int dl\, l^4 \langle z^\pm(x+l) z^\pm(x) \rangle$ (Galtier *et al.*, 1997). However, here, too, the invariance is questionable. Both closure theory (Lesieur & Schertzer, 1978) and direct numerical simulations (see, e.g., Hossain *et al.*, 1995) indicate $E \sim t^{-1}$. It is interesting to note that the transition from $E \sim t^{-1}$ to $E \sim t^{-1/2}$ is found to occur at relatively small values of the magnetic helicity, $H/H_{\max} \sim 0.2$–0.3, hence the latter decay law should be more typical.

5.4.3 Spatial scaling properties

Theoretical concepts Turbulence is characterized by a broad range of spatial scales exhibiting certain scaling properties. The scaling range, or inertial range, is defined by

$$l_0 > l > l_d, \tag{5.94}$$

or in k-space

$$k_0 < k < k_d, \tag{5.95}$$

where $l_0 = k_0^{-1}$ is the integral scale-length of the turbulence defined in (5.85) or (5.90). On the small-scale side the inertial range is limited by $l_d = k_d^{-1}$, where dissipation becomes important. For a normal turbulence cascade l_0 is of the order of the injection scale l_{in}, at which the turbulence is excited. In the presence of an inverse cascade, however, l_0 is in general much larger than l_{in}, such that there may be two different inertial ranges,

$l_0 > l > l_{in}$ and $l_{in} > l > l_d$. Here we only discuss recent results concerning the direct cascade.

The scaling properties are conveniently described by the statistics of the increments of the turbulent fields, $\delta v_l = v(x + l) - v(x)$ etc., which for homogeneous turbulence is independent of x. (The vector character of $\mathbf{v}$, $\mathbf{x}$, $\mathbf{l}$ is not important at this point.) For stationary conditions the energy decay rate ϵ equals the energy injection rate ϵ_{in}, which is also equal to the energy transfer rate ϵ_t across the inertial range and to the energy dissipation rate ϵ_d. Hence we denote all these quantities by ϵ. The classical turbulence concept introduced by Kolmogorov (1941), which is (approximately) valid for hydrodynamic turbulence, rests on the assumption that the cascade process is local (in k-space), meaning that δv_l depends only on the energy transfer rate ϵ and the scale l, thus by simple dimensional arguments

$$\delta v_l \sim \epsilon^{1/3} l^{1/3}. \tag{5.96}$$

(A more physical argument is that $\epsilon \sim \delta v_l^2 / \tau_l$, where $\tau_l \sim l/\delta v_l$ is the local eddy-turnover or distortion time, which leads to the same result.)

In MHD turbulence the assumption of a local cascade process is not a priori justified, as noted by Iroshnikov (1964) and Kraichnan (1965). The idea is that the magnetic field B_0 in the large energy-containing eddies[*] should exert a strong influence on the small-scale eddies, coupling velocity and magnetic field fluctuations, such that $\delta v_l \simeq \pm \delta \mathbf{B}_l$, forming Alfvén waves propagating along the local guide field B_0 with the Alfvén velocity v_A. As seen directly from (3.12), only wave-packets moving in opposite directions interact, whose crossing time $\tau_A \sim l/v_A$ is much shorter than the nonmagnetic eddy distortion time $\tau_l \sim l/\delta v_l$. Hence many "collisions" are required to change the amplitude significantly. The energy transfer time is longer by the factor τ_l/τ_A,

$$\tau_l^{IK} \sim \tau_l^2 / \tau_A.$$

Hence the energy transfer rate is $\epsilon \sim \delta v_l^2 / \tau_l^{IK} \sim \delta v_l^4 / l v_A$, or

$$\delta v_l \sim \delta B_l \sim \delta z_l^{\pm} \sim (\epsilon v_A)^{1/4} l^{1/4}. \tag{5.97}$$

This is called Iroshnikov–Kraichnan (IK) phenomenology, which is based on the Alfvén-wave effect. (In the case of strong alignment, where one field, z^+ or z^-, dominates, δz_l^+ and δz_l^- have different scaling properties, for a review see Biskamp, 1993a.)

---

[*] Turbulent structures are generally called eddies, though their shape may be quite different from the eddies in a hydrodynamic flow.

The energy spectrum $E_k{}^\dagger$ is essentially the Fourier transform of the second-order moment $\langle \delta v_l^2 \rangle$.‡ In the case of the Kolmogorov scaling, (5.96), one obtains the famous Kolmogorov spectrum,

$$E_k \sim \epsilon^{2/3} k^{-5/3}, \qquad (5.98)$$

while the IK scaling, (5.97), gives the Iroshnikov–Kraichnan spectrum

$$E_k \sim (\epsilon v_A)^{1/2} k^{-3/2}. \qquad (5.99)$$

The self-similarity of the turbulence expressed by the scaling relations (5.96) and (5.97) is, however, not strictly valid due to the phenomenon of *intermittency*. This means that with decreasing l the spatial distribution of δv_l becomes more and more sparse or intermittent, which is particularly evident in the very nonuniform distribution of the dissipative structures. Hence the turbulence looks different on different scales. A complete statistical picture of the spatial distribution of the inertial-range turbulent structures is obtained from the set of scaling exponents ζ_p, $p = 0, 1, 2, ...,$ of the structure functions $S_p(l)$, the moments of δv_l,

$$S_p(l) \equiv \langle (\delta v_l)^p \rangle \sim l^{\zeta_p}. \qquad (5.100)$$

One usually considers the longitudinal structure functions, the moments of the longitudinal velocity increments $\delta v_l = [\mathbf{v}(\mathbf{x}+\mathbf{l}) - \mathbf{v}(\mathbf{x})] \cdot \mathbf{l}/l$. For these there is an exact relation in hydrodynamic turbulence, $\langle (\delta v_l)^3 \rangle = -(4/5)\epsilon l$, known as Kolmogorov's 4/5 relation (see, e.g., Monin & Yaglom, 1975, Vol. II). Hence the exponents satisfy the relations $\zeta_3 = 1$ and, of course, $\zeta_0 = 0$. In addition, basic statistical arguments require that $d\zeta_p/dp > 0$ and $d^2\zeta_p/dp^2 < 0$. $\zeta_{p0} = p/3$ is the non-intermittent limiting case, so that $\zeta_p \geq \zeta_{p0}$ for $p \leq 3$ and $\zeta_p \leq \zeta_{p0}$ for $p > 3$. The larger the deviations of ζ_p from the linear relation ζ_{p0}, the more intermittent the turbulence is said to be. From a phenomenological model for a hierarchy of structures, She & Leveque (1994) derived the relation

$$\zeta_p^{\text{SL}} = p/9 + 2\left[1 - (2/3)^{p/3}\right], \qquad (5.101)$$

which satisfies $\zeta_3 = 1$ and fits the experimental results surprisingly well.

The She–Leveque relation (5.101) contains no adjustable parameters. In reality, however, there are mainly two parameters in the model, for which

---

† In turbulence theory one usually considers the angle-integrated spectrum $E_k = \int d\Omega_k E_\mathbf{k}$, such that $E = \int_0^\infty E_k dk$, while in turbulence experiments and observations the one-dimensional spectrum, e.g., $E_{k_x} = \int dk_y dk_z E_\mathbf{k}$, is more easily accessible.

‡ To derive the scaling of E_k simply form $(\delta v_l)^2|_{l=k^{-1}}$ and divide by k.

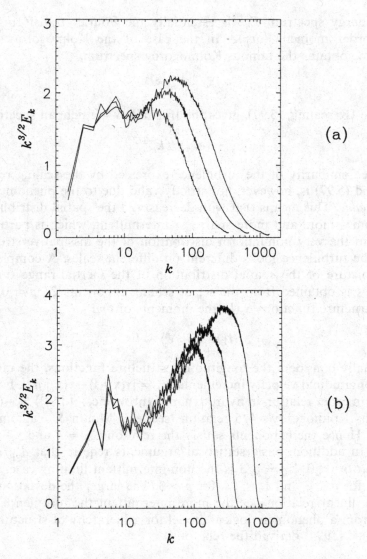

Fig. 5.19.   Nonlocal bottleneck effect in 2D MHD turbulence: energy spectrum compensated by $k^{3/2}$ plotted on a linear–log scale. (a) Normal diffusion; (b) high-order hyperdiffusion. The three curves in each chart correspond to resolutions of 1024^2, 2048^2, and 4096^2 modes, respectively (after Biskamp *et al.*, 1998b).

She and Leveque make physical assumptions. The general relation reads (Politano & Pouquet, 1995):

$$\zeta_p = \frac{p}{g}\left(1 - \frac{2}{g}\right) + C_0\left[1 - \left(1 - \frac{2}{gC_0}\right)^{p/g}\right]. \qquad (5.102)$$

Here g is related to the fundamental scaling, $\delta v_l \sim l^{1/g}$, and C_0 to the dimension D of the dissipative eddies, $C_0 = 3 - D$. Hydrodynamic turbulence is governed by the Kolmogorov scaling, $g = 3$, and the dissipative eddies are vorticity filaments, hence $C_0 = 2$, from which results (5.101). For MHD turbulence, by contrast, the IK phenomenology predicts $g = 4$ and the dissipative eddies are micro-sheets, i.e., $C_0 = 1$, hence (Grauer *et al.*, 1994)

$$\zeta_p^{\text{IK}} = p/8 + 1 - (1/2)^{p/4}, \tag{5.103}$$

which implies $\zeta_4 = 1$.

Simulation results Previous numerical studies of 2D MHD turbulence seemed to support the IK scaling, in particular the $k^{-3/2}$ energy spectrum (Biskamp & Welter, 1989) and the scaling exponents of the structure functions (Grauer & Marliani, 1995). Recent developments, however, tend to cast some doubt on the general validity of the IK phenomenology. 2D simulations show that the scaling exponents ζ_p differ significantly from the IK relation, (5.103), (Politano *et al.*, 1998). Simulations at still higher resolution reveal an anomalous behavior (Biskamp *et al.*, 1998b; Biskamp & Schwarz, 2000), which indicates that the previous results are not asymptotic. It is found that the energy spectrum is enhanced toward the high-k edge by a nonlocal bottleneck effect[*] (see fig. 5.19) and that the scaling exponents of the structure functions, instead of approaching asymptotic values, decrease with increasing Reynolds number. Only the relative values, for instance $\xi_{3,p} = \zeta_p/\zeta_3$, are constant.

The scaling properties in 3D MHD turbulence have recently been studied by numerical simulations of decaying turbulence with resolutions of up to 512^3 modes (Müller & Biskamp, 2000). Figure 5.20 shows a scatter plot of the normalized energy spectrum compensated by $k^{5/3}$ obtained from an extended period of fully developed turbulence.[†] The

---

[*] The bottleneck effect designates the hump in the energy spectrum observed at the transition from the inertial to the dissipation range, where the energy flux from the former encounters some kind of a soft barrier, a "bottleneck", which the energy must cross before being dissipated. Mathematically the effect is caused by an overshoot in the Fourier transform resulting from the rather abrupt dissipative fall-off of the fluctuation amplitudes. While in 3D hydrodynamic turbulence this hump has a finite spectral extent, which is independent of the Reynolds number (see, e.g., Lohse & Müller-Groeling, 1995), in 2D MHD it is found to affect an increasingly larger part of the inertial range.

[†] In decaying turbulence one cannot simply average over the spectrum at different times because of the secular change of the system. Hence the spectrum must be normalized to eliminate the time variation of the global quantities, $E_k(t) \rightarrow \widehat{E}(\widehat{k})$, where $\widehat{k}$ is normalized to the dissipation scale l_d, $\widehat{k} = k l_d$. In the Kolmogorov case $l_d = l_K = (\mu^3/\epsilon)^{1/4}$, the Kolmogorov length, and normalization of the energy spectrum (5.98) results from simple dimensional analysis:

$$\widehat{E}(\widehat{k}) = E_k/(\epsilon\mu^5)^{1/4} = CF(\widehat{k})\widehat{k}^{-5/3}, \tag{5.104}$$

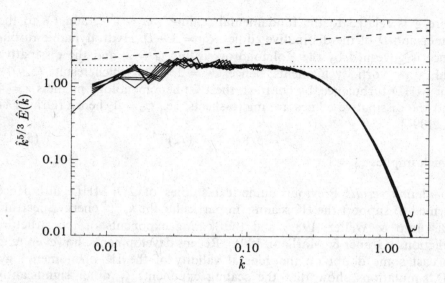

Fig. 5.20. Scatter plot of the normalized energy spectrum compensated by $k^{5/3}$ from a simulation run with 512^3 modes, $H/H_{max} = 0.6$, $K = 0$, using normal diffusion. The dashed line indicates the IK spectrum $\propto \hat{k}^{-3/2}$, the dotted line the spectrum $C\hat{k}^{-5/3}$ with $C = 1.7$ (after Müller & Biskamp, 2000).

spectrum exhibits a clear scaling behavior over almost one decade, which is close to $k^{-5/3}$, definitely steeper than the IK spectrum $k^{-3/2}$ indicated by the dashed line. The dotted line corresponds to $C\hat{k}^{-5/3}$ with $C = 1.7$.

In contrast to the energy spectrum in fig. 5.20, the structure functions from this simulation run do not exhibit a clear scaling range. (It is generally observed, both in experiments and in numerical simulations, that the structure functions have a shorter scaling range than the energy spectrum, in particular for higher orders, $p > 2$.) However, using hyperdiffusion, i.e., replacing the operator $\eta \nabla^2$ by $-\eta_2 (\nabla^2)^2$ and similarly for the viscosity, concentrates dissipation more strongly at the smallest scales and results in a longer inertial range. In fig. 5.21 the normalized time-averaged structure functions $\hat{S}_p^+(\hat{l}) = \langle |\delta z_l^+|^p \rangle / E^{p/2}$, $\hat{l} = l/l_d$, are plotted for $p = 3$ and 4. We

---

where F is a dimensionless function, $F(0) = 1$, which describes the spectrum in the dissipation range $\hat{k} \gtrsim 1$, and C is the Kolmogorov constant.

Though the energy spectrum in 3D MHD seems to follow a $k^{-5/3}$ law, the standard normalization, (5.104), must be generalized to account for the dependence on H. A suitable form is (see Müller & Biskamp, 2000)

$$\hat{E}(\hat{k}) = E_k \Big/ \left[(\epsilon \mu^5)^{1/4} (v_A^5/\epsilon H)^{3/8} \right] = C F(\hat{k}) \hat{k}^{-5/3}, \qquad (5.105)$$

where the dissipation scale is slightly changed compared with the Kolmogorov case, $l_d = l_K (v_A^5/\epsilon H)^{1/8}$.

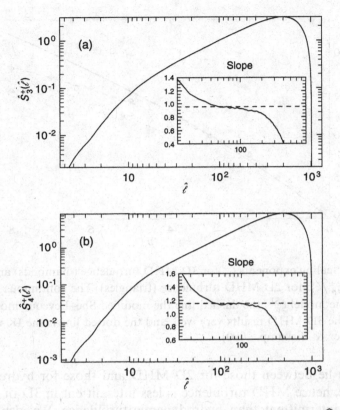

Fig. 5.21. Log–log plots of the normalized structure functions $\widehat{S}_p^+(\widehat{l})$ for (a) $p = 3$, (b) $p = 4$, from a simulation run with hyperdiffusion but otherwise the same parameters as in fig. 5.20. The insets show the logarithmic derivative, the horizontal dashed lines indicating the most probable values of the scaling exponents (from Müller & Biskamp, 2000).

find in particular that ζ_3 is close to the Kolmogorov value $\zeta_3 = 1$, a result which has recently also been derived analytically for certain triple products of δz_l in MHD turbulence (Politano & Pouquet, 1998a,b).

Assuming $\zeta_3 = 1$ to hold exactly, allows us to obtain rather accurate values for the remaining scaling exponents by using the property of extended self-similarity (Benzi *et al.*, 1993). Here the structure functions S_p are plotted not as functions of l but of S_3, $l = l(S_3)$, which yields a significantly extended scaling range. Figure 5.22, summarizing the results, gives ζ_p for MHD turbulence in 3D (diamonds) and 2D (triangles), compared with the She–Leveque model ζ_p^{SL} for hydrodynamic turbulence (continuous line) and the ζ_p^{IK}, (5.103) (dotted line). (Note that the 2D values are only relative, ζ_p/ζ_3, since here ζ_3 is smaller than unity, as mentioned above.) The figure shows that the scaling exponents for 3D MHD

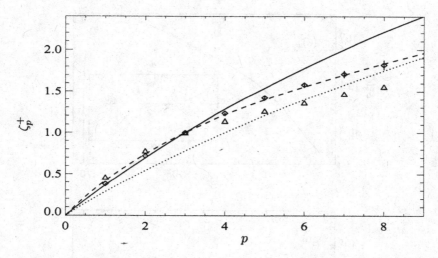

Fig. 5.22.   Scaling exponents ζ_p^+ for 3D MHD turbulence (diamonds) and relative exponents ζ_p^+/ζ_3^+ for 2D MHD turbulence (triangles). The continuous line is the She–Leveque model ζ_p^{SL}, the dashed line the modified She–Leveque model ζ_p^{MHD}, which fits the 3D MHD results very well, and the dotted line is the IK model ζ_p^{IK} (from Müller & Biskamp, 2000).

turbulence lie between those for 2D MHD and those for hydrodynamic turbulence, hence MHD turbulence is less intermittent in 3D than in 2D, but more intermittent than hydrodynamic turbulence. We also see that the simulation results differ basically from the IK prediction ζ_p^{IK}.

Let us now see, how the numerically observed values for 3D MHD turbulence fit into the framework of the generalized She–Leveque formalism (5.102). The numerical results indicate that the fundamental scaling is Kolmogorov-like, i.e., $g = 3$. Visualizations of the small-scale turbulent structures clearly demonstrate the sheet-like character of the dissipative eddies, i.e., $C_0 = 1$. We thus obtain the intermittency model for 3D MHD turbulence

$$\zeta_p^{MHD} = p/9 + 1 - (1/3)^{p/3}, \tag{5.106}$$

which in fact agrees well with the numerically observed scaling exponents, as seen in fig. 5.22.

These findings change the conventional interpretation of the turbulence dynamics in MHD. Although the Alfvén-wave effect is present, as clearly seen from the fact that the kinetic and magnetic energy spectra come close at small scales, the interaction of Alfvén waves propagating along the local field does not seem to govern the dynamics, which is dominated instead by cross-field eddy-like motions, much as in hydrodynamic turbulence. Thus the Alfvén-wave interaction, which is basically a weak turbulence process, is only a subdominant effect, which taken separately would lead to a steeper energy spectrum $E_k \sim k^{-2}$ (Galtier *et al.*, 1999). The Iroshnikov–Kraichnan mechanism giving rise to the $k^{-3/2}$ spectrum does not seem to be active in a physical system.

It is interesting to note the difference between 2D and fully 3D MHD turbulence. Though the asymptotic behavior in 2D is difficult to assess, as mentioned above, at least in the Reynolds number range considered in the 3D studies the 2D results seem to be consistent with the IK phenomenology, showing in particular a clear $k^{-3/2}$ energy spectrum (see, e.g., Biskamp & Welter, 1989). The difference between the 2D and the 3D behavior is not implausible. There are two competing dynamical effects in MHD turbulence – eddy motions and Alfvén waves. Since in 2D hydrodynamic turbulence the eddy dynamics is weak, resulting in the steep k^{-3} energy spectrum, Alfvén waves are likely to play an important role. In 3D, by contrast, eddy dynamics is much more vigorous and may thus dominate over the Alfvén-wave effect.

5.4.4 Homogenous turbulent dynamo

The statistical properties just described refer to turbulent systems in which the magnetic energy is of the same order as or larger than the kinetic energy, assuming, in particular, that the magnetic field is smooth, but its derivative, the current density, is distributed sparsely or intermittently. An example in nature is the turbulence in the solar wind, where the magnetic field has simply expanded from the space-filling quasi-static field in the corona. Conditions are, however, quite different for systems in which the turbulence is primarily kinetic, stirred by mechanical or thermal forces. Here the magnetic field is convected with the fluid and possibly amplified by dynamo action. Such a field is not space-filling but tends to be concentrated in thin flux tubes, similar to vortex tubes in hydrodynamic turbulence. This behavior can be understood from the structure of the equations. If the magnetic field is weak, the Lorentz force is negligible, such that vorticity and magnetic field follow the equations

$$\partial_t \mathbf{B} + \mathbf{v} \cdot \nabla \mathbf{B} = \mathbf{B} \cdot \nabla \mathbf{v} + \eta \nabla^2 \mathbf{B}, \tag{5.107}$$

$$\partial_t \mathbf{w} + \mathbf{v} \cdot \nabla \mathbf{w} = \mathbf{w} \cdot \nabla \mathbf{v} + \mu \nabla^2 \mathbf{w}, \tag{5.108}$$

which are formally identical for magnetic Prandtl number $Pr_M = \mu/\eta = 1$. Hence a similar structure of the solutions can be expected. For the inertial range, the analogy between $\mathbf{w}$ and $\mathbf{B}$ implies the magnetic energy spectrum

$$E_k^M \propto k^2 E_k^K \sim k^{1/3}, \tag{5.109}$$

assuming a Kolmogorov kinetic energy spectrum. We thus see that E_k^M increases with k, up to the dissipative cutoff, which implies that the magnetic field is concentrated in small-scale structures. The argument, (5.109), leading the the spectrum $E_k^M \sim k^{1/3}$ should, however, be considered with some caution. Since the magnetic energy is not an ideal invariant, the spectrum cannot be inferred from a cascade argument but is only based on the analogy with the vorticity.

Nor does the analogy give information about the *magnitude* of E^M. Will a weak seed field decay or grow in time? Starting from a smooth field distribution, the magnetic energy will always increase initially due to field line stretching, twisting and folding, as idealized in the Vainshtein–Zeldovich process, where the magnetic field is rapidly concentrated in rope-like structures of decreasing diameter d, until the thinning is limited by resistive diffusion. Whether, subsequently, the field energy continues to grow or decays depends on the balance between convective driving and diffusion, which is measured by the magnetic Reynolds number Re_M or, for given kinetic Reynolds number, on the magnetic Prandtl number, $Re_M = Pr_M Re$. Only if $Re_M > Re_{M,c}$ (or $Pr_M > Pr_{M,c}$) does true dynamo action occur, where $Re_{M,c} \sim 20$–40 typically.

Saturation is caused by the Lorentz force, but no clear answer has yet been given about the final magnetic energy level and the spectral distribution. Three-dimensional numerical simulations (Kida *et al.*, 1991) indicate that for the most interesting case of $Pr_M \gg 1$ the magnetic spectrum is essentially flat at small k where $E_k^K > E_k^M$, while for large k where $E_k^K \ll E_k^M$, the magnetic field is passively convected by the large-scale motions, hence $E_k^M \sim k^{-1}$ up to the resistive cutoff. In this regime the energy dissipation is mainly ohmic, in spite of the fact that $\eta \ll \mu$.

6

Noncollisional reconnection processes

We have seen in the previous chapters that, in the framework of resistive MHD, reconnection rates are in general low ($\sim \eta^{1/2}$) becoming independent of η only when the system is fully turbulent. The resistive MHD approximation is, however, only valid for relatively dense and cool plasmas. In hot plasmas, which are of special interest in laboratory fusion devices and in space, resistivity is no longer the most important nonideal effect in Ohm's law. Instead $\mathbf{R}$ in (2.1) is dominated by noncollisional terms, which are associated with different intrinsic microscopic length-scales: the ion and electron inertial lengths c/ω_{pj}, $j = i,e$; the ion and electron Larmor radii $\rho_j = v_{tj}/\Omega_j$; and the Debye length $\lambda_D = v_{te}/\omega_{pe}$. Here the plasma frequencies ω_{pj}, the gyrofrequencies Ω_j, and the thermal velocities v_{tj} are defined by

$$\omega_{pj} = \sqrt{4\pi n e^2/m_j}, \quad \Omega_j = eB/m_j c, \quad v_{tj} = \sqrt{k_B T_j/m_j}. \quad (6.1)$$

In a low-β plasma we have typically the ordering

$$c/\omega_{pi} \gg c/\omega_{pe} \gtrsim \rho_i \gg \rho_e \gtrsim \lambda_D. \quad (6.2)$$

The particle density n is, properly speaking, the electron density n_e or the ion density n_i, respectively. However, deviations from charge neutrality $n_e \neq n_i$ occur only on spatial scales of order λ_D. Since most plasmas of interest in the present context are sufficiently dense, such that λ_D is small and ω_{pe}^{-1} short compared to all other spatial and time-scales, we have $(n_i - n_e)/n_e \ll 1$, so that we can set $n_e = n_i = n$. This approximation is called quasi-neutrality, since the net electric charge density is small but not identically zero. There is still an electrostatic field, which is, however, no longer calculated from Poisson's equation but from the quasi-neutrality condition $\nabla \cdot \mathbf{j} = 0$. Only in chapter 7 will we discuss certain electrostatic microinstabilities, which involve fluctuations on the Debye scale. Effects associated with the intrinsic scales in (6.2) become more important than resistivity, when the resistive scale-length $\delta_\eta \sim (\eta \tau_A)^{1/2}$, the width of a resistive current sheet, becomes smaller than one of these scale-lengths. For instance, electron inertia dominates over resistive diffusion in a current sheet, if

$$\frac{\delta_\eta}{c/\omega_{pe}} \sim \left(\frac{\tau_A}{\tau_e}\right)^{1/2} < 1, \tag{6.3}$$

where $\tau_A = L/v_A$ is the Alfvén time associated with a typical macroscopic length L and $\tau_e = v_{ei}^{-1}$ is the electron collision time.

This chapter is focused on nonresistive, i.e., nondissipative reconnection, which we treat in the framework of two-fluid theory. Since a fluid description in general requires some dissipation to prevent solutions from developing singularities, application of a fluid model to collisionless systems might appear doubtful. The processes investigated here are therefore not truely collisionless – we call them quasi-collisionless or noncollisional – but the reconnection dynamics is found to be independent of the residual values of the dissipation coefficients.

Starting from Braginskii's theory (Braginskii, 1965), a simplified two-fluid model is developed in section 6.1, which serves as the framework for this chapter. Section 6.2 considers the high-β regime, where at small scales $l < c/\omega_{pi}$ the dynamics is dominated by the Hall term, which leads to a decoupling of the electrons from the ions introducing a dispersive mode, the whistler. In this scale range the dynamics is carried only by the electron and is described by the equations of electron magnetohydrodynamics (EMHD), a subset of our two-fluid model. We investigate whistler-dominated reconnection, where the reconnection rate is found to be independent of the electron physics and is controlled only by the ion dynamics.

Section 6.3 treats the low-β regime, which is described by a reduced set of two-fluid equations, the four-field model or, neglecting the sound wave, the three-field model. Here, the most efficient reconnection process results from a combination of the Hall term with the parallel electron compressibility, which introduces a different dispersive mode, the kinetic Alfvén wave valid at scales below the ion-sound gyroradius $\rho_s = c_s/\Omega_i$, $c_s = \sqrt{k_B(T_i + T_e)/m_i}$. Since this regime is of particular relevance for tokamak plasmas, we consider first the linear stability theory in some detail. Nonlinear reconnection dynamics is studied in section 6.4 for the internal kink mode, the basic mechanism of the sawtooth collapse, which is an important reconnection process in a tokamak. Recent experimental results on the sawtooth oscillation and their theoretical interpretation are presented in section 6.5. Lastly, in section 6.6, we discuss briefly several plasma experiments built specifically to study reconnection physics in well-defined and controlled configurations.

6.1 Two-fluid theory

The fluid description of a plasma results from taking moments of the kinetic equations for each plasma component. As in most of this book we restrict consideration to simple plasmas consisting of electrons and a singly charged ion species, which we usually take to be hydrogen. Inclusion of multiple ion species and ionization states would be straightforward, but since it is not essential for the reconnection physics it would only complicate the formalism.

The lowest-order moment equations for $n(= n_i = n_e), \mathbf{v}_{i,e}, p_{i,e}$ are

$$\partial_t n = -\nabla \cdot n\mathbf{v}_i = -\nabla \cdot n\mathbf{v}_e, \tag{6.4}$$

$$m_e n(\partial_t \mathbf{v}_e + \mathbf{v}_e \cdot \nabla \mathbf{v}_e) = -\nabla p_e - \nabla \cdot \boldsymbol{\pi}_e - en\left(\mathbf{E} + \frac{1}{c}\mathbf{v}_e \times \mathbf{B}\right) + \boldsymbol{\Gamma}, \tag{6.5}$$

$$m_i n(\partial_t \mathbf{v}_i + \mathbf{v}_i \cdot \nabla \mathbf{v}_i) = -\nabla p_i - \nabla \cdot \boldsymbol{\pi}_i + en\left(\mathbf{E} + \frac{1}{c}\mathbf{v}_i \times \mathbf{B}\right) - \boldsymbol{\Gamma}, \tag{6.6}$$

$$\frac{3}{2}(\partial_t p_e + \mathbf{v}_e \cdot \nabla p_e) + \frac{5}{2} p_e \nabla \cdot \mathbf{v}_e = -\nabla \cdot \mathbf{q}_e - \boldsymbol{\pi}_e : \nabla \mathbf{v}_e + Q_e, \tag{6.7}$$

$$\frac{3}{2}(\partial_t p_i + \mathbf{v}_i \cdot \nabla p_i) + \frac{5}{2} p_i \nabla \cdot \mathbf{v}_i = -\nabla \cdot \mathbf{q}_i - \boldsymbol{\pi}_i : \nabla \mathbf{v}_i + Q_i. \tag{6.8}$$

As before, we assume that the plasma components follow the ideal gas law with the ratio of specific heats $\gamma = 5/3$, which is well satisfied for the dilute plasmas of interest here,

$$p_{i,e} = nT_{i,e}, \tag{6.9}$$

using the customary shorthand notation $k_B T \rightarrow T$. (Major deviations from pressure isotropy may become important in collisionless plasmas, where often $T_\perp \neq T_\parallel$ and where even nondiagonal elements in the pressure tensor may be important. Such effects will be discussed in chapter 7.) $\boldsymbol{\Gamma}$ is the friction force between electrons and ions, which must appear with opposite sign in the (6.5) and (6.6) to insure momentum conservation, $\boldsymbol{\pi}_j$ are the stress tensors, $\mathbf{q}_j$ the heat fluxes and Q_j the heat sources. Equations (6.4)–(6.8) are generally valid, but to be useful the transport quantities $\boldsymbol{\Gamma}, \boldsymbol{\pi}_j, \mathbf{q}_j, Q_j$ must be specified, which have been calculated by Braginskii (1965) for a collisional plasma.

In such a plasma the microscopic velocity distributions are nearly Maxwellian, i.e., the system is close to local thermodynamic equilibrium for each plasma component. The conditions are that the time-scale of the fluid dynamics is long compared with the collision times τ_j, $\tau_j\partial_t \ll 1$, and spatial scales long compared with the mean free paths λ_j, $\lambda_j\nabla \ll 1$, where $\lambda_j = v_{tj}\tau_j$. In a magnetized plasma the latter condition is, however, only relevant along the field, $\lambda_j\nabla_{\parallel} \ll 1$, since for $\Omega_j\tau_j = \lambda_j/\rho_j \gg 1$, the mean free path in perpendicular direction is replaced by the Larmor radius, i.e., we have the condition $\rho_j\nabla_{\perp} \ll 1$.

The exact expressions of the collision times τ_j or the collision frequencies $v_j = \tau_j^{-1}$, are not important in this context, but it is useful to have an order-of-magnitude estimate, in particular to know their relative sizes. One has to distinguish collisions between like particle and different particle species using the notation that, e.g., v_{ei} refers to electrons being scattered by ions:

$$v_{ee} \sim v_{ei} \sim \frac{\omega_{pe}}{n\lambda_D^3} \sim \frac{n}{T_e^{3/2}m_e^{1/2}}, \tag{6.10}$$

$$v_{ii} \sim \frac{\omega_{pi}}{n\lambda_D^3} \sim \left(\frac{m_e}{m_i}\right)^{1/2} v_{ee}, \tag{6.11}$$

where $n\lambda_D^3$, the number of particles in the Debye cell, is the measure of the plasma collisionality, $n\lambda_D^3 \to \infty$ meaning the collisionless limit. The effect of electron collisions on ions is weak, the energy transfer rate being

$$v_{ie} \sim \frac{m_e}{m_i}v_{ei}. \tag{6.12}$$

This relation can easily be understood. The velocity change of an ion by a collision with a thermal electron is $\Delta v \sim (m_e/m_i)v_{te}$, corresponding to a change of the energy $\Delta E = m_i(v_i + \Delta v)^2 - m_iv_i^2 \sim (m_e/m_i)^{1/2}E$. Since the process is diffusive – collisions may increase or decrease the energy – the time required to change the ion energy by its own order of magnitude is $(E/\Delta E)^2v_{ei}^{-1} \sim (m_i/m_e)v_{ei}^{-1}$. Because of the slow energy transfer, ions and electrons may have substantially different temperatures, either $T_i \ll T_e$ or $T_i \gg T_e$, while each component is nearly in thermal equilibrium.

In the following it will be sufficient to use Braginskii's expressions for the transport quantities in a simplified form. The basis for these approximations are weak collisionality $\Omega_j\tau_j \gg 1$ and high parallel electron mobility, in particular $\nabla_{\parallel}T_e = 0$. The friction term Γ has two contributions, $\Gamma = \Gamma_u + \Gamma_T$, where Γ_u is the usual resistive term

$$\Gamma_u = \frac{m_e}{e\tau_e}(\mathbf{j}_{\perp} + 0.51\mathbf{b}j_{\parallel}) \simeq en\eta\mathbf{j}, \tag{6.13}$$

while Γ_T is connected with the electron temperature gradient

$$\Gamma_T = -0.71n\nabla_{\parallel}T_e - \frac{3}{2}\frac{n}{\Omega_e\tau_e}\mathbf{b} \times \nabla T_e. \tag{6.14}$$

The first term in $\mathbf{\Gamma}_T$ vanishes because of our assumption $\nabla_\parallel T_e = 0$, the second term is of the order $\beta\mathbf{\Gamma}_u$, hence we have for not too high β

$$\mathbf{\Gamma} \simeq ne\eta\mathbf{j}. \tag{6.15}$$

The approximations in the stress tensors $\boldsymbol{\pi}_j$ are even less systematic. The stress tensor consists of three groups of terms connected with the parallel viscosity $\mu_{\parallel j} \sim v_{tj}^2\tau_j$, the perpendicular viscosity $\mu_{\perp j} \sim \rho_j^2/\tau_j = v_{tj}^2/(\Omega_j^2\tau_j)$ and the nondissipative gyroviscosity $\mu_{0j} \sim v_{tj}^2/\Omega_j$, which is independent of the collision time. Though $\mu_\parallel$ is much larger than $\mu_\perp$, this is compensated by the difference in the derivatives. To estimate the ratio of the ion viscous effects we use $\nabla_\parallel \sim L_\parallel^{-1}$, $\nabla_\perp \sim L_\perp^{-1} \lesssim \rho_i^{-1}$, since the perpendicular scales may be as small as the ion gyroradius,

$$\frac{\mu_{\parallel i}\nabla_\parallel^2}{\mu_{\perp i}\nabla_\perp^2} \sim \frac{v_{ti}^2\tau_i^2}{\rho_i^2}\frac{L_\perp^2}{L_\parallel^2} \sim \left(\frac{v_{ti}\tau_i}{L_\parallel}\right)^2 < 1, \tag{6.16}$$

if we satisfy the requirement $\lambda_i\nabla_\parallel \sim v_{ti}\tau_i/L_\parallel < 1$ for the validity of the fluid approach. Moreover, dynamical studies show that the parallel flow is usually only weakly excited. We will therefore neglect the effect of parallel ion viscosity. The gyroviscosity is formally much larger than the perpendicular viscosity and can therefore play an important role in a weakly collisional plasma. The effect comes from the nondiagonal part of $\boldsymbol{\pi}$ in the plane perpendicular to $\mathbf{B}$, π_{xy} (assuming $\mathbf{B}$ in the z-direction) describing the transfer of x-momentum in y-direction and vice versa,

$$\pi_{xy} = \pi_{yx} = nm_i\mu_0(\partial_x v_x - \partial_y v_y). \tag{6.17}$$

Compared with the perpendicular viscosity, the gyroviscosity μ_0 has the collision frequency replaced by the gyrofrequency, hence $\mu_0 \gg \mu_\perp$ for $\Omega\tau \gg 1$.[*] Neglecting the smaller mixed parallel–perpendicular terms π_{xz}, the stress tensor contribution in the ion momentum equation (6.7) assumes the form:

$$-(\nabla \cdot \boldsymbol{\pi})_\perp = nm_i\mu_{i\perp}\nabla^2\mathbf{v}_{i\perp} - nm_i\mu_0\mathbf{b} \times \nabla^2\mathbf{v}_i. \tag{6.18}$$

In the electron momentum equation (6.5) the stress tensor is usually neglected, since it is proportional to the electron mass. Though the parallel viscosity could still be formally larger than the friction term $\mathbf{\Gamma}_\parallel$, the latter is more important since it describes the momentum exchange between electrons and ions, i.e., resistivity, instead of a mere relaxation of the parallel electron momentum gradient. However the perpendicular electron

---

[*] It is worth mentioning that the expression $\rho_i^2\Omega_i(\simeq \rho_e^2\Omega_e) \simeq cT/eB$ has a certain reputation or notoriety as the Bohm diffusion coefficient (originally including a factor $1/16$) in the theory of plasma transport.

viscosity $\mu_{e\perp}$ is often kept, since it may be effectively enhanced compared with the negligibly small collisional value $\rho_e^2 \nu_e$ in the presence of braided magnetic field lines (Furth *et al.*, 1973; Kaw *et al.*, 1979). The electron momentum equation (6.5) is usually written in the form of Ohm's law. With the approximations on π_e and Γ and substituting $ne\mathbf{v}_e = ne\mathbf{v}_i - \mathbf{j}$, the generalized Ohm's law reads

$$\mathbf{E} = -\frac{1}{c}\mathbf{v}_i \times \mathbf{B} + \frac{1}{nec}\mathbf{j} \times \mathbf{B} - \frac{1}{ne}\nabla p_e$$

$$-\frac{m_e}{ne}(\partial_t n\mathbf{v}_e + \nabla \cdot n\mathbf{v}_e\mathbf{v}_e) + \eta\mathbf{j} + \frac{\mu_{e\perp}m_e}{e}\nabla_\perp^2\mathbf{v}_e. \qquad (6.19)$$

Compared with the simple form of Ohm's law in resistive MHD, $\mathbf{E} = -(1/c)\mathbf{v} \times \mathbf{B} + \eta\mathbf{j}$, with $\mathbf{v} = \mathbf{v}_i$, several additional terms appear in two-fluid theory: the Hall term ($\propto \mathbf{j} \times \mathbf{B}$); the electron pressure gradient; the electron inertia; and electron viscosity, which are nondissipative except the last one. Electron inertia and viscosity are only important in cases where the electron flux $n\mathbf{v}_e$ is large, which occurs on scales small compared to the smallest ion scales, hence we can substitute $\mathbf{v}_e \simeq -\mathbf{j}/ne$ in these terms. To illucidate the order of magnitude of the non-MHD effects let us write (6.19) in non-dimensional form using the normalisations $x/L \to x, v/v_A \to v, B/B_0 \to B, n/n_0 \to n, cE/v_AB_0 \to E, p_e/(B_0^2/4\pi) \to p_e, j/(cB_0/4\pi L) \to j$:

$$\mathbf{E} = -\mathbf{v}_i \times \mathbf{B} + \frac{d_i}{n}(\mathbf{j} \times \mathbf{B} - \nabla p_e) + \frac{d_e^2}{n}\frac{d\mathbf{j}}{dt} + \eta\mathbf{j} - \eta_2\nabla^2\mathbf{j}, \qquad (6.20)$$

where

$$d_i = \frac{c}{\omega_{pi}L}, \quad d_e = \frac{c}{\omega_{pe}L}, \qquad (6.21)$$

$$\eta = \frac{\eta c^2}{4\pi v_A L} = \frac{1}{S}, \quad \eta_2 = d_e^2\frac{\mu_{e\perp}}{v_A L}, \qquad (6.22)$$

and η_2 is also called hyperresistivity. While the dissipative terms are known to cause reconnection, this is, in general, not obvious for the nondissipative effects. For instance the Hall term itself conserves field line topology, see section 2.2, the field being simply convected with the electron flow $\mathbf{v}_e$, but it may drastically enhance the efficiency of a weak dissipative reconnection process as will be discussed in detail in the following section. The electron pressure contribution in Ohm's law is usually small. Inserting $\mathbf{E}$ in Faraday's law gives a term $\propto \nabla n \times \nabla p_e$, which is only important in a dense plasma, while it vanishes for a polytropic gas law $p = p(n)$. (Here the underlying assumption is that collisions, though rare, are still frequent enough to make the pressure tensor isotropic.) The electron pressure gradient will, however, play an crucial role through the diamagnetic drift in a low-β plasma.

Substituting $\mathbf{E}$ in the ion momentum equation, neglecting electron inertia, and using the approximation (6.18) and $ne(\mathbf{v}_i - \mathbf{v}_e) = \mathbf{j}$, gives

$$nm_i(\partial_t \mathbf{v}_i + \mathbf{v}_i \cdot \nabla \mathbf{v}_i) = -\nabla(p_i + p_e) + \frac{1}{c}\mathbf{j} \times \mathbf{B}$$
$$+ nm_i(\mu_{i\perp}\nabla_\perp^2 \mathbf{v}_i - \mu_0 \mathbf{b} \times \nabla^2 \mathbf{v}_i). \qquad (6.23)$$

Since magnetic reconnection processes are rather insensitive to the details of the evolution of the temperatures T_i, T_e, it is sufficient to treat the transport terms on the r.h.s. of (6.7) and (6.8) in a crude way. We neglect all heat fluxes except for the parallel electron one $q_{e\parallel}$, which we take to be infinite such that $\nabla_\parallel T_e = 0$. Also, the terms involving the stress tensors are neglected. Q_e consists mainly of the Ohmic heating, while Q_i contains the viscous dissipation.

Summarizing our two-fluid model written in non-dimensional form with the notation $\mathbf{v} = \mathbf{v}_i, \mu = \mu_{\perp i}, \mathbf{v}_e = \mathbf{v} - d_i \mathbf{j}/n$, we have:

$$\partial_t \mathbf{v} + \mathbf{v} \cdot \nabla \mathbf{v} = -\frac{1}{n}(\nabla(p_i + p_e) - \mathbf{j} \times \mathbf{B}) + \mu \nabla_\perp^2 \mathbf{v} - \mu_0 \mathbf{b} \times \nabla_\perp^2 \mathbf{v}, \qquad (6.24)$$

$$\mathbf{E} = -\mathbf{v} \times \mathbf{B} + \frac{d_i}{n}(\mathbf{j} \times \mathbf{B} - \nabla p_e) + \frac{d_e^2}{n}\frac{d\mathbf{j}}{dt} + \eta \mathbf{j} - \eta_2 \nabla_\perp^2 \mathbf{j}, \qquad (6.25)$$

$$\frac{3}{2}(\partial_t p_e + \mathbf{v}_e \cdot \nabla p_e) + \frac{5}{2}p_e \nabla \cdot \mathbf{v}_e = -\eta j^2 - \eta_2(\nabla \mathbf{j})^2, \qquad (6.26)$$

$$\frac{3}{2}(\partial_t p_i + \mathbf{v} \cdot \nabla p_i) + \frac{5}{2}p_i \nabla \cdot \mathbf{v} = -\mu(\nabla_\perp \mathbf{v})^2. \qquad (6.27)$$

In addition, the continuity equation for n and Faraday's law have to be satisfied. This set of equations conserves energy in the form (neglecting dissipation and boundary effects):

$$W = \frac{1}{2}\int\left(nv^2 + B^2 + \frac{d_e^2}{n}j^2 + 3(p_i + p_e)\right)dV = const. \qquad (6.28)$$

The energy expression differs from the (compressible) MHD expression, (3.32), only by the electron inertia term. What happens to the other ideal MHD invariants in the two-fluid framework, most notably the magnetic helicity $H = \int \mathbf{A} \cdot \mathbf{B} dV$? We find up to resistive and boundary effects

$$\frac{dH}{dt} = -2d_e^2\int \mathbf{B} \cdot \frac{1}{n}\frac{d\mathbf{j}}{dt}dV = O(m_e). \qquad (6.29)$$

Hence also in two-fluid theory magnetic helicity is conserved except for the usually very weak electron inertia effect, since the contribution from the electron pressure term in (6.25) vanishes on the assumption $\nabla_\| T_e = 0$. On very small scales $\lesssim c/\omega_{pe}$, which are dominated by electron inertia, we introduce, in section 6.2.1, a generalized helicity expression, which is conserved in the framework of electron magnetohydrodynamics.

6.2 High-β whistler-mediated reconnection

We first consider noncollisional reconnection in high-β plasmas, $\beta \gtrsim 1$, which is typical for plasma parameters in the solar wind and the geomagnetic tail. In this case the two-fluid equations can be simplified by assuming the ion motion to be incompressible. (A rigorous justification of this approximation requires $\beta \gg 1$, such that the sound wave is much faster than, and hence decouples from, the Alfvén wave governing the reconnection time-scale. However, compressibility effects are found to be weak already at moderately high $\beta \gtrsim 1$). If $\nabla \cdot \mathbf{v}_i = 0$, one can choose uniform density, whence the quasi-neutrality condition $\nabla \cdot \mathbf{j} = 0$ requires that also the electrons are incompressible $\nabla \cdot \mathbf{v}_e = 0$. We can now eliminate the pressure by taking the curl of (6.24), which leaves us with two equations for the vorticity $\mathbf{w} = \nabla \times \mathbf{v}$ and the magnetic field $\mathbf{B}$, generalizing the incompressible MHD equations (3.9), (3.10):

$$\partial_t \mathbf{w} + \mathbf{v} \cdot \nabla \mathbf{w} - \mathbf{w} \cdot \nabla \mathbf{v} = \mathbf{B} \cdot \nabla \mathbf{j} - \mathbf{j} \cdot \nabla \mathbf{B} - \nabla \times (\nabla \cdot \pi_i), \tag{6.30}$$

$$\partial_t \mathbf{B} + \mathbf{v} \cdot \nabla \mathbf{B} - \mathbf{B} \cdot \nabla \mathbf{v} = -d_i (\mathbf{B} \cdot \nabla \mathbf{j} - \mathbf{j} \cdot \nabla \mathbf{B})$$
$$- d_e^2 \nabla \times (d\mathbf{j}/dt) - \nabla \times (\eta \mathbf{j} - \eta_2 \nabla^2 \mathbf{j}), \tag{6.31}$$

where the last term in (6.30) comprises the viscous effects. While the $\mathbf{w}$ equation remains essentially unchanged, there are two new non-ideal terms in the induction equation (6.31), originating from the Hall effect and electron inertia, which are associated with the scales d_i and d_e, respectively.

To study the linear mode characteristics, we linearize these equations about a homogeneous static equilibrium neglecting dissipation and making the usual Fourier ansatz for the perturbations $\tilde{\mathbf{B}} = \mathbf{B}_1 \exp\{-i\omega t + i\mathbf{k} \cdot \mathbf{x}\}$ etc.,

$$\omega \mathbf{v}_1 + k_\| B \mathbf{B}_1 = 0, \tag{6.32}$$

$$\omega(1 + k^2 d_e^2)\mathbf{B}_1 + k_\| B \mathbf{v}_1 = i d_i k_\| B \mathbf{k} \times \mathbf{B}_1, \tag{6.33}$$

$$\mathbf{k} \cdot \mathbf{v}_1 = \mathbf{k} \cdot \mathbf{B}_1 = 0,$$

which gives the dispersion relation (writing v_A^2 instead of B^2 for clarity)

$$\left[\omega^2(1 + d_e^2 k^2) - k_\|^2 v_A^2\right]^2 - \omega^2 d_i^2 k^2 k_\|^2 v_A^2 = 0. \tag{6.34}$$

While for long wavelength $kd_i < 1$ one recovers the shear Alfvén wave $\omega^2 = k_\parallel^2 v_A^2$, in the short-wavelength regime $kd_i > 1$ the mode is dispersive

$$\omega^2 = \frac{d_i^2 k^2 k_\parallel^2 v_A^2}{(1 + d_e^2 k^2)^2}, \qquad (6.35)$$

called whistler.* In this regime the frequency no longer depends on the ion mass, since

$$v_A c / \omega_{pi} = \Omega_i c^2 / \omega_{pi}^2 = \Omega_e c^2 / \omega_{pe}^2 = cB/4\pi ne.$$

The dispersion relation (6.35) has two different regimes depending on the magnitude of $d_e k$. Written in dimensional form, we have

$$\omega \simeq \Omega_e \frac{c^2}{\omega_{pe}^2} k_\parallel k, \qquad \frac{\omega_{pi}}{c} < k < \frac{\omega_{pe}}{c}, \qquad (6.36)$$

the whistler proper, and

$$\omega \simeq \Omega_e \frac{k_\parallel}{k}, \qquad k > \frac{\omega_{pe}}{c}, \qquad (6.37)$$

the electron cyclotron wave. The coupling of the two components of $\mathbf{B}_1$ (and of $\mathbf{v}_1$) by the cross-product on the r.h.s. of (6.33) makes the whistler circularly polarized,

$$\frac{B_{1x}}{i B_{1y}} = \pm 1, \qquad (6.38)$$

right-hand for wave propagation in the direction of $\mathbf{B}_0$, left-hand for propagation in the opposite direction. The sense of rotation is that of electron gyration.

6.2.1 The EMHD approximation

While on large scales $l > c/\omega_{pi}$ ions and electrons move essentially together, $\mathbf{v}_i \simeq \mathbf{v}_e = O(v_A)$, $\mathbf{v}_i - \mathbf{v}_e = \mathbf{j}/ne = O(v_A c/\omega_{pi} l)$, at small scales $l < c/\omega_{pi}$ the ions are too heavy to follow the electrons, hence $\mathbf{v}_i \ll \mathbf{v}_e \simeq -\mathbf{j}/ne$. Neglecting the ion velocity in this regime, the system is determined by the induction equation (6.31) depending only on $\mathbf{B}$ and $\mathbf{j} = \nabla \times \mathbf{B}$. With

---

*The name results from the properties of such disturbances in the reception of radio waves. Whistlers are audio-frequency electromagnetic bursts originating from lightning flashes, which propagate along the ionospheric magnetic field. Since the group velocity increases with increasing frequency $\partial\omega/\partial k \propto \sqrt{\omega}$, (6.36), the received signal exhibits a descending audio tone.

the usual normalization to the Alfvén time $\tau_A = L/v_A$ the electron inertia term reads

$$d_e^2 \nabla \times \frac{d\mathbf{j}}{dt} = d_e^2 \nabla \times (\partial_t \mathbf{j} - d_i \mathbf{j} \cdot \nabla \mathbf{j}) = -d_e^2 \left[\partial_t \nabla^2 \mathbf{B} - d_i \nabla \times (\mathbf{j} \times \nabla^2 \mathbf{B}) \right]. \quad (6.39)$$

If, instead of τ_A, we choose as unit time the characteristic time of the whistler $\tau_W = \tau_A/d_i$, which can also be written as

$$\tau_W = L^2 \omega_{pi}/(cv_A) = (\Omega_e d_e^2)^{-1} \quad (6.40)$$

(note that τ_W is independent of both m_i and m_e), (6.31) assumes the form

$$\partial_t(\mathbf{B} - d_e^2 \nabla^2 \mathbf{B}) + \nabla \times \left[\mathbf{j} \times (\mathbf{B} - d_e^2 \nabla^2 \mathbf{B}) \right] = \eta \nabla^2 \mathbf{B} - \eta_2 (\nabla^2)^2 \mathbf{B}. \quad (6.41)$$

This equation is called electron magnetohydrodynamics (EMHD), since the electrons play a role similar to that played by the ions in ordinary MHD.[*] EMHD is simpler than MHD, consisting of only one vector equation, but it contains an intrinsic scale-length d_e. There are two ideal invariants in EMHD, the energy

$$W = \tfrac{1}{2} \int (B^2 + d_e^2 j^2) d^3 x \quad (6.42)$$

and a generalized helicity

$$H_e = \int (\mathbf{A} - d_e^2 \mathbf{j}) \cdot (\mathbf{B} - d_e^2 \nabla^2 \mathbf{B}) d^3 x. \quad (6.43)$$

6.2.2 *Properties of the reconnection region*

As in the resistive case, we consider reconnection processes in the two-dimensional approximation $\partial_z = 0$, where magnetic field and current density can be written in the form

$$\mathbf{B} = \mathbf{e}_z \times \nabla \psi + \mathbf{e}_z(B_{z0} + b), \quad (6.44)$$

$$\mathbf{j} = \nabla b \times \mathbf{e}_z + \mathbf{e}_z \nabla^2 \psi, \quad (6.45)$$

splitting the axial field into a mean field B_{z0} and a fluctuating part b. Reconnection occurs in a region of small $\mathbf{B}_\perp$, which for stationary

---

[*] EMHD has previously been discussed mostly in the context of fast plasma processes such as plasma switches and Z-pinches, see the reviews by Kingsep *et al.*, (1990) and Gordeev *et al.*, (1994). Since in such processes the density is strongly inhomogeneous, the constant-density approximation implicit in (6.41) assuming $\mathbf{v}_e = -\mathbf{j}/en_0$ is not justified, as the term $\nabla(1/n) \times (\mathbf{j} \times \mathbf{B})$ can play an important role. In the context of magnetic reconnection, however, electron inertia is usually more important than density inhomogeneity, hence (6.41) is appropriate, see, e.g., Bulanov *et al.*, (1992).

conditions has the form of an X-point configuration. In the region $|x| < d_i$ around the X-point, where the ion velocity can be neglected, the system is described by the EMHD equation (6.41), which in 2D reads

$$\partial_t(\psi - d_e^2 j) + \mathbf{v}_e \cdot \nabla(\psi - d_e^2 j) = \eta j - \eta_2 \nabla^2 j, \tag{6.46}$$

$$\partial_t(b - d_e^2 w_e) + \mathbf{v}_e \cdot \nabla(b - d_e^2 w_e) + \mathbf{B} \cdot \nabla j = \eta w_e - \eta_2 \nabla^2 w_e, \tag{6.47}$$

$$\mathbf{v}_e = -\mathbf{j}_\perp = \mathbf{e}_z \times \nabla b, \ j = j_z = \nabla^2 \psi, \ w_e = \nabla^2 b, \tag{}$$

hence the axial field fluctuation b plays the role of an electron stream function, which we therefore denote by ϕ_e in the following, $b = \phi_e$. Equation (6.46) expresses the conservation of $F = \psi - d_e^2 j$, the electron canonical momentum in the direction of the ignorable coordinate, cf. (7.7).

We also assume stationarity, which is valid as long as the flow remains stable. Instability and resulting turbulence will be treated in section 6.2.4. As in section 3.3.3, the coordinate system is chosen such that $\pm x$ defines the inflow and $\pm y$ the outflow directions. For $|x| > d_e$ (6.46) and (6.47) reduce to

$$E + \mathbf{v}_e \cdot \nabla\psi = 0, \tag{6.48}$$

$$\mathbf{B} \cdot \nabla\nabla^2\psi = 0, \tag{6.49}$$

where $E = \partial_t\psi$ is the reconnection rate. The equations have the similarity solution

$$\psi = \frac{1}{2}(x^2 - a^2 y^2), \tag{6.50}$$

$$\phi_e = \frac{E}{2} \ln\left|\frac{x + ay}{x - ay}\right|, \tag{6.51}$$

which is identical with the corresponding MHD solution, (3.142) and (3.143), illustrated in fig. 3.15. Finite resistivity or hyperresistivity is only needed to smooth the solution on the separatrix $x = \pm y$. The scale parameter a allows a finite uniform current density $j_0 = 1 - a^2$, so that the separatrix branches may intersect at any angle.

We now show (Biskamp *et al.*, 1997) that the similarity solution valid for $|x| > d_e$ can be matched to the solution in the electron inertia-dominated layer, or simply the inertial layer, $|x| < d_e$. In the limit of small resistivity, the current sheet which forms in this region exhibits a complicated multi-scale structure, in particular the current density develops a cusp-like singularity, resulting from continuous acceleration at the X-point, the stagnation point of the perpendicular electron flow. However, the singularity does not give a finite contribution to the integrated current (see also the discussion in section 6.3.3). Because of the cusp-like current profile the hyperresistivity η_2 associated with electron viscosity can be a more efficient damping and dissipation mechanism than resistivity even for very

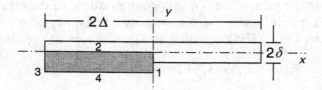

Fig. 6.1.   Integration domain in (6.53).

small values of η_2. In the following order-of-magnitude discussion these substructures are irrelevant and will be neglected.

The inertial layer consists of a current sheet of width δ and length Δ. From the convection term in (6.46), it is seen that inertia becomes important if

$$\partial_x \psi = B_y \sim d_e^2 \partial_x j \sim d_e^2 B_y / \delta^2,$$

hence the layer width is

$$\delta \sim d_e. \tag{6.52}$$

Integrating (6.47),

$$d_e^2 \mathbf{v}_e \cdot \nabla w_e = \mathbf{B} \cdot \nabla j, \tag{6.53}$$

over a quadrant of the current layer and using Gauss' theorem to transform the surface integrals into line integrals over the boundary we have

$$d_e^2 \oint v_{en} w_e \, dl = \oint B_n j \, dl,$$

as illustrated in fig. 6.1. Since $w_e = 0$ on the paths 1 and 2 (exactly) and 4 (approximately), only 3 contributes to the integral on the left side with $w_e \simeq \partial_x^2 b = \partial_x v_{ey}$, while the major contribution to the right side comes from path 2, hence

$$d_e^2 \int_3 v_{ey} \partial_x v_{ey} \, dx = \int_2 j \partial_y \psi \, dy = d_e^2 \int_2 j \partial_y j \, dy, \tag{6.54}$$

where we have used the property that $F = \psi - d_e^2 j$ is constant along the layer. (This property is clearly visible in fig. 6.2, resulting from a numerical simulation, which will be discussed in more detail in section 6.2.3.) Equation (6.54) gives the outflow velocity v,

$$v \sim j. \tag{6.55}$$

This relation results from the gyration of the out-of-plane current $j \, (= j_z)$ into the outflow (y) direction. The current layer can also be interpreted as a (finite amplitude) whistler perturbation obeying the relation (6.38), or in the present notation,

$$b_k = \pm k \psi_k, \tag{6.56}$$

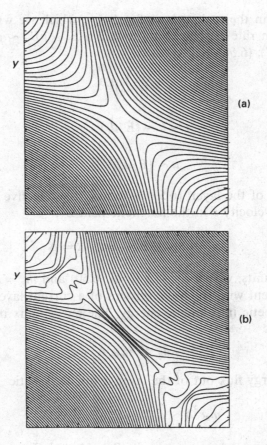

Fig. 6.2. Structure of (a) ψ and (b) $F = \psi - d_e^2 j$ in the reconnection region.

which is valid for the whistler satisfying the dispersion relation (6.35) independently of the value of kd_e. Relation (6.56) is essentially equivalent to (6.55). Integration of the continuity equation $\nabla \cdot \mathbf{v}_e = 0$ over a quadrant of the current layer gives

$$u\Delta \sim v\delta, \qquad (6.57)$$

where u is the inflow velocity.

Finally, to derive an equation for the layer length Δ we use directly the conservation of $F = \psi - d_e^2 j$ along the current sheet, $F|_{y=\Delta} = F|_{y=0}$. To determine the variation of ψ we note that, while B_y is changed by the presence of the current sheet, B_x is not, $\partial_y B_x = -\partial_y^2 \psi \sim 1$ from (6.50). (This property can be checked a posteriori using the scaling laws for j and Δ.) Hence the constancy of F gives

$$\psi(\Delta) - \psi(0) \sim \Delta^2 \sim d_e^2 j. \qquad (6.58)$$

We now obtain the scaling laws for Δ, j, u, which we write in terms of the reconnection rate E. Use of the relation $E \sim uB_y \sim ujd_e$ and (6.52) and (6.55), (6.57), (6.58) yields

$$\Delta \sim (Ed_e^2)^{1/3}, \tag{6.59}$$

$$j \sim v \sim (E/d_e)^{2/3}, \tag{6.60}$$

$$u \sim (E/d_e)^{1/3}. \tag{6.61}$$

The length Δ of the layer, (6.59), is simply the effective Larmor radius of electrons of velocity v in the magnetic field B_x,

$$\rho_e \sim vm_ec/eB_x \sim vd_e^2/\Delta = \Delta$$

in our present units, where $B_x \sim \Delta$ from (6.50) and $m_e = d_e^2$. The scaling laws are consistent with the basic physics of a current layer, the transformation of magnetic into kinetic energy. The energy flux into the layer is mainly magnetic

$$uB_y^2\Delta \sim d_e \gg m_eu^3\Delta \sim d_e^{5/3},$$

whereas the energy flux out of the layer is mainly kinetic

$$m_ev^3\delta \sim d_e \gg vB_x^2\delta \sim d_e^{5/3},$$

using again that in our units $m_e = d_e^2$, $B_x \sim \Delta$ and $B_y \sim jd_e$. We also see that the energy fluxes into and out of the layer are of the same order, which differs from the behavior in resistive MHD, (3.156), where the main part of the energy flux into the layer is dissipated.

Relation (6.59) implies that $\Delta \to 0$ for $d_e \to 0$. Since $B_y \sim d_e^{1/3} \to 0$, there is no flux pile-up in front of the layer. *The reconnection rate E is independent of the small parameter d_e.* In the region $|x| < d_i$, E is a free parameter of the electron flow, determined only by the free energy of the global configuration.

This behavior differs markedly from the scaling of resistive current-sheet reconnection. It is true that outside the current layer a solution equivalent to (6.50) and (6.51) is permitted also in resistive MHD, but it cannot be matched to the diffusion region, which has macroscopic length $\Delta \sim L$. The basic difference is that in resistive MHD the outflow (= downstream) velocity equals the upstream Alfvén velocity, which remains finite, such that the reconnection rate is slow, $u \sim v_A\delta/\Delta \ll v_A$, depending strongly on the small parameter η, $E \sim \eta^{1/2}$.

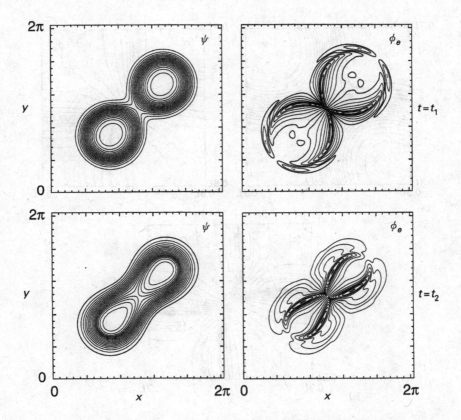

Fig. 6.3. Coalescence of two flux bundles in EMHD: contours of ψ and ϕ_e at two different times, t_1, t_2; $t_2 > t_1$.

6.2.3 Coalescence of EMHD flux bundles

The EMHD scaling predictions have been confirmed by a series of numerical simulations (Biskamp *et al.*, 1995). We consider the coalescence of two flux bundles located on the diagonal in a square box of edge size $L = 2\pi$,

$$\psi(x, y, t = 0) = \sum_{j=1,2} \exp\{-r_j^4/4\}, \qquad (6.62)$$

where $r_j^2 = (x-x_j)^2 + (y-y_j)^2$, $x_1 = y_1 = (\pi/2)+0.6$, $x_2 = y_2 = (3\pi/2)-0.6$. Figure 6.3 gives contour plots of ψ and ϕ_e at two times during the reconnection process. The conspicuous feature is that the flux surfaces seem to be pulled into the reconnection region instead of being pushed against it as in the MHD case. This property is due to the flow pattern with streamlines converging (i.e., velocity increasing) toward the X-point, which agrees with the similarity solution (6.51). In resistive reconnection

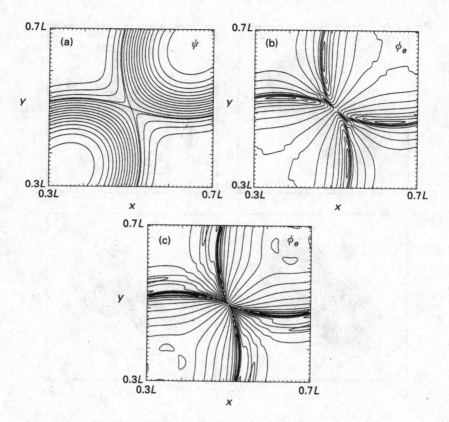

Fig. 6.4.   Reconnection region of EMHD flux bundle coalescence: (a) ψ; (b) ϕ_e for $d_e = 0.06$; and (c) for $d_e = 0.03$. ψ is essentially identical in both cases (from Biskamp *et al.*, 1995).

the flow is rather uniform, even diverging in front of the macroscopic current sheet, see fig. 3.16.

The structure of the region around the X-point and its dependence on d_e are illustrated in fig. 6.4. The inertial layer appears as a small layer of high electron velocity (note that the strong flows along the separatrix are a consequence of the similarity solution (6.51)), but does not show up in the flux function ψ, which exhibits a structureless X-point behavior. The layer shrinks both in length and width with decreasing d_e, which is found to be in quantitative agreement with the scaling laws derived above.

A further blow-up of the reconnection region is shown in fig. 6.2, which demonstrates an important feature of noncollisional electron dynamics. Equation (6.46) implies that $F = \psi - d_e^2 j$ is convected with the electron fluid. Since symmetry requires a stagnation point of the flow at the X-point, F would simply pile up in front of the X-point and would remain topologically invariant in the absence of dissipation. This pile-up of F is clearly seen in fig. 6.2(b) making F nearly constant along the layer. The distortion of the contours of F in the outflow region, however, indicates that reconnection is taking place due to the small dissipation included in the computation. Since the current density j assumes a quasi-singular cusp-like behavior at the X-point (see section 6.2.4), which produces a similar behavior of F, F may change topology even in the limit of vanishingly small dissipation.

6.2.4 Electron Kelvin–Helmholtz instability of the current layer

In the derivation of the scaling laws in section 6.2.2 we have ignored the substructure of the current density in the inertial layer d_e. Instead of the bell-shaped cross-layer current profile typical of resistive MHD (see section 3.2.1), noncollisional current layers, because of free electron acceleration at the X-point, develop a cusp-like profile. Hence transverse gradient scales become very short, $l_s \ll d_e$, such that $w_e \gg j/d_e$, the vorticity terms dominate in (6.47), $d_e^2 \mathbf{v}_e \cdot \nabla w_e \gg \mathbf{B} \cdot \nabla j$, and the equation reduces to the 2D Euler equation for the electron flow,

$$\partial_t w_e + \mathbf{v}_e \cdot \nabla w_e = 0.$$

As shown in section 4.6.2, the strong flow along the inertial layer may be Kelvin–Helmholtz unstable, if the parallel magnetic field is sufficiently weak. While a resistive current sheet is Kelvin–Helmholtz stable since the outflow velocity is sub-Alfvénic, $v \leq v_A$, the magnetic field at an EMHD current layer is weak as discussed above, hence shear flow instability is expected to occur for sufficiently narrow structure l_s, i.e., for sufficiently small dissipation coefficients.

Figure 6.5 illustrates the onset of the electron Kelvin–Helmholtz instability (EKHI) and the generation of EMHD turbulence. Since the largest flow velocity occurs at the layer edges, turbulence appears mainly in these regions. (For still smaller dissipation coefficients than used in the simulation of fig. 6.5, which further reduce the sublayer width l_s and hence increase the velocity shear, the entire layer breaks up leading to large regions of fully developed EMHD turbulence, which is found to exhibit many interesting properties, e.g., a Kolmogorov $k^{-5/3}$ energy spectrum, see Biskamp *et al.*, 1996, 1999).

To this point, our consideration of the EMHD dynamics has been confined to 2D geometry. While this approximation is justified for a stable laminar flow configuration, it is no longer valid when the system is susceptible to the EKHI, where not only the azimuthal flow v but also the axial current density j_z in the central part of the layer will

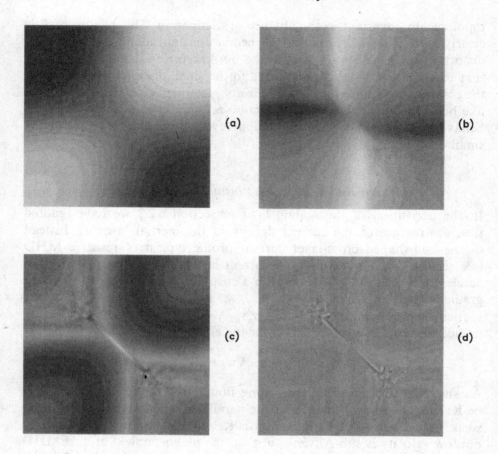

Fig. 6.5.   Simulation of electron Kelvin–Helmholtz instability giving rise EMHD to turbulence in the outflow regions of coalescing flux bundles. (a) ψ, (b) ϕ_e, (c) j, (d) w_e.

drive the instability. Since the extent of the layer in the axial direction is macroscopic, much larger than in the azimuthal plane, there is no finite-length effect limiting the excitation of turbulence due to convection out of the unstable region, which implies that in 3D the layer should become turbulent more readily.

The first 3D EMHD simulations of the break-up of a current layer were performed by Drake *et al.*, (1994). Here, we discuss some more recent numerical studies, where the current layer is generated self-consistently during the coalescence of two flux bundles (Drake *et al.*, 1997). This behavior is illustrated in fig. 6.6, showing a numerical simulation with the same initial conditions as in the 2D simulations of figs. 6.3, 6.4 but for somewhat larger d_e than used in the 2D runs. Shown are greyscale plots of

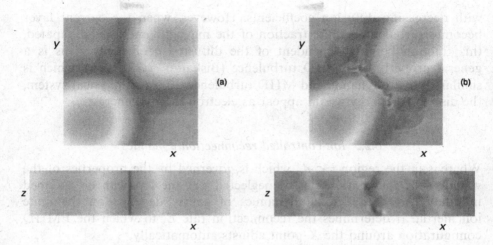

Fig. 6.6. 3D simulation of flux bundle coalescence illustrating the development with time of 3D EMHD turbulence for $d_e = 0.2$. Grayscale plots of axial current density j_z in the azimuthal (x, y) plane and the (x, z) plane along the diagonal $x = y$ across the current sheet. (a) $t = 0.8$, (b) $t = 1.75$ (from Drake *et al.*, 1997).

the axial current density in the x, y and the x, z plane. At time $t = 0.8$ (in units of the whistler time, (6.40)) the coalescence process is still laminar and practically two-dimensional, characterized by a smooth narrow current sheet between the two flux bundles. Subsequently turbulence sets in, generated mainly by EKHI of the axial current, which leads to a turbulent broadening of the entire layer as shown at $t = 1.75$. A 2D control run with the same parameters and initial conditions remains laminar – excitation of 2D turbulence as in fig. 6.5 would require a smaller dissipation coefficient – which demonstrates the importance of fully 3D geometry.

The 3D simulation just discussed has been performed assuming that the mean axial field B_{z0} vanishes. Since the EKHI is stabilized by a parallel magnetic field, an axial field of finite strength $B_{z0} > 1$ strongly reduces the effect of the instability, as shown in section 4.6.2.

It might be expected that the turbulent break-up of the current layer would result in a strongly enhanced reconnection rate. This is, however, not the case. The reason is that, even in a laminar state, the electron current layer does not control the reconnection speed, as we have shown in the preceding section. Electron inertia is needed only to break up the magnetic topology, but does not lead to magnetic flux pile-up which could slow down the reconnection process. The main effect of the turbulence is to *enhance the energy dissipation rate*. In the absence of EKHI the magnetic energy is mainly converted into kinetic energy of the electrons in the form of high-speed flows, while the dissipation rate is small, decreasing with decreasing diffusion coefficients. However, when the current layer becomes turbulent, a finite fraction of the magnetic energy is dissipated, this fraction being independent of the diffusion coefficients. This is a general property of EMHD turbulence (Biskamp *et al.*, 1996), which is similar to hydrodynamic and MHD turbulence. In a real physical system, the dissipated energy would appear as electron thermal energy.

6.2.5 Ion-controlled reconnection dynamics

Whereas in the region $l < d_i$, which is governed by the properties of the whistler, the ion motion can be neglected, i.e., the ions can be assumed infinitely heavy, the global dynamics of course depends on the finite ion inertia. It determines the reconnection rate E, to which the EMHD configuration around the X-point adjusts automatically.

The first simulations of whistler-mediated reconnection in the coalescence of two flux bundles were performed by Mandt *et al.* (1994), where the ions were treated as macro-particles. Here, we discuss results from two-fluid studies (Biskamp *et al.*, 1995), which allow us to use higher spatial resolution. Figure 6.7 illustrates a typical state during the coalescence process, where both $d_e (= 0.015)$ and $d_i (= 0.1)$ (corresponding to mass ratio $m_i/m_e = 44$) are small compared with the diameter of a flux bundle $L \sim 2$, in contrast to the EMHD coalescence studies in section 6.2.3, where $L < d_i$ is implied. The difference of the flow patterns in the vicinity of the X-point in clearly recognizable, electron streamlines converging toward the reconnection point, where ion streamlines are already deflected into the downstream cone. The magnitude of the inflow velocities along the main diagonal in fig. 6.7 are plotted in fig. 6.8. While on the MHD scale $l > d_i$ both ions and electrons move closely together, the ions are slowed down on a scale $l \sim d_i$, whereas the electrons are further accelerated reaching a maximum speed at a much smaller distance $l \sim d_e$.

A series of computer runs has been performed with different values of d_i and d_e (Biskamp *et al.*, 1997). The main result is that the maximum reconnection rate reached during the coalescence is almost independent of d_i, in other words, the reconnection speed is quasi-Alfvénic. Consistent with this fast process, the ion flow does not form an extended layer. In fact, the ion velocity changes abruptly in a shock-like way from the lower inflow to the higher outflow values, a behavior reminiscent of Petschek's shock configuration (see section 3.3.4). Recent simulations of the tearing mode (Shay *et al.*, 1998), which have been performed for substantially larger system size as used before, seem to confirm the Alfvénic scaling. We will come back to these studies in section 7.6.3.

The results of the incompressible theory are corroborated by numerical solutions of the fully compressible equations (6.24)–(6.27) in the case of

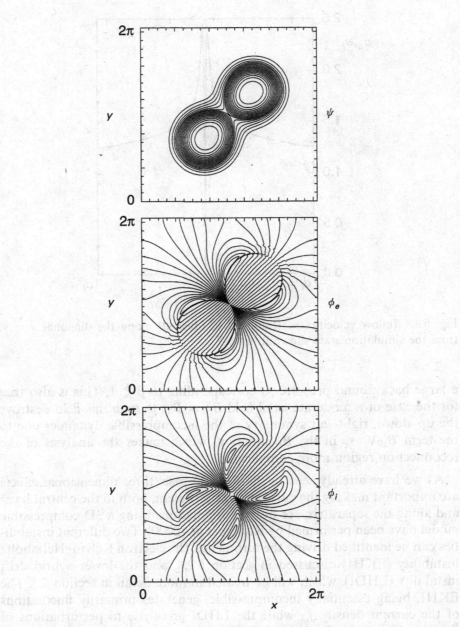

Fig. 6.7. Simulation of flux bundle coalescence with $d_i = 0.1$ and $d_e = 0.015$. Plotted are contours of ψ, the electron stream function ϕ_e and the ion stream function ϕ_i.

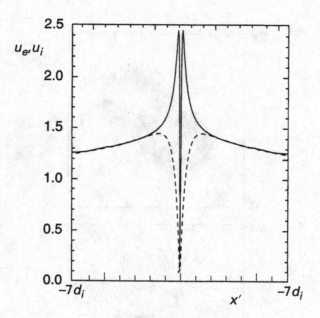

Fig. 6.8.   Inflow velocities u_e (full) and u_i (dashed) along the diagonal $x = y$, from the simulation state shown in fig. 6.7.

a large background pressure p_0 corresponding to $\beta \gtrsim 1$. This is also true for the case of a moderate axial field $B_{z0} \sim B_\perp$, though this field destroys the up–down, right–left symmetry of the incompressible dynamics due to the term $B_{z0}\nabla \cdot \mathbf{v}_e$ in the B_z equation, which makes the analysis of the reconnection region more complicated.

As we have already seen in the EMHD case, three-dimensional effects are important making the current sheet turbulent, both in the central layer and along the separatrix, see fig. 6.9. Simulations using a 3D compressible model have been performed by Rogers *et al.* (2000). Two different instabilities can be identified driving the turbulence, the electron Kelvin–Helmholtz instability (EKHI) discussed in section 6.2.4, and the lower-hybrid drift instability (LHDI), which will be treated in more detail in section 7.3. The EKHI, being essentially incompressible, generates primarily fluctuations of the current density j_z, while the LHDI gives rise to perturbations of the plasma density. Both instabilities remain active in the presence of a moderate axial field. The turbulence does not further enhance the reconnection speed, which is already quasi-Alfvénic in the 2D laminar case, but increases both ion and electron dissipation. It should, however, be kept in mind that dissipation in these high-β cases is, properly speaking, a kinetic process, since Larmor radii are large and particle orbits are

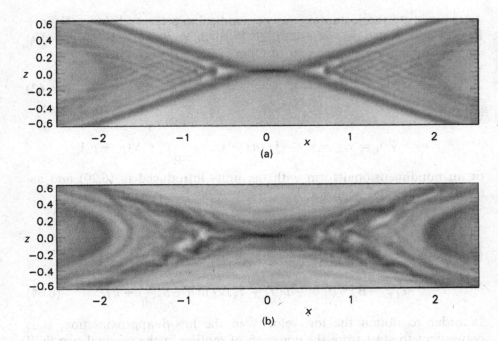

Fig. 6.9.   Contours of current density j_z in (a) a 2D simulation, (b) a fully 3D simulation (from Rogers *et al.*, 2000).

nonadiabatic in the reconnection region. As will be seen in section 7.6, finite particle excursions broaden the sheet structure, in particular for the electrons.

6.3 Low-β noncollisional reconnection

6.3.1 The four-field and three-field models

Most laboratory plasmas are embedded in a strong, nearly constant magnetic field $\mathbf{B}_0$. As already discussed for the reduced-MHD approximation in section 3.1.4, in a low-β plasma only the field $\mathbf{B}_\perp$ generated by currents along $\mathbf{B}_0$ is an independent dynamic variable. To determine the small change of the main field $\delta B_\parallel$, one uses the fact that for low β the magnetosonic wave is the fastest mode, even faster than the whistler. Let us compare typical time-scales of both waves, $\tau_A = (v_A k_\perp)^{-1}$, (3.24), and $\tau_W \lesssim (\Omega_e k_\parallel / k_\perp)^{-1}$ for $k_\perp d_e \sim 1$, (6.37):

$$\tau_A / \tau_W = \sqrt{m_i / m_e} \, k_\parallel / k_\perp \ll 1,$$

since in the vicinity of a reconnection region $k_\parallel / k_\perp$ will be very small, e.g., $k_\parallel / k_\perp \sim c / \omega_{pe} R \sim 10^{-4}$ in a typical tokamak plasma. Hence $\delta B_\parallel$ is determined by the instantaneous perpendicular pressure balance $\nabla_\perp(p + B^2/8\pi) = 0$, hence $\delta B_\parallel = -4\pi\delta p/B$. With $\mathbf{B}_0 = B\mathbf{e}_z$ and $\mathbf{B}_\perp = \mathbf{e}_z \times \nabla\psi$ and $E_z = \partial_t\psi$ the z-component of Ohm's law (6.19) becomes

$$\partial_t\psi - \mathbf{B} \cdot \nabla\phi_e = d_e^2\frac{dj}{dt} + \eta j, \qquad (6.63)$$

where

$$\mathbf{e}_z \times \nabla\phi_e = \mathbf{v}_{e\perp} = \mathbf{v}_{i\perp} - \mathbf{j}_\perp/ne = \mathbf{v}_{i\perp} - \frac{c}{nB}\mathbf{e}_z \times \nabla(p_i + p_e)$$

or, in nondimensional form with the units introduced in (6.20) and assuming isothermal behavior,

$$\mathbf{v}_{e\perp} = \mathbf{v}_{i\perp} - \frac{d_i}{nB}(T_i + T_e)\mathbf{e}_z \times \nabla n.$$

Insertion into (6.63) gives, with $\mathbf{v}_{i\perp} = \mathbf{e}_z \times \nabla\phi_i$,

$$\partial_t\psi - \mathbf{B} \cdot \nabla\phi_i = -d_i(T_i + T_e)\nabla_\parallel \ln n + d_e^2\frac{dj}{dt} + \eta j. \qquad (6.64)$$

In order to obtain the ion velocity in the low-β approximation, it is convenient to start from the equation of motion in the original two-fluid theory, (6.6). Since we are interested in time-scales of the order of the Alfvén time $\tau_A = L/v_A$, which is much longer than the ion cyclotron period Ω_i^{-1}, the inertia term is $(\Omega_i\tau_A)^{-1}$ times smaller than the $\mathbf{v}_i \times \mathbf{B}$ term. The perpendicular velocity $\mathbf{v}_{i\perp}$ is therefore calculated in the so-called drift approximation, which corresponds to an expansion in terms of $(\Omega_i\tau_A)^{-1} \propto m_i$, such that in lowest order the inertia term, the stress tensor, and Γ are neglected. Multiplication of (6.6) by $\mathbf{B}\times$ gives the zeroth-order ion drift velocity

$$\mathbf{v}_{i\perp}^{(0)} = \mathbf{v}_E + \mathbf{v}_{*i}, \qquad (6.65)$$

where

$$\mathbf{v}_E = c\frac{\mathbf{E} \times \mathbf{B}}{B^2} = \frac{c}{B}\mathbf{b} \times \nabla\varphi \qquad (6.66)$$

is called the E$\times$B drift, see (3.46), and

$$\mathbf{v}_{*i} = c\frac{\mathbf{b} \times \nabla p_i}{neB} \qquad (6.67)$$

the ion diamagnetic drift. (The terminology derives from the fact that $en(\mathbf{v}_{*i} - \mathbf{v}_{*e}) = (c/B)\mathbf{b} \times \nabla p = \mathbf{j}_\perp$, the perpendicular current density, generating the diamagnetic field reduction in the plasma.) The first-order correction is obtained by inserting $\mathbf{v}_{i\perp}^{(0)}$ into the inertia term and the stress tensor,

$$\mathbf{v}_{i\perp}^{(1)} = \frac{1}{\Omega_i}\mathbf{b} \times (\partial_t + \mathbf{v}_E \cdot \nabla)\mathbf{v}_{i\perp}^{(0)} = \mathbf{v}_p, \qquad (6.68)$$

called the (ion) polarization drift, where the contribution from the gy-roviscosity, the μ_0 part of the stress tensor, just cancels the diamagnetic advection term $\mathbf{v}_{*i} \cdot \nabla \mathbf{v}_{i\perp}$ in (6.68) (Hinton & Horton, 1971). Here we have omitted effects of magnetic curvature. (A general set of reduced two-fluid equations has been derived by Zeiler *et al.*, 1997.)

For the electrons, the zeroth-order drift velocity is similar to that of the ions,

$$\mathbf{v}_{e\perp}^{(0)} = \mathbf{v}_E + \mathbf{v}_{*e}, \qquad (6.69)$$

where the electron diamagnetic drift is

$$\mathbf{v}_{*e} = -c \frac{\mathbf{b} \times \nabla p_e}{neB}. \qquad (6.70)$$

Whereas the electron polarization drift ($\propto m_e$) can be neglected, the electron parallel velocity is important

$$v_{e\|} = v_{i\|} - j_\|/en. \qquad (6.71)$$

Both the ion and the electron velocities are constrained by the quasi-neutrality condition

$$\nabla \cdot \mathbf{j} = e\nabla \cdot n(\mathbf{v}_i - \mathbf{v}_e) = e\nabla \cdot n\mathbf{v}_p + \nabla_\| j_\| = 0. \qquad (6.72)$$

Though $\mathbf{v}_p$ is small compared with $\mathbf{v}_{i\perp}^{(0)}$, it contributes in the quasi-neutrality equation, since the E×B drifts cancel and the diamagnetic contributions vanish, $\nabla \cdot n\mathbf{v}_{*j} = 0$, assuming, as before, constant $\mathbf{B}$. Since $\mathbf{v}_p$ contains the time derivative of $\mathbf{v}_{i\perp}^{(0)}$, (6.72) actually constitutes a dynamical equation for $\mathbf{v}_{i\perp}^{(0)}$, replacing the original ion equation of motion, (6.6). To make the formalism more transparent, we introduce a further convenient approximation extending the assumption $\nabla_\| T_e = 0$ to T_e (and T_i) constant throughout. In this approximation $\mathbf{v}_{i\perp}^{(0)}$, henceforth denoted simply by $\mathbf{v}_\perp$, is incompressible, $\nabla \cdot \mathbf{v}_\perp = 0$, such that it can be written in terms of a generalized stream function ϕ,

$$\mathbf{v}_\perp = \mathbf{b} \times \nabla\phi, \quad \phi = \frac{c}{B}\left(\varphi + \frac{T_i}{e}\ln n\right), \qquad (6.73)$$

where φ is the electrostatic potential. Since the spatial variation of n is usually weak compared with that of $\mathbf{v}_p$, it is neglected in (6.72), $\nabla \cdot n\mathbf{v}_p \simeq n\nabla \cdot \mathbf{v}_p$, which is similar to the approximation in reduced MHD, (3.58). In this way and with the normalizations introduced before, (6.72) takes the form of a generalized vorticity equation for $w = \nabla^2\phi$

$$\partial_t w + \mathbf{v}_\perp \cdot \nabla w = \nabla \cdot (\mathbf{v}_{*i} \cdot \nabla\nabla\phi) + \mathbf{B} \cdot \nabla j + \mu_\perp \nabla^2 w, \qquad (6.74)$$

which, apart from the $\mathbf{v}_{*i}$ term, is identical with the vorticity equation (3.65) of reduced MHD.

The low-β form of the continuity equation is conveniently derived for the electron density $n_e = n$ using (6.71),

$$\partial_t n + \mathbf{v}_\perp \cdot \nabla n = -\nabla_\parallel(v_\parallel - d_i j), \tag{6.75}$$

since the perpendicular electron drift is incompressible $\nabla \cdot \mathbf{v}_{e\perp} = 0$ and $\mathbf{v}_{*e} \cdot \nabla n = 0$.

To derive the equation for the parallel plasma velocity $v_\parallel$ we add the parallel components of (6.5) and (6.6). As in the polarization drift the ion stress tensor contribution cancels the ion diamagnetic part in the advection term, hence

$$\partial_t v_\parallel + (\mathbf{v}_\perp - \mathbf{v}_{*i}) \cdot \nabla v_\parallel = -(T_i + T_e)\nabla_\parallel \ln n + \mu_\parallel \nabla^2 v_\parallel. \tag{6.76}$$

Equations (6.64), (6.74), (6.75), (6.76) for $\psi, w, n, v_\parallel$ constitute the four-field model developed by Hazeltine et al. (1985), which serves as a convenient tool for the study of reconnection processes in noncollisional low-β plasmas.

Since the parallel ion dynamics is only weakly excited in reconnection processes, as will be seen in section 6.4.2, one can further simplify this model by neglecting $v_\parallel$, which leaves us essentially with the three-field model discussed by Hsu et al. (1986):

$$\partial_t \psi - \mathbf{B} \cdot \nabla\phi = -d_i(T_i + T_e)\nabla_\parallel \ln n + d_e^2 dj/dt + \eta j, \tag{6.77}$$

$$\partial_t w + \mathbf{v}_\perp \cdot \nabla w = \nabla \cdot (\mathbf{v}_{*i} \cdot \nabla\nabla\phi) + \mathbf{B} \cdot \nabla j + \mu_\perp \nabla^2 w, \tag{6.78}$$

$$\partial_t n + \mathbf{v}_\perp \cdot \nabla n = d_i \nabla_\parallel j. \tag{6.79}$$

6.3.2 Linear stability theory

The linear stability theory of reconnecting modes, in particular of the internal kink mode, has received much attention in the literature. A general classification of the different collisionality regimes was developed by Drake & Lee (1977). For the simplified treatment presented here it suffices to distinguish between the collisional regime if the resistive diffusion layer exceeds the electron inertial length, $\delta_\eta = \sqrt{\eta/\omega} > d_e$, and the collisionless regime in the opposite case $\delta_\eta < d_e$. Moreover, one has to distinguish between very low plasma pressure $\beta < 2m_e/m_i$, where $\rho_s < d_e$, and relatively high pressure $\beta > 2m_e/m_i$, where $\rho_s > d_e$. Here $\rho_s = c_s/\Omega_i$ and $c_s = \sqrt{(T_i + T_e)/m_i}$ is the isothermal sound speed. Though a proper treatment for large ion gyroradius seems to require a kinetic approach, which we postpone to chapter 7, the fluid theory turns out to be at least semi-quantitatively correct.

We start by deriving the dispersion relation for linear modes in a homogeneous plasma embedded in a uniform field $\mathbf{B} = Be_z$. Linearizing the four-field equations (6.63), (6.64), (6.75), (6.76) neglecting dissipation and making the Fourier ansatz $\tilde{\psi} = \psi_1 \exp\{-i\omega t + \mathbf{k} \cdot \mathbf{x}\}$ etc.,

$$\omega(1 + d_e^2 k_\perp^2)\psi_1 + k_\parallel B\phi_1 - d_i k_\parallel (T_i + T_e)n_1 = 0, \qquad (6.80)$$

$$\omega\phi_1 + k_\parallel B\psi_1 = 0, \qquad (6.81)$$

$$\omega n_1 - k_\parallel v_{\parallel 1} - d_i k_\parallel k_\perp^2 \psi_1 = 0, \qquad (6.82)$$

$$\omega v_{\parallel 1} - (T_i + T_e)k_\parallel n_1 = 0, \qquad (6.83)$$

gives the dispersion relation

$$(\omega^2 - k_\parallel^2 c_s^2)\left[\omega^2(1 + k_\perp^2 d_e^2) - k_\parallel^2 v_A^2\right] - \omega^2 k_\parallel^2 v_A^2 k_\perp^2 \rho_s^2 = 0, \qquad (6.84)$$

where we have reintroduced the dimensional notation to elucidate the physics, $B \rightarrow v_A$, $T_i + T_e \rightarrow c_s^2$, $d_i^2(T_i + T_e) \rightarrow \rho_s^2$. For long wavelength $k_\perp \rho_s \ll 1$ the last term in (6.84) is negligible and we recover the two MHD modes, important in a low-β plasma, the sound wave $\omega_s^2 = k_\parallel^2 c_s^2$ and the shear Alfvén wave $\omega_A^2 = k_\parallel^2 v_A^2$, which are very disparate, $\omega_s^2 \sim \beta\omega_A^2$. Since in reconnection theory interest is mainly in the Alfvén branch, we neglect the c_s^2 term in (6.84), which means neglecting the ion parallel dynamics. In this case the dispersion relation is

$$\omega^2 = \frac{(1 + k_\perp^2 \rho_s^2) k_\parallel^2 v_A^2}{1 + k_\perp^2 d_e^2}. \qquad (6.85)$$

In the short-wavelength range $k_\perp \rho_s > 1$ and for $\rho_s > d_e$ the mode $\omega^2 \simeq k_\perp^2 \rho_s^2 k_\parallel^2 v_A^2$ is called the kinetic Alfvén wave. It is generated by parallel electron compressibility, the r.h.s. in (6.79), which will be shown to act as a powerful reconnection mechanism.

In the next step we include a density gradient, $n_0' = -n_0/L_n$,[*] in the x-direction and hence also pressure gradients $p_{0j}' = T_j n_0'$. Since we are primarily interested in small scales $kL_n \gg 1$, the change of the equilibrium density over a wavelength is negligible, i.e., when not differentiated, n_0 can be taken constant. The density gradient introduces the diamagnetic drift frequencies ω_{*j}, which lead to important modifications of the dispersion relation. Neglecting the sound-wave branch as indicated above, we restrict consideration to the three-field model equations (6.77)–(6.79), which are

---

[*] Since in a plasma column the radial density gradient is usually negative, it is convenient to define L_n with the minus sign.

linearized about an equilibrium state with vanishing electrostatic field $\mathbf{v}_E = 0$ and hence with the equilibrium mass flow $m_i n v_{*i}$,

$$(\omega - \omega_{*e})\psi_1 + k_\parallel B \phi_1 - k_\parallel (T_i + T_e)d_i\, n_1 = -k_\perp^2(i\eta + \omega d_e^2)\psi_1, \quad (6.86)$$

$$\omega\phi_1 + k_\parallel B\psi_1 = 0, \quad (6.87)$$

$$(\omega - \omega_{*i})n_1 + k_y n_0' \phi_1 = k_\parallel k_\perp^2 d_i \psi_1, \quad (6.88)$$

where in our isothermal model

$$\omega_{*i} = k_y \frac{T_i d_i}{B} n_0', \quad \omega_{*e} = -k_y \frac{T_e d_i}{B} n_0', \quad (6.89)$$

or in dimensional form

$$\omega_{*i} = -k_y \frac{cT_i}{eBL_n}, \quad \omega_{*e} = k_y \frac{cT_e}{eBL_n}. \quad (6.90)$$

Thus by including diamagnetic effects and resistivity the dispersion relation (6.85) is generalized

$$1 - \frac{k_\parallel^2 v_A^2}{\omega(\omega - \omega_{*i})}\left(1 + \frac{\omega}{\omega - \omega_{*e}}k_\perp^2 \rho_s^2\right) + \frac{\omega}{\omega - \omega_{*e}}k_\perp^2\left(d_e^2 + i\frac{\eta}{\omega}\right) = 0, \quad (6.91)$$

which shows that for $\omega_{*i}, \omega_{*e} \gtrsim k_\parallel v_A$ the mode frequency is strongly affected by the diamagnetic drifts.

We now proceed to investigate the general inhomogeneous case, where magnetic shear introduces a resonant surface, $k_\parallel(x_s) = 0$, which may destabilize a macroscopic MHD mode. The width δ of the resonant layer is determined by the physics in the layer. Similarly to the treatment in chapter 4, we assume that the nonideal processes are sufficiently weak, such that δ is small compared to the macroscopic scale L. As discussed in sections 4.1 and 4.2, outside this layer, where $k_\parallel$ is finite, nonideal as well as inertia effects can be neglected, since the system relaxes on the Alfvén time-scale $(k_\parallel v_A)^{-1}$, such that the perturbation is quasi-stationary obeying the linearized equilibrium equation. Following the analysis of section 4.2.1, the external solution in the vicinity of the resonant surface $x = x_s = 0$ is given by (4.70) for the displacement $\xi = \xi_x$ depending on the parameter λ_H, which characterizes the free energy of the mode. In the resonant layer we can use $\partial_x^2 \gg k_y^2$, such that $j_1 = -k_\perp^2 \psi_1 \to \psi_1''$, $w_1 \to \phi_1'' = (\omega/k)\xi''$, while $k_\parallel B \to xkB_0'$ with the notation $B_y' = B_0'$, $k_y = k$. Since in the linearized equations (6.86)–(6.88) n occurs only algebraically, it can be eliminated, which leaves us with two second-order differential equations, a generalized form of (4.65) for ψ_1 and (4.64) for ξ,

$$\omega^2 \xi'' = xk^2 B_0' \psi_1'', \tag{6.92}$$

$$(\omega - \omega_{*e}) \left(\psi_1 + \frac{\omega}{\omega - \omega_{*i}} x B_0' \xi \right) = \left(i\eta + \omega d_e^2 - \frac{x^2 (kB_0')^2 \rho_s^2}{\omega - \omega_{*i}} \right) \psi_1''. \tag{6.93}$$

The two-fluid result, (6.93), differs from the MHD result, (4.65), by (a) the presence of the drift frequencies, which in general provide a stabilizing effect; (b) the ρ_s^2 term, which arises from the combination of the Hall term and parallel electron compressibility; and (c) the electron inertia effect. The latter two terms are, in general, destabilizing. To simplify the formalism and the discussion, we first neglect diamagnetic effects, which will be considered subsequently. In the very low-β regime, where $\rho_s < d_e$, we can omit the ρ_s^2 term. Hence the only change compared with the resistive case is to substitute $\eta \to \eta + \gamma d_e^2, \gamma = -i\omega$, in the dispersion relation (4.66). While for strong ideal instability $\hat{\lambda}_H \gg 1$ we recover the ideal growth rate $\gamma = \lambda_H k B_0'$, for near-marginal ideal stability $\hat{\lambda}_H \simeq 0$ the dispersion relation reduces to $\hat{\gamma} = 1$ or

$$\gamma = (\eta + \gamma d_e^2)^{1/3} (kB_0')^{2/3}. \tag{6.94}$$

In the collisional regime the growth rate of the resistive mode, (4.68), is reproduced, whereas in the collisionless regime the growth rate becomes (Hazeltine & Strauss, 1978; Basu & Coppi, 1981)

$$\gamma = d_e k B_0'. \tag{6.95}$$

For deeply stable ideal conditions $-\hat{\lambda}_H \gg 1$, (4.69) gives the generalized tearing mode

$$\gamma = \left(\frac{\Gamma(\frac{1}{4})}{2\Gamma(\frac{3}{4})} \right)^{4/5} (\eta + \gamma d_e^2)^{3/5} (kB_0')^{2/5} |\lambda_H|^{-4/5}, \tag{6.96}$$

whence in the collisionless regime

$$\gamma = 0.22 d_e^3 k B_0' \Delta'^2 \tag{6.97}$$

using the relation $\lambda_H = -\pi/\Delta'$, (4.72). Note that the condition for the validity of collisionless behavior, $d_e^2 > \eta/\gamma$, is more stringent for smaller growth rate, such that even in high-temperature plasmas the tearing mode is often resistive, while the kink mode is collisionless.

In the case of relatively high β, $\beta > 2m_e/m_i$, where $\rho_s > d_e$, which is typical for hot tokamak plasmas, the structure of (6.93) is different from that of the resistive equation, leading to a more complicated dispersion relation than (4.66) (Aydemir, 1991, for $\lambda_H = 0$ but including the sound wave; Zakharov & Rogers, 1992). We only discuss the case $\lambda_H = 0$, where one obtains

$$\gamma^3 = \frac{2}{\pi}(kB'_0)^3 \rho_s^2 \left(\frac{\eta}{\gamma} + d_e^2\right)^{1/2}, \qquad (6.98)$$

hence for collisionless conditions

$$\gamma = \left(\frac{2}{\pi}\right)^{1/3} kB'_0 \rho_s^{2/3} d_e^{1/3}. \qquad (6.99)$$

For $\rho_s \gg d_e$ the growth rate is much larger than in the inertia-dominated regime (6.95). However, we also notice that the instability, as in the whistler-dominated high-β case, requires a genuine reconnection mechanism, either resistivity or electron inertia, since the Hall term itself does not allow reconnection. (Nonlinearly the dynamics does not depend on these dissipation processes, as we shall see.)

Conventional wisdom is that for conditions such that the ion Larmor radius $\rho_i > d_e$, finite Larmor radius (FLR) effects have to be taken into account, which would require a kinetic treatment. It is therefore noteworthy that, at least for the linear instability, these effects, which are not included in the two-fluid theory, are not important, as found by Zakharov and Rogers, who compare the growth rates obtained in the two-fluid approach with several different kinetic studies (Drake, 1978; Porcelli, 1991; Coppi & Detragiache, 1992). The results differ by less than ten percent, less than the difference between adiabatic and isothermal electrons in the fluid framework. The origin of this insensitivity to FLR effects is that, due to parallel electron compressibility, the sheet width is broadened to ρ_s, which exceeds the ion Larmor radius ρ_i, $\rho_s = \sqrt{2}\rho_i$ for $T_e = T_i$.

Let us now include diamagnetic drift effects omitted in the preceding discussion, which provide an efficient damping by phase mixing of ψ and ϕ, if $\omega_* > \gamma_0$, where γ_0 is the growth rate for $\omega_* = 0$, (6.99). A simple estimate gives

$$\frac{\omega_*}{\gamma_0} \sim \frac{L_s}{L_n} \beta^{1/2}, \qquad (6.100)$$

using the growth rate, (6.99), but neglecting a factor $(\rho_s/d_e)^{1/3} \sim (\beta m_e/m_i)^{1/6}$. Here L_s is the magnetic shear length defined by $B'_0 = B_0/L_s = B_\theta \hat{s}/r$ in a cylindrical configuration, (4.78). Hence ω_* has to be included for sufficiently high β. We only give the generalized dispersion relation for marginal ideal stability $\lambda_H = 0$:

$$\omega(\omega - \omega_{*e})(\omega - \omega_{*i}) = -k^2 B'^2_0 (i\eta + \omega d_e^2) \quad \text{for} \quad d_e > \rho_s, \qquad (6.101)$$

$$\omega(\omega - \omega_{*e})^3(\omega - \omega_{*i})^3 = -\frac{4}{\pi^2}(kB'_0)^6 \rho_s^4 (i\eta + \omega d_e^2) \quad \text{for} \quad \rho_s > d_e. \qquad (6.102)$$

Solving these equations shows that $\omega_* \gtrsim \gamma_0$ stabilizes the internal kink mode, which has been invoked as the mechanism for suppression of the sawtooth oscillation in tokamak discharges (Levinton *et al.*, 1994). We shall come back to this point in section 6.5.

6.4 Nonlinear noncollisional kink mode

The linear theory of the kink mode outlined in the preceding section indicates that, in the absence of a dissipative process in Ohm's law, electron inertia may act as a genuine reconnection mechanism. We consider now the nonlinear evolution, which will show a faster behavior than in the resistive case discussed in section 4.3.2. The nonlinear characteristics play a crucial role in understanding the rapid collapse in the sawtooth oscillation.

6.4.1 Electron inertia-dominated reconnection

Consider first the very low-β regime $\beta < m_e/m_i$, where $\rho_s < d_e$, such that the effect of the pressure term in Ohm's law, (6.64), can be neglected. Different from the electron-dominated reconnection dynamics in the high-β whistler regime, the perpendicular motions of ions and electrons are now tightly coupled. The system is described by the vorticity equation (6.78) and (6.77), which in the low-β limit can be written in terms of $F = \psi - d_e^2 j$, the canonical electron momentum in the direction of the ignorable coordinate, (7.7),

$$\partial_t w + \mathbf{v} \cdot \nabla w = \mathbf{B} \cdot \nabla j, \tag{6.103}$$

$$\partial_t F + \mathbf{v} \cdot \nabla F = 0. \tag{6.104}$$

The structure of these equations is similar to that of 2D EMHD for $kd_e > 1$, but here the l.h.s. of the vorticity equation is the ion inertia term ($\propto d_i^2$ in the normalization chosen in section 6.2), and $\mathbf{v}$ is the ion velocity. Hence the ions follow the electrons into the narrow electron layer d_e and are accelerated along the sheet, basically as in a resistive current sheet with d_e replacing δ_η. The vorticity equation gives again the relation that the outflow velocity v_0, the plasma velocity at the edge of the current sheet, equals the Alfvén speed in the magnetic field immediately in front of the sheet, $v_0 \simeq v_A = B_0$, (3.106), while from the continuity equation $\nabla \cdot \mathbf{v} = 0$ we obtain

$$u_0 \Delta \sim v_0 d_e, \tag{6.105}$$

such that

$$\frac{u_0}{v_A} \equiv M_0 \sim \frac{d_e}{\Delta}, \tag{6.106}$$

see Wesson (1990). As in the resistive kink mode, reconnection occurs in
a macroscopic current sheet $\Delta \sim r_1$. Proceeding as in section 4.3.2, we
combine the continuity equation (4.97) with the expression (4.98) for the
helical field at the position of the sheet and use the relation $\dot{w}_I \simeq 2\dot{\xi} = 2u_0$,
which gives (ignoring factors of order unity)

$$\dot{w}_I \simeq d_e(\psi_0''/r_1)\, w_I = \gamma_0 w_I, \qquad\qquad (6.107)$$

where γ_0 is the linear growth rate, (6.95). Since the layer width d_e is
constant, different from the resistive case where δ_η shrinks as the inflow
speed increases, (4.99), the evolution is exponential,

$$w_I \simeq w_{I0} \exp\{\gamma_0 t\}, \qquad\qquad (6.108)$$

i.e., the behavior of the linear instability continues into the nonlinear
regime, which is faster than the algebraic growth, (4.101), $w_I \propto t^2$, of the
resistive kink mode.

In the preceding discussion quasi-stationary reconnection is implied, in
particular $E + \mathbf{v} \cdot \nabla F = 0$ from (6.104). However, this equation cannot be
valid at the X-point, where the second term vanishes; an argument which is
sometimes used against electron inertia as a true reconnection mechanism.
In fact, (6.104) indicates, that the dynamics at the X-point is governed by
electron acceleration along the z-direction. Reconnection occurs, but not
in a quasi-stationary way. The process has been analyzed by Ottaviani &
Porcelli (1993). The authors consider the double tearing mode in a plane
configuration, specifically the periodic equilibrium $\psi_0 = \cos x$, similar to
(4.74), but the results apply, *mutatis mutandis*, also to the cylindrical kink
mode, the dynamics being very similar, as discussed in sections 4.2 and
4.3. The starting point is the assumption that the outflow is confined to a
layer of constant width d_e and macroscopic length $\Delta \sim k^{-1} \gg d_e$, where
k is the wavenumber of the mode, hence one can make the simple ansatz
for the stream function

$$\phi(x, y, t) = -v_0(t)g(x)k^{-1}\sin ky \qquad\qquad (6.109)$$

with

$$g(x) = \begin{cases} 1, & x < -d_e \\ x/d_e, & -d_e < x < 0 \end{cases} \qquad\qquad (6.110)$$

and $g(-x) = -g(x)$. This corresponds to a uniform flow in the x-direction
into the flow layer. With the velocity thus prescribed, we can determine
the flux function ψ from the conservation of F along the orbit of a
fluid element $F(x, y, t) = F(x_0, -\infty)$, where $x_0(x, y, t) = x - \xi(x, y, t)$ is the
initial position of the fluid element located at (x, y) at time t and ξ is the
displacement along the x-direction, defined by $d\xi/dt = v_x$, $\xi(t = -\infty) = 0$.

Consider the motion along the line $y = 0$, where $v_y = 0$ and $v_x = -\partial_y \phi = v_0(t)g(x)$, $v_x > 0$ for $x < 0$. The following integral is independent of x:

$$\int_{-\infty}^{t} \frac{v_x}{g} dt' = \int_{x_0}^{x} \frac{dx'}{g(x')} = \xi_0(t), \tag{6.111}$$

which equals the absolute value of the uniform shift in the region outside the layer, as seen by choosing x_0, x in this region, where $|g| = 1$. The nonlinear regime is characterized by $\xi_0 > d_e$. Different from the flow layer d_e, the current layer width $\delta(t)$ shrinks with time which is defined by

$$\int_{d_e}^{\delta} \frac{dx'}{g(x')} = -d_e \ln \frac{\delta}{d_e} = \xi_0,$$

whence

$$\delta(t) = d_e \exp\{-\xi_0(t)/d_e\}. \tag{6.112}$$

This implies that the initially shallow $F(x_0)$ profile inside d_e is squeezed into a narrow sublayer of width δ, forming much stronger gradients. But this only affects the current density, whereas the flux function ψ remains smooth, varying only on the scale d_e, as can be seen by integrating the equation

$$d_e^2 \nabla^2 \psi - \psi \simeq d_e^2 \partial_x^2 \psi - \psi = -F$$

using Green's function $-(1/2d_e)\exp\{-|x - x'|/d_e\}$,

$$\psi(x, y, t) \simeq \frac{1}{2d_e} \int_{-\infty}^{\infty} e^{-|x-x'|/d_e} F(x', y, t)dx', \tag{6.113}$$

which shows that ψ is smoothed out over the flow layer width d_e. Hence in the resonant layer $|x| \lesssim d_e$ one has

$$F(x_0) \simeq F_0 - \tfrac{1}{2}x_0^2 F_0'' = F_0 - \tfrac{1}{2}(x - \xi_0)^2 F_0'', \tag{6.114}$$

$$F(x_0) \simeq \psi(x_0) \simeq \psi_0 - \tfrac{1}{2}(x - \xi_0)^2 \psi_0'', \tag{6.115}$$

since in the nonlinear regime $\xi \simeq \xi_0 > d_e$ and $j(x_0)$ is smooth, such that $j'' d_e^2 \ll \psi_0''$.

Along the layer, $F \simeq const$ (see also fig. 6.2(b)), hence the variation of the current density along the layer is

$$\delta j = \delta \psi / d_e^2 \simeq -\tfrac{1}{2}(\xi_0/d_e)^2 \psi_0''. \tag{6.116}$$

To derive an equation for ξ_0 we make use of the vorticity equation, which we integrate over a quadrant of the layer, as indicated in fig. 6.1:

$$\frac{d}{dt} \int w d^2x = \int \mathbf{B} \cdot \nabla j d^2x. \tag{6.117}$$

Here we are neglecting the vorticity advection term, which is small compared with $\partial_t w$ in this strongly nonstationary phase. The l.h.s. of (6.117) becomes approximately

$$\frac{d}{dt} \int \nabla^2 \phi d^2 x = \frac{d}{dt} \oint \partial_n \phi ds \simeq \frac{d}{dt} \int_0^\Delta \partial_x \phi dy \simeq \frac{1}{d_e k^2} \ddot{\xi}_0 \qquad (6.118)$$

using (6.109) and (6.110) and $k\Delta \sim 1$. Similarly, the r.h.s. of (6.117) becomes

$$\int \mathbf{B} \cdot \nabla j d^2 x = \oint j \partial_s \psi ds \simeq \frac{1}{2} \frac{(\delta \psi)^2}{d_e^2} \simeq \frac{1}{8} \xi_0^4 \frac{\psi_0''^2}{d_e^2}, \qquad (6.119)$$

inserting (6.114) and (6.116). Combining the two expressions gives the desired equation for the nonlinear evolution of ξ_0

$$d^2 \hat{\xi}_0 / d\hat{t}^2 = c \hat{\xi}_0^4, \qquad (6.120)$$

where $\hat{\xi}_0 = \xi_0/d_e$, $\hat{t} = t\gamma_0$, $\gamma_0 = d_e k \psi_0''$ (cf. (6.95)), and c is a numerical factor of order unity. Equation (6.120) has the similarity solution

$$\hat{\xi}_0 = \frac{a}{(\hat{t}_0 - \hat{t})^{2/3}} \qquad (6.121)$$

indicating explosive (= faster than exponential) growth with a singularity at the finite time $t_0 = O(\gamma_0^{-1})$.

The solution (6.121) can, however, only be valid in the very early nonlinear stage, where $\xi_0 \gtrsim d_e$, since further increase of ξ_0 would lead to extremely narrow widths δ given by (6.112), which should easily be smoothed out by residual resistivity or, more importantly, by electron viscosity. Hence the substructure in the current layer will not be important in the main nonlinear phase, much in the same way it has been neglected in the EMHD analysis in section 6.2. Instead of the explosive behavior, the dynamics is governed by the quasi-stationary reconnection process leading to exponential island growth, (6.108). Figure 6.10 gives the current density from a numerical simulation of the electron inertia-driven kink mode with the parameters $d_e = 0.01$, $\eta = 10^{-6}$. The value of the resistivity is sufficiently small such that $d_e \gg \delta_\eta$, i.e., the system is in the collisionless regime, where the growth rate is given by (6.95) and the sheet width is d_e. The current-density distribution resembles that in the resistive kink mode, see fig. 4.14(a).

6.4.2 Kinetic Alfvén-wave-mediated reconnection

In most plasmas of interest β is sufficiently high, $\beta > m_e/m_i$, such that the pressure term in (6.64) is more important than electron inertia, $\rho_s > d_e$. In this case the structure of the reconnection region is quite different

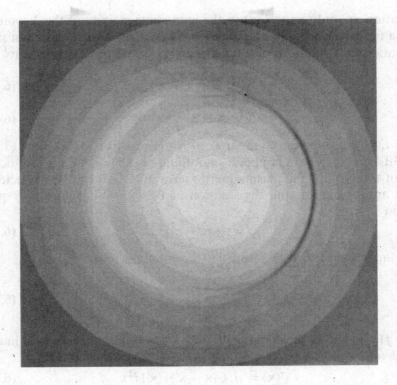

Fig. 6.10. Grayscale plot of simulation of the current density in the nonlinear development of the electron inertia-driven kink mode showing an extended current sheet of width d_e.

from the current sheet of macroscopic length characteristic of the low-β electron inertia-dominated dynamics. Since the pressure term requires a perpendicular field component, $\mathbf{B} \cdot \nabla n \simeq B_x \partial_x n$, it is more effective, leading to faster reconnection, if B_x is large, i.e., if ψ forms an X-point.

To discuss the structure of the reconnection region it is useful to simplify (6.77) by neglecting the ion temperature and assuming that the equilibrium pressure gradient are small, such that diamagnetic effects can be neglected. Writing $\nabla_\| j = B_z^{-1} \mathbf{B} \cdot \nabla j$ makes clear that in this case n obeys the same equation as $(d_i/B_z)w$. Since in the evolution of an instability the memory of the initial perturbation is rapidly lost, we can substitute n in Ohm's law, which leaves us with only two equations for ψ and w,

$$\partial_t \psi - \mathbf{B} \cdot \nabla \phi = -\rho_s^2 \mathbf{B} \cdot \nabla w + d_e^2 dj/dt, \qquad (6.122)$$

$$\partial_t w + \mathbf{v} \cdot \nabla w = \mathbf{B} \cdot \nabla j, \qquad (6.123)$$

since $d_i^2 T_e / B_z^2 = \rho_s^2$ in the units introduced in section 6.1. The ρ_s^2 effect gives rise to the kinetic Alfvén wave, (6.85). (Electron inertia and residual dissipation are only important in the immediate vicinity of the X-point to allow a finite reconnection rate $\partial_t \psi$, since here the pressure term vanishes.) For stationary conditions and $d_e \ll \rho_s$, (6.122) and (6.123) reduce to

$$E - \mathbf{B} \cdot \nabla(\phi - \rho_s^2 \nabla^2 \phi) = 0, \tag{6.124}$$

$$\mathbf{v} \cdot \nabla \nabla^2 \phi - \mathbf{B} \cdot \nabla j = 0, \tag{6.125}$$

where $E = \partial_t \psi = const$. Compared with the resistive MHD case, (3.140) and (3.141), the ρ_s^2 term implies a smoothing of the flow over a distance ρ_s. For not too large E the plasma inertia term in (6.125) can be neglected to lowest approximation and the solution of $\mathbf{B} \cdot \nabla j = 0$ around the X-point is again

$$\psi = \tfrac{1}{2}(x^2 - a^2 y^2), \tag{6.126}$$

while the stream function ϕ satisfies the equation

$$\phi - \rho_s^2 \nabla^2 \phi = \frac{E}{2} \ln \left| \frac{x + ay}{x - ay} \right| \equiv f(x, y), \tag{6.127}$$

where $f(x, y)$ solves the equation $E - \mathbf{B} \cdot \nabla f = 0$. Equation (6.127) has the formal solution

$$\phi(\mathbf{x}) = \int G(\mathbf{x} - \mathbf{x}') f(\mathbf{x}') d^2 x' \tag{6.128}$$

with Green's function

$$G(\mathbf{x}) = \int d^2 k \frac{e^{i\mathbf{k} \cdot \mathbf{x}}}{1 + k^2 \rho_s^2}. \tag{6.129}$$

A numerical solution of (6.127) is plotted in fig. 6.11. For finite ρ_s the singularity of ϕ on the separatrix $x = \pm ay$ is eliminated, and there is only a weak logarithmic singularity of $w = \nabla^2 \phi$, which is smoothed by residual resistivity or electron viscosity. Hence the effect of plasma inertia, the first term in (6.125), remains small, which allows the X-point configuration, (6.126), to persist also for finite reconnection rate, in contrast to the resistive case, where the inertia term enforces the formation of a macroscopic current sheet. It is worth mentioning that for $\rho_s = 0$ the electron inertia term in (6.122) modifies only the current profile, but does not smooth the singular flow pattern. Hence in this case the plasma inertia term in (6.125) is not negligible and an X-point configuration is not possible. Instead a macroscopic current sheet is formed, see the preceding section, much in the same way as in the resistive case.

Numerical simulation studies of pressure-mediated reconnection have been performed for the coalescence of two flux bundles (Kleva *et al.*, 1995),

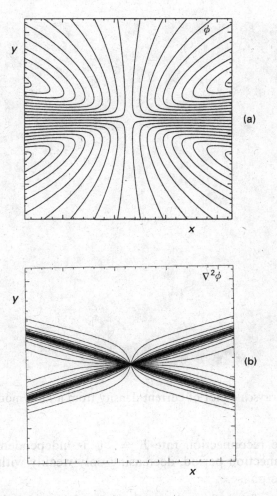

Fig. 6.11. Solution of (6.127): (a) flow pattern ϕ; (b) vorticity $\nabla^2\phi$.

the cylindrical kink mode (Aydemir, 1992), and the double tearing mode, which differ only by the geometry of the system, the reconnection dynamics being very similar. Simulations confirm that the magnetic configuration exhibits an X-point with a downstream cone of finite angle. In the limit $d_e^2 \ll \rho_s^2$, vorticity and current density form shock-like structures along the magnetic separatrix just as in the solution of (6.124), see fig. 6.11(b). Figure 6.12 illustrates a simulation run with $d_e = 0.01$, $\rho_s = 0.02$ and $d_i = 0.2$, i.e., $m_i/m_e = 400$, which demonstrates the change of the current-density distribution compared with the corresponding electron inertia-driven case with $\rho_s = 0$ shown in fig. 6.10. Reconnection now occurs in a micro-current sheet. Within the range of ρ_s/L values amenable in the

Fig. 6.12.   Grayscale plot of current density from a kink mode simulation.

simulations the reconnection rate $E = \partial_t \psi$ is independent of ρ_s during the early reconnection period, decreasing only weakly with ρ_s during the fastest dynamic phase.

It is found that for sufficiently small values of the resistivity the structure of the reconnection region and the reconnection rate are independent of η. When d_e is increased, the length of the small central current sheet increases, but the same basic configuration persists even for $d_e \gtrsim \rho_s$. Only if $d_e \gg \rho_s$ does electron inertia dominate and a macroscopic current sheet is formed.

The fast reconnection dynamics in the pressure-dominated regime is reflected in the nonlinear evolution of the kink mode. Whereas in the electron inertia regime the kink mode effectively continues to grow exponentially in the nonlinear phase, the pressure-dominated mode exhibits an explosive time behavior, as was first observed numerically by Aydemir (1992). This behavior can be interpreted by a simple model. Since reconnection does not depend on the smallness parameters of the system ρ_s, d_e, η, i.e., it is essentially Alfvénic, the angle α of the outflow cone is determined purely geometrically,

$$r_1\alpha \simeq \xi, \tag{6.130}$$

where ξ is the helical shift of the of the main magnetic axis, and the island width is $w_I \simeq 2\xi$, see section 4.3.2. In addition, we use the continuity equation

$$u_0 r_1 \simeq v_0\delta(\alpha) \simeq v_0 r_1\alpha. \tag{6.131}$$

As in the resistive MHD case the outflow velocity is determined by pressure balance and hence equals the upstream Alfvén speed

$$v_0 = v_A \simeq |\psi_0''|\xi/2, \tag{6.132}$$

while the inflow velocity is given by the geometry of the system $u_0 = \dot{\xi}/2$. Combining these relations we find

$$\frac{d\hat{\xi}}{d\tau} \simeq \hat{\xi}^2, \tag{6.133}$$

with $\hat{\xi} = \xi/r_1$ and $\tau = t|\psi_0''| = t/\tau_A$. Hence the helical shift grows explosively

$$\hat{\xi} = \frac{1}{\tau_0 - \tau}. \tag{6.134}$$

Note that we did not make use of Ohm's law. The only physical assumption regards the pressure balance in the layer, i.e., $v_0 = v_A$, which is justified for processes slow compared to the compressional Alfvén time-scale. The finite-time singularity in (6.134) should not be taken literally, since this self-similar solution is only valid in the early nonlinear phase of the kink mode evolution, much as the self-similar solutions (6.108) and (4.101). At large amplitude $\xi \sim r_1$ the change of the radius of the plasma core and of the helical flux distribution must be considered, which leads to saturation and final decay of the growth rate.

6.4.3 Influence of diamagnetic effects

In the previous subsection the equilibrium pressure gradient has been neglected, thus omitting diamagnetic rotation. In fact, the electron pressure term in the ψ-equation contains two distinctly different effects, which can clearly be seen when linearizing the term

$$\delta(\mathbf{B} \cdot \nabla p_e) = \mathbf{B}_0 \cdot \nabla\delta p_e + \delta B_r p_{e0}'. \tag{6.135}$$

The first term on the r.h.s., which gives rise to the kinetic Alfvén wave, is responsible for the nonlinear acceleration of the growth, (6.134), discussed above. The second term introduces the diamagnetic effect, which implies that a perturbation propagates in the poloidal direction with the drift velocity $v_{*e} = cT_e/(eB_zL_n) = \rho_s c_s/L_n$, which constitutes an efficient damping process by phase mixing ψ and ϕ, as discussed in section 6.3.2, stabilizing the linear kink mode if $\omega_* > \gamma_0$. For $\omega_* \sim \gamma_0$ the kink mode is still unstable, but is found to saturate at a finite island width. Let us therefore study in more detail the nonlinear evolution in this regime. Because of the n'_0 term in the continuity equation (6.79) we can no longer replace n by $(d_i/B_z)w$, but have to regard n and w as independent dynamic variables, i.e., go back to the full three-field equations (6.77)–(6.79), which we write in a slightly different form, neglecting ion temperature and introducing electron viscosity instead of resistivity:

$$\partial_t F + \mathbf{v} \cdot \nabla F = -\alpha_i \beta_p \mathbf{B} \cdot \nabla n - \mu_e \nabla^2 j, \qquad (6.136)$$

$$\partial_t w + \mathbf{v} \cdot \nabla w = \mathbf{B} \cdot \nabla j + \mu \nabla^2 w, \qquad (6.137)$$

$$\partial_t n + \mathbf{v} \cdot \nabla n = \alpha_i \mathbf{B} \cdot \nabla j + D \nabla^2 n, \qquad (6.138)$$

where $F = \psi - d_e^2 j$, $\alpha_i = d_i B_0/B_z = c/\omega_{pi}R$, since $q = r_1 B_z/RB_0 = 1$, $\beta_p = 4\pi n_0 T_e/B_0^2$, R = major torus radius, B_0 = poloidal field at $r = r_1$. We discuss a numerical simulation study with the following parameters: $d_e = 10^{-2}$, $\alpha_i = 5 \times 10^{-2}$, $\mu_e = 10^{-9}$, $\mu = D = 10^{-6}$; β_p and L_n are varied, where L_n is the gradient scale of a parabolic density profile $n_0(r) = 1 - (r/L_n)^2$. These parameter values are chosen more for numerical convenience than for a realistic tokamak simulation. We only want to demonstrate the qualitative behavior.

If L_n is sufficiently large, $L_n > 1$, i.e., for weak diamagnetic effects, the pressure term is found to be destabilizing nonlinearly. Figure 6.13(a) gives the growth rate defined by the kinetic energy, $\gamma(t) = (dE^K/dt)/2E^K$, for $L_n = 10$ and several values of β_p. While according to (6.99) the linear growth rate γ_0 changes only weakly with β_p, $\gamma_0 \propto \beta_p^{1/3}$, the most conspicuous feature is the accelerated growth in the nonlinear phase, first noted by Aydemir (1992), which corresponds to the explosive solution (6.134). We have therefore plotted γ/γ_0 as function of $\gamma_0 t$, which shows that the nonlinear enhancement of the growth rate increases with β_p. If, on the other hand, L_n is small, $L_n < 1$, where the diamagnetic effects dominate, increasing β_p reduces the nonlinear growth rate, fig. 6.13(b), which for sufficiently high β_p results in finite-amplitude saturation, $\beta_p \gtrsim 2$ in this case. Saturation occurs when the helical shift ξ of the plasma core no longer points toward the X-point but is rotated out of phase, such that reconnection is no longer driven.

From a series of simulation runs one obtains a semi-quantitative condition for saturation:

$$\omega_* > \gamma_{max}, \qquad (6.139)$$

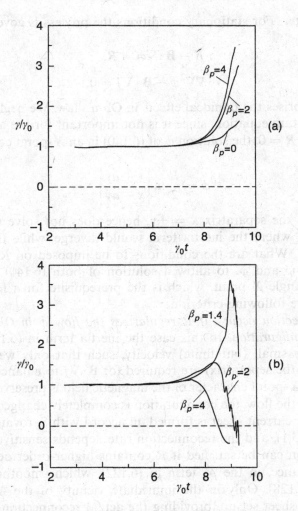

Fig. 6.13. Normalized growth rate of the nonlinear kink mode as function of normalized time: (a) $L_n = 10$; (b) $L_n < 1$ (from Biskamp & Sato, 1997).

where ω_* is the linear mode frequency and $\gamma_{\max}$ the maximum (nonlinear) growth rate. For $\omega_* < \gamma_{\max}$ full reconnection occurs, even if ω_* exceeds the linear growth rate. This condition generalizes the linear stability criterion, which relates ω_* to γ_0.

6.4.4 Criterion for fast reconnection

In this subsection we want to generalize the discussion following (6.124) and (6.125), formulating a criterion to decide whether reconnection is fast, i.e., basically Alfvénic, or depends more or less strongly on the respective small parameter. For stationary conditions the process is governed by the equations

$$E - \mathbf{B} \cdot \nabla \phi = R, \tag{6.140}$$

$$\mathbf{v} \cdot \nabla \nabla^2 \phi - \mathbf{B} \cdot \nabla j = 0, \tag{6.141}$$

where R comprises the nonideal effects in Ohm's law. We neglect viscosity in the momentum equation, since it is not important for the argument. In the ideal limit $R = 0$, the solution ϕ of (6.140) in an X-point configuration, (6.126),

$$\phi = \frac{E}{2} \ln \left| \frac{x + ay}{x - ay} \right| \tag{6.142}$$

is singular on the separatrix $x = \pm y$, hence does not solve the momentum equation, where the inertia term would diverge, while the magnetic term vanishes. What are the conditions to be imposed on R, which is a functional of ψ and ϕ, to allow a solution of both (6.140) and (6.141) with a finite-angle X-point, which is the prerequisite for a fast process? We suggest the following criterion:·

Fast reconnection occurs, if R regularizes the flow ϕ in Ohm's law for an X-point configuration. In this case the inertia term in (6.141) is small for sufficiently small (but finite) velocity, such that only weak currents (essentially at the separatrix) are required for $\mathbf{B} \cdot \nabla j$ to balance the inertia term, and the X-point character of the magnetic field is preserved. If R does not regularize the flow, the configuration is completely changed. In general a macroscopic current sheet is formed, in accord with Syrovatskii's theory (see section 3.3.1), and the reconnection rate depends sensitively on R.

The criterion can be satisfied if R contains higher-order derivatives of ϕ as, for instance, in the ρ_s^2 term in (6.122), which smoothes the basic singularity, (6.128). Only in the immediate vicinity of the X-point is a micro-current sheet set up, providing the actual reconnection mechanism due to residual resistivity or electron inertia. The length of this sheet remains small, shrinking with η or d_e.

By contrast, the electron inertia term $R = d_e^2 \mathbf{v} \cdot \nabla j$ alone does not affect the singularity of ϕ, which arises from the inversion of the operator $\mathbf{B} \cdot \nabla$, nor does a purely resistive $R = \eta j$. In both cases extended current sheets are formed as discussed in detail in sections 6.4.1 and 3.3.3, respectively. It is interesting to consider the relative importance of the ρ_s^2 and the d_e^2 terms in (6.122). The reconnection process is fast, dominated by the ρ_s^2 term even for $\rho_s < d_e$. Only if $\rho_s \ll d_e$ does the d_e^2 term control the dynamics with a macroscopic current sheet.

In the high-β case, where the ions are immobile at scales $< d_i$, the ion or plasma stream function ϕ in Ohm's law is replaced by ϕ_e in this region. Here electron inertia (section 6.2.2) and non-gyrotropic terms in the electron pressure tensor (section 7.6.2) have a regularizing effect on ϕ_e similar to that exerted by the ρ_s^2 term on ϕ in the low-β case, which allows an X-point configuration to be established.

The X-point solution, (6.126), (6.142), or the regularized flow, (6.128), are symmetric with respect to inflow and outflow, the angle of the downstream cone being a free parameter. This is true for sufficiently low velocity, where the inertia term in the momentum equation is unimportant. At higher velocities the term introduces an inflow–outflow asymmetry. Inertia limits the outflow velocity to the Alfvén speed, as shown in Petschek's model, the basic configuration in the presence of a fast reconnection mechanism, which now couples the downstream angle to the inflow speed.

6.5 Sawtooth oscillation in tokamak plasmas

The sawtooth oscillation is the classical paradigm of a reconnection process in a tokamak plasma. In my previous book (Biskamp, 1993a) I have given an extended review of the historical developments during the 20 years following the first observations by von Goeler *et al.* (1974) and the first theoretical model by Kadomtsev (1975). Here, I will therefore concentrate mainly on the results obtained during the 1990s, which have contributed essential new pieces to the puzzle posed by this intriguing phenomenon.

6.5.1 Basic experimental observations

The sawtooth oscillation is a relaxation process in a tokamak plasma corresponding to a periodic flattening of the temperature and the density profiles in the core of the discharge column. The conventional diagnostic tool for studying this phenomenon is an array of soft X-ray (SX) diodes measuring the emission along different chords across the plasma, as indicated schematically in fig. 6.14. The signal along a chord, passing through the central part of the plasma column is modulated by a periodic oscillation consisting of a slowly rising part, the rise phase, followed by a rapid drop, the sawtooth collapse or crash. The signal has the shape of a sawtooth, whence the name. For a sufficiently off-centered chord, the signal is inverted, with a sudden rise coinciding with the sawtooth collapse and a gradual subsequent decay. The cross-over from one type of signal to the other occurs at the radius where the safety factor q passes through unity, which is called the inversion radius.

Sawtooth oscillations, or simply sawteeth, occur in practically all tokamak devices under various operational conditions. While the standard diagnostics is still via SX emission, which allows the highest time resolution but provides only an indirect measure of the electron temperature,

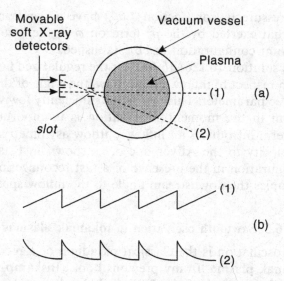

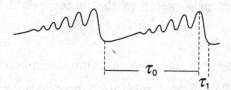

Fig. 6.14. Observation of sawtooth oscillations by soft-X-ray diagnostics. Schematic drawings of (a) the experimental set-up, (b) the SX signals along chords (1) and (2).

Fig. 6.15. Schematic drawing of the $m = n = 1$ precursor oscillation superimposed on the $m = n = 0$ sawtooth signal. τ_0 is the sawtooth period, τ_1 the collapse time-scale.

these observations are now supplemented by direct measurements of (i) the electron temperature using electron cyclotron emission (ECE), and (ii) the electron density using microwave interferometry.

An indication of the basic mechanism responsible for the sawtooth relaxation is obtained from a precursor oscillation, which in many cases is observed to precede the collapse, fig. 6.15. While the main sawtooth signal is symmetric with an $m = n = 0$ mode signature, the precursor corresponds to a helical shift of the central part of the plasma column, i.e., an $m = n = 1$ mode, which gives rise to the observed oscillating signal owing to diamagnetic and plasma rotation.

In contrast to the simple regular sawteeth characteristic of earlier small tokamak experiments with minor radius $a \sim 10$ cm, the saw-

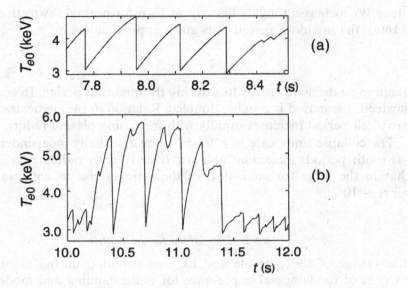

Fig. 6.16. ECE measurements of the central electron temperature for different types of sawtooth oscillations obtained in the JET tokamak. (a) Normal sawteeth from an ohmic discharge (from Edwards *et al.*, 1986); (b) so-called giant sawteeth from an ion-cyclotron-heated plasma (after Campbell *et al.*, 1987).

tooth phenomenon exhibits a more complex behavior in large-diameter tokamak plasmas with minor radius $a \sim 10^2$ cm. Depending on the operational mode – pure ohmic, neutral beam, or radio-frequency (mainly ion-cyclotron resonance) heating – the period, amplitude and shape of the SX or ECE signals may vary considerably, examples being given in fig. 6.16. Sawteeth often show a compound structure, where roughly half-way through the rise phase a minor or partial collapse occurs, as seen in fig. 6.16(b), which affects mainly the region close to the $q = 1$ radius, while the central energy loss is small.

The sawtooth period is directly related to the confinement time τ_{E1} of the energy W_1 inside the $q = 1$ radius defined by

$$\frac{W_1}{\tau_{E1}} = P, \qquad (6.143)$$

where P is the heating power deposited inside this radius. The sawtooth collapse results in a certain fraction ΔW_1 of the energy being ejected. This generates an imbalance between heating power and energy transport, from which the plasma recovers during the rise phase,

$$\frac{dW_1}{dt} \simeq P - \frac{W_1 - \Delta W_1}{\tau_{E1}} = \frac{\Delta W_1}{\tau_{E1}}. \qquad (6.144)$$

Since W_1 increases roughly linearly, at least for normal sawteeth as in fig. 6.16(a), the sawtooth period τ_0 is given approximately by

$$\tau_0 \simeq \tau_{E1} \propto r_1^2 \tag{6.145}$$

assuming a diffusive energy loss during the quiescent period. In compound sawteeth the period is roughly doubled. Relation (6.145) indicates that the sawtooth period increases rapidly with increasing plasma radius.

The collapse time-scale τ_1 is found to be essentially independent of the sawtooth period, increasing at most linearly with radius $\tau_1 \propto r_1$, such that in the large tokamak devices the ratio of rise to collapse time is $\tau_0/\tau_1 \sim 10^3$.

6.5.2 The safety factor profile

Knowledge of the q-profile and its time variation during sawtooth activity is of fundamental importance for understanding and modeling the sawtooth phenomenon. Since the signature of the precursor as well as the fast dynamics during the collapse, which will be discussed below, clearly point to the kink mode as the basic agent responsible for the collapse, the central value $q_0 = q(0)$ should be smaller than unity, giving rise to a resonant surface $q(r_1) = 1$ at some finite radius r_1. On the other hand, the resistive kink mode itself leads to complete reconnection of the internal helical flux, which relaxes the plasma into a state with $q \geq 1$ as outlined in section 4.3.3. Since in typical hot and large-diameter plasmas the resistive skin time is much longer than the sawtooth period, one could expect that after the first sawtooth the q-profile remains essentially flat and close to unity inside the inversion radius.

In earlier tokamak experiments the current profile and hence the q-profile were inferred from the condition of resistive equilibrium $\eta j = const$ using the measured electron temperature profile and a reasonable guess of the impurity ion distribution, which resulted in rather inaccurate q-values in the crucial center part. During the 1990s, however, several diagnostic methods have been developed to measure the poloidal magnetic field directly using either the Faraday rotation effect (Soltwisch et al., 1987) or the motional Stark effect (see, e.g., Levinton et al., 1993), which allow determination of the q-profile with an error of less than 5%. Averaging over many equivalent sawteeth, making use of the near-periodicity of the sawtooth oscillation, yields accurate information about the time behavior of q_0, the central q value, see fig. 6.17. The surprising and now firmly established result is that q_0 remains distinctly below unity, typically jumping up from 0.7 to 0.8 during the collapse and slowly decaying back to 0.7 during the sawtooth rise phase. The q-profile

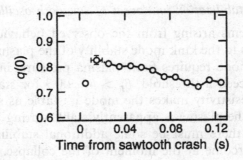

Fig. 6.17. Average time behavior of the q-value on the magnetic axis during a sawtooth period (from Levinton *et al.*, 1993).

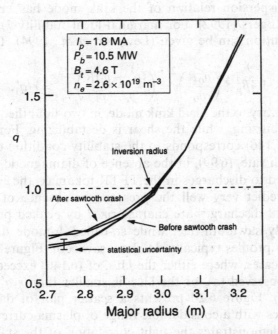

Fig. 6.18. The q-profile measured before and after the sawtooth collapse for two separate sawteeth (after Levinton *et al.*, 1993).

before and after the collapse is shown in fig. 6.18, which differs fundamentally from the behavior of the resistive kink mode, see fig. 4.17(b). These experimental findings impose severe constraints on the theoretical interpretation of the sawtooth collapse dynamics, as we will see later.

6.5.3 Stabilization and onset of sawtooth oscillations

One of the problems arising from the observed behavior of the central safety factor refers to the kink mode stability of the plasma column. While the ideal MHD mode requires finite plasma pressure, in particular must the poloidal β exceed a threshold, $\beta_p > \beta_c \simeq 0.3$ for stability (Bussac et al., 1975), finite resistivity makes the mode unstable as soon as q_0 drops below unity. But the plasma is apparently stable during the the sawtooth rise phase, hence there must be some additional stabilizing mechanism, which is only overcome at the moment of the collapse. Moreover, there are plasma conditions which do not show sawtooth activity at all, though q_0 is measured to be clearly below unity.

The simplest of such effects arise in two-fluid theory. As discussed in section 6.3.2, diamagnetic rotation has a distinctly stabilizing influence. The general dispersion relation of the kink mode has been derived by Zakharov & Rogers (1992). For marginal ideal stability $\lambda_H = 0$ a simple threshold condition can be given (Levinton et al., 1994). The kink mode is stable, if

$$1.4\beta_p^{2/3} \left(\frac{n'_0 R}{n_0}\right)^{2/3} \left(\frac{(p'_e + p'_i)R}{p_e + p_i}\right)^{1/3} > r_1 q'_1. \qquad (6.146)$$

Note that, contrary to the ideal kink mode, in two-fluid theory the pressure gradient is stabilizing, while the shear is destabilizing. For $T'_e = T'_i = 0$ the criterion (6.146) corresponds to the stability condition $\omega_* > \gamma_0$, where γ_0 is the growth rate, (6.99), in the absence of diamagnetic rotation.[*]

When applied to discharges in the TFTR tokamak, the criterion (6.146) is found to predict very well the presence or absence of sawteeth. So-called supershot discharges are characterized by peaked pressure profiles and are usually sawtooth-free, while so-called L-mode discharges with flatter pressure profiles typically do show sawteeth. Figure 6.19 gives two representative cases, where either the l.h.s. of (6.146) exceeds $r_1 q'_1$ and no sawteeth are observed (a), or the l.h.s. is smaller than $r_1 q'_1$ and sawteeth are present (b). Figure 6.20 presents a scatter plot of data comprising many discharges with a considerable range of plasma current and heating power, which demonstrates the tight correlation of the stability criterion (6.146) with the appearance of sawteeth.

The success of the criterion (6.146) is surprising and somewhat counterintuitive since, by general physical arguments, the pressure gradient constituting a free-energy reservoir should be destabilizing, whereas the shear obstructs plasma motions and should have a stabilizing influence.

---

[*] If $\lambda_H > 0$, the linear kink mode is always unstable. However, analyzing the nonlinear behavior, it is found that the mode saturates at low amplitude if the diamagnetic frequency is somewhat above the $\lambda_H = 0$ threshold value (Rogers & Zakharov, 1995), such that the criterion remains, practically speaking, valid.

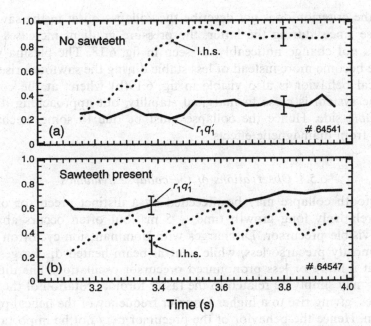

Fig. 6.19. The l.h.s. of (6.146) and the shear $r_1 q_1'$ for: (a) a supershot discharge with out sawteeth: (b) an L-mode discharge with sawteeth (from Levinton *et al.*, 1994).

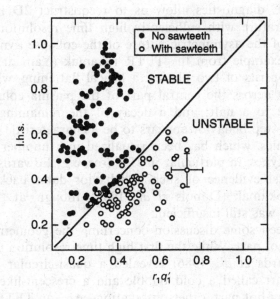

Fig. 6.20. Scatter plot of discharge data in the l.h.s. of (6.146) vs $r_1 q_1'$ plane. Cases where sawteeth are present have open circles, those without sawteeth full circles (from Levinton *et al.*, 1994).

In fact, the criterion does *not* describe the collapse onset in the sawtooth rise phase since, during the latter, the pressure gradient increases while $r_1 q_1'$ does not change noticeably, as seen in fig. 6.18. The plasma would therefore become more instead of less stable during the sawtooth rise. This unphysical behavior is also visible in fig. 6.19(b) where, at the collapse point, the system is close to marginal stability, but approaching it from the *unstable* side. Hence the collapse must be due to some mechanism different from diamagnetic effects.

6.5.4 Observations of the collapse dynamics

The sawtooth collapse may be preceded by a distinct precursor oscillation of relatively long growth time ~ 5 ms, but often occurs abruptly without visible precursor. Discharges with dominant ion-cyclotron heating are mostly precursorless, while neutral beam-heated discharges tend to exhibit a more or less pronounced precursor oscillation. The different behavior may simply be related to the faster toroidal rotation of the latter discharges, giving rise to a higher Doppler frequency of the helical plasma distortion. Hence the behavior of the precursor may not be important for the final collapse itself, which shows a similar behavior in both cases and occurs on time-scales much shorter than the precursor growth time, $\tau_1 \sim$ 100–200 μs.

SX and ECE diagnostics allow us to reconstruct 2D images of the electron temperature with sufficiently high time resolution to obtain a direct picture of the asymmetric nature of the collapse dynamics. Figure 6.21 gives an example from the TFTR tokamak (Yamada *et al.*, 1994). The process consists of two phases: a helical flattening, which proceeds about half-way across the central part of the plasma column, where it seems to come to a halt; and a decay of the remaining temperature peak. This two-step behavior appears to be a characteristic feature of the collapse dynamics, which has been visualized in a number of tokamaks and discharge types, in particular in JET (see, e.g., Edwards *et al.*, 1986). In fact, the first evidence of such a behavior dates back many years to the TFR tokamak (Dubois *et al.*, 1983), though, at that time, the time-resolution was still insufficient.

There has been some discussion concerning the geometrical shape of the flattened cool part. While the first high time-resolution measurements on JET (Edwards *et al.*, 1986) revealed a quasi-circular shape of the cool plasma part called a cold bubble and a crescent-like deformation of the remaining hot part, other observations, e.g., on TFTR (Nagayama *et al.*, 1991) find just the reverse, a crescent-like region of cooler plasma and a circular hot core. It appears, however, that this morphological difference should not be overemphasized. As a rule of thumb, a circular

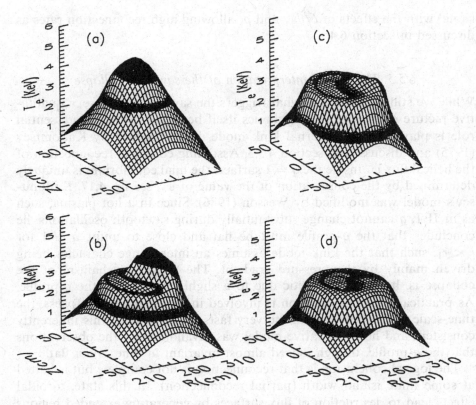

Fig. 6.21. Development of the $T_e(r, \theta)$ profile during a sawtooth collapse. The time interval between each figure is about 120 μs (from Yamada *et al.*, 1994).

cold bubble tends to appear in very fast collapse events, while a circular hot core is more typical for slower events. The difference may to a certain extent also depend on the method of tomographic reconstruction of the profiles.

A more crucial point, which has led to intense debates, has been the question of the collapse time-scale τ_1. Observations show a rather broad span of collapse times $\tau_1 \sim 50$–500 μs for similar sizes of the inversion radius $r_1 \sim 0.5$ m in JET or TFTR, depending on the discharge type. Since the collapse almost inevitably involves magnetic reconnection, the time-scale should reflect in some way the dominant reconnection mechanism. Assuming reconnection to occur in a resistive current sheet, the reconnection time would be much longer, typically 10 ms, even if only a certain fraction of the helical flux is reconnected (partial reconnection) as may be inferred from the q-profile measurements. Considering the plasma parameters of hot tokamak discharges it is, however, clear that resistive diffusion is negligible and the reconnection process should be noncollisional with the effects of c/ω_{pe} and ρ_s allowing high reconnection rates as discussed in section 6.4.

6.5.5 Theoretical interpretation of the sawtooth collapse

While we still do not know what triggers the sawtooth collapse, a qualitative picture of the collapse dynamics itself becomes visible. The essential role is played by the internal kink mode, first proposed by Kadomtsev (1975) and discussed in section 4.3.3. Assuming complete reconnection of the helical flux ψ_* inside the $q = 1$ surface, the final equilibrium is uniquely determined by the conservation of the value of ψ_*, see fig. 4.17. Kadomtsev's model was modified by Wesson (1986). Since in a hot plasma, such as in JET, q cannot change substantially during sawtooth oscillations, he concludes that the q-profile must be flat and close to unity, $q \simeq 1$ for $r < r_1$, such that the kink mode assumes an interchange character being driven mainly by the pressure gradient. The trigger mechanism of the collapse is, however, magnetic due to a slight inversion of the q-profile. As practically no reconnection is involved in an interchange process, the time-scale of the collapse can be very fast. Unfortunately, this inherently consistent and hence attrative model was invalidated by the observations that the q-profile, though indeed almost invariant in time, is *not* flat.

The logical conclusion is that reconnection is not complete but is halted at some finite island width (partial reconnection). In this state, toroidal effects lead to destruction of flux surfaces by generating extended regions of stochastic field-line behavior, which allow rapid radial electron energy transport along field lines (though this is not true to the same extent for the ion energy and the density). One is tempted to argue that due to the flattening of the pressure gradient the driving agent of the kink mode is turned off and hence the mode saturates. Extensive numerical simulations using the full compressible MHD equations in toroidal geometry have, however, shown that such saturation does not occur. Instead the kink mode, once started, always continues until complete reconnection and hence $q \geq 1$ (Aydemir et al., 1989; Baty et al., 1992). Diamagnetic effects, which in principle may result in a saturated state as discussed in section 6.4.3, should not be important for very fast collapse events, where $\gamma > \omega_*$.

An alternative model of the sawtooth collapse has been suggested by Biskamp & Drake (1994). The basic idea is that in a fast reconnection process, such as that mediated by the noncollisional c/ω_{pe} and ρ_s effects, plasma inertia associated with the rapid kink motion will drive the system back to a state with $q < 1$. The dynamics, which is demonstrated by numerical simulations, consists of a three-stage process: (i) the classical kink mode evolution to complete reconnection, which is followed by (ii) the reintroduction of the helical flux "through the backdoor" and

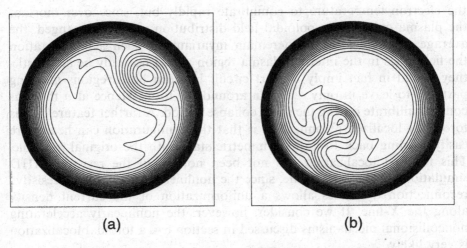

(a)   (b)

Fig. 6.22. Contour plots of the helical flux function ψ_*, illustrating the two reconnection stages in the plasma inertia-driven collapse model. (a) $t = 1652$, (b) $t = 1727$ (from Biskamp & Drake, 1996).

(iii) a secondary reconnection process leading to the final state similar to the initial one. Figure 6.22 illustrates the two reconnection phases. However, the efficiency of this process depends crucially on the weakness of diamagnetic effects. These distort the flow and tend to impede the second stage of flux reversal. Indeed subsequent measurements on the TEXTOR tokamak with sufficient time-resolution show that the change of the poloidal field remains small throughout the sawtooth collapse.

The dynamical models which we have just discussed have in common the restriction to helical geometry, i.e., they are two-dimensional. On the other, hand we have already pointed out that toroidal geometry, which breaks the helical symmetry, may play an important role by providing an alternative mechanism of fast electron transport other than cross-field plasma convection. Recent experimental studies of the sawtooth collapse reveal a strong asymmetry between the inside of the torus (= high-field) and the outside (= low-field) (Nagayama et al., 1996). Reconnection appears to occur mainly on the outside, i.e., reconnection is concentrated in a small section of the torus where the X-line, winding around the plasma column, is located on the outside, and so reconnection is not uniform along the X-line as implied in the helical model. Across this narrow toroidal section all field lines from inside the $q = 1$ surface may connect to outside ones without strongly disturbing the poloidal field in the main remaining part of the plasma column. (A similar toroidal focusing has been found in numerical studies of the nonlinear development of ballooning modes by Park et al., 1995.) In this sense complete reconnection does occur, allowing the electron temperature to equilibrate rapidly but, since over most of the plasma column the poloidal field distribution is hardly changed, the average q-profile appears to remain invariant. In such a configuration the field lines in the internal plasma region would not only be stochastic, they would in fact imply a short circuit between both regions allowing particles to leave in only one turn around the torus. Hence also the ions could equilibrate during the short collapse period. A further feature of this toroidally localized reconnection is that the configuration can heal more easily, leading back to an axisymmetric state with the original q-profile. This toroidal localization has not been noticed in the resistive MHD simulations of the kink mode, since the nonlinear slowing of the resistive reconnection dynamics allows a uniformization of the current density along the X-line. If we consider, however, the nonlinearly accelerating noncollisional mechanisms discussed in section 6.4, a toroidal localization is very likely.

At relatively high β, MHD ballooning modes may also play an important role, which are destabilized by the steep pressure gradient generated in the region around the X-line at finite island size. These will enhance the process of field-line stochastization of the plasma (for a recent study see Nishimura et al., 1999).

We still have not discussed the mechanism by which the kink motion is halted before the entire plasma column has been turned inside out, i.e., why the kink mode saturates, since the MHD simulations have shown that, in spite of the flattening of the pressure profile, the process continues until the safety-factor profile also has become flat with $q \simeq 1$. Here a new element is likely to play a role, the strong (positive) radial electric field due to charge separation generated by the rapid (parallel) electron energy loss. This field necessarily appears during the collapse and leads to rotation of the internal plasma, such that the kink motion no longer points toward the X-line, an effect which is similar to that of diamagnetic rotation. The onset of such rotation has recently been observed in TEXTOR. It thus appears that kink mode saturation is a necessary consequence of the thermal transport processes during the collapse. These ideas should be corroborated by suitable numerical simulations, which is, however, not a simple task since, in addition to toroidal geometry, noncollisional processes and transport effects would have to be included.

6.6 Laboratory reconnection experiments

Though reconnection processes play a certain role in most plasma devices, besides the tokamak the reversed-field pinch being the most notable example (for an introduction, see Biskamp, 1993a, chapter 9), several experiments have been constructed with the explicit goal of studying magnetic reconnection in a well-defined and controlled configuration. Here, I would like to discuss three of these in some detail: the University of California (UCLA) reconnection experiments, which were performed during the 1980s; and two more recent studies: the magnetic reconnection experiment of the Princeton Plasma Physics Laboratory (PPPL), and the spherical torus experiment at the Tokyo University. It is true that plasma parameters in these devices are rather modest in terms of Lundquist numbers S or the ratio of scale-lengths, e.g., $L/(c/\omega_{pe})$, but they nevertheless allow reproduction of essential features of the reconnection physics and also reveal unexpected effects which show that real plasmas are more complex than their theoretical counterparts.

6.6.1 *The UCLA magnetic reconnection experiment*

The experiments on magnetic field-line reconnection by Stenzel and Gekelman, which were reported in a series of publications in the 1980s (joined also by other authors), were the first major attempt to study reconnection physics in the laboratory under well-controlled conditions, in particular the diffusion region (Stenzel & Gekelman, 1981). The setup was a relatively large linear device of length $L = 2$ m and diameter 1 m. The vessel contained two parallel conducting plates 32 cm apart. The space between the plates was filled by a nearly homogeneous plasma of density $n \simeq 10^{-12}$ cm^{-3} and temperatures $T_e \simeq T_i \simeq 5$ eV, such that the plasma conditions are nearly collisionless. The plasma was confined by an axial field $B_z = 20$ G. Parallel axial currents along the plates were switched on for a period long compared with the Alfvén transit time but short compared with the resistive diffusion time, generating a poloidal field with $B_{p,\text{max}} \sim 100$ G, which generates an X-type neutral-line configuration in vacuum. The device was designed for optimal diagnostic access. Plasma parameters allow the insertion of probes and magnetic loops without significant perturbation of the plasma or damage to the probes, which measure plasma density and magnetic and electric fields with good spatial resolution. Also measured are the particle fluxes and energies in all three directions.

In the presence of a plasma between the plates, the frozen-in property implies that when increasing the plate currents the expansion of the poloidal field drives a plasma flow toward the mid-plane, where a narrow current sheet is formed with a width of the order of the electron inertial length c/ω_{pe}. Since only the electrons are strongly magnetized, while the ion gyroradius is of the order of the plate spacing, this is not a true MHD plasma but, more properly speaking, an EMHD system, and the Lundquist number is rather small $S \sim 20$. Nevertheless important features of an MHD current sheet are recovered, in particular the fast plasma jetting out of the sheet edges approaching the Alfvén speed (Gekelman *et al.*, 1982). Using data from comprehensive measurements of fields, flows, density, and temperature in the generalized Ohm's law, the effective resistivity has been found to be spatially inhomogeneous with values exceeding the classical resistivity by one to two orders of magnitude. Hence binary collisions do not play an important role. Energy dissipation is dominated by electron heating, but T_e is largest at the sheet edges rather than at the sheet center. While macroscopically the current sheet is stable, it reveals a broad spectrum of high-frequency fluctuations which have been identified mainly as whistler waves (as expected for an EMHD system) but also as ion-sound waves. These could be responsible for the observed anomalous resistivity. Only when the current density in the sheet exceeds a certain threshold does the sheet disrupt, leading to the formation of a potential double layer in the plasma, which gives rise to particle acceleration up to energies much higher than expected for steady-state electric fields in the plasma (Stenzel *et al.*, 1983).

6.6.2 *The PPPL magnetic reconnection experiment*

This experiment (with the acronym MRX) differs substantially from the UCLA experiment, since plasma conditions are now in the MHD regime, in particular $S \sim 10^3 \gg 1$, $\rho_i/L \ll 1$. The device consists of two annular flux cores of diameter nearly 1 m suspended vertically in a vacuum vessel (see Yamada *et al.*, 1997). The flux cores contain toroidal windings and a helical toroidal solenoid producing a toroidal magnetic flux inside the core. Toroidal currents in the cores generate an axisymmetric quadrupole vacuum field, fig. 6.23(a). The plasma is produced by a poloidal electric field induced by changing the toroidal flux in the cores. The poloidal plasma currents also give rise to a toroidal magnetic field component in the plasma. Reconnection is driven by either increasing or decreasing the poloidal field currents I_{PF} in the cores, called "push reconnection" or "pull reconnection", respectively, which corresponds to a vertical current sheet (fig. 6.23(b)) or a horizontal one (fig. 6.23(c)). Typical plasma conditions are $n_e \sim 10^{14}$ cm^{-3}, $T_e \sim 10$–30 eV, $B_p \lesssim 500$ G. The relatively low temperature and the short plasma pulse duration (< 1 ms) allow the use of internal probes to measure local plasma parameters. A 2D array of magnetic probes gives the poloidal field distribution. The reconnection rate is measured from the inflow plasma velocity u or directly from the change of the poloidal flux $\partial_t \psi = E_t$, results from both methods being in good agreement.

In the classical Sweet–Parker model considered in chapter 3, the magnetic field component in the third orthogonal direction does not affect the reconnection dynamics because of the incompressibility assumption.

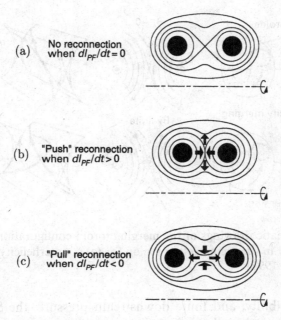

Fig. 6.23. Poloidal cross-section of the MRX configuration. (a) Vacuum field; (b) "push reconnection", when $dI_{PF}/dt > 0$; and (c) "pull reconnection", when $dI_{PF}/dt < 0$. I_{PF} is the toroidal current in the cores. The dashed horizontal line indicates the torus axis (from Yamada *et al.*, 1997).

In MRX, however, the toroidal field B_t has a significant effect, the current sheet being narrower (of the order of ρ_i) and reconnection proceeding faster for $B_t \simeq 0$, where the reconnecting field is anti-parallel, than for finite B_t, which indicates that plasma compressibility is important. As expected, compressibility effects are strongest for $B_t \simeq 0$, where the density in the sheet center is found to be twice as large as in the upstream state, while for sizable B_t the central density increase is weak. The main difference compared with the Sweet–Parker model arises, because the plasma pressure p_1 in the downstream state at the sheet edges p_1 is significantly higher than the upstream pressure p_0. Hence it follows from the parallel momentum balance, that the outflow velocity v is smaller than the Alfvén speed, typically $v \lesssim 0.2 v_A$. As in the UCLA experiment, the effective resistivity η_{eff} determined from Ohm's law, $E_t + u B_p = \eta_{\text{eff}} j_t$, is found to be enhanced compared with the classical value η_c. The ratio η_{eff}/η_c grows strongly with decreasing collisionality parameter δ/λ_e, the ratio of sheet thickness to mean free path. Since δ decreases with decreasing B_t, the anomaly of the resistivity is particularly large for $B_t \simeq 0$. The increase of the effective resistivity can be related to the observation of high-frequency fluctuations. It is true that, by incorporating effective resis-

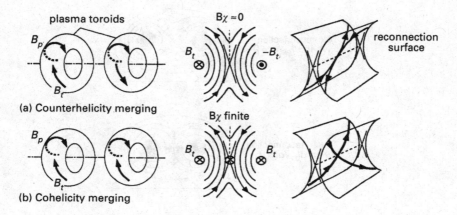

Fig. 6.24. Schematic drawings of the merging toroid configurations in TS-3 and the magnetic field in the reconnection region for (a) counterhelicity, (b) cohelicity (from Ono *et al.*, 1997).

tivity, compressibility, and finite downstream pressure, the Sweet–Parker model can formally be salvaged; defining an effective Lundquist number S_{eff}, the observed reconnection rate is found to satisfy the Sweet–Parker relation $u/v_A \propto S_{eff}^{-1/2}$ (Ji *et al.*, 1999). Nevertheless the experimental results indicate that it is surprisingly difficult to produce plasma conditions corresponding to simple resistive MHD reconnection, as collisionless effects seem to dominate even for $\lambda_e \lesssim \delta$. On the other hand in the case of finite B_t, which should be closest to the MHD regime, the current sheet becomes rather broad and the reconnection process seems to be affected by the limited size of the plasma volume.

6.6.3 *The Tokyo University reconnection experiment*

This experiment (acronym TS-3) is, in a sense, a sister experiment of MRX described in the previous subsection. The device is a cylindrical vacuum vessel of roughly 1 m length and 1 m diameter, inside which two pairs of poloidal field coils are suspended vertically (similar to MRX). The initial plasma configuration consists of two axisymmetric plasma columns or toroids carrying both poloidal and toroidal magnetic fields. An axial coil along the cylinder axis can provide an additional external toroidal field. The plasma toroids are accelerated toward each other by reversed currents in the poloidal field coils, which drives the merging into a single toroid. Depending on the magnitude and direction of the toroidal fields in the plasma columns one distinguishes between counterhelicity and cohelicity reconnection as sketched in fig. 6.24. The configuration is thus characterized by the field B_X along the X-line. In the counterhelicity case $B_X = 0$, the reconnecting field being strictly antiparallel, while in the cohelicity case B_X is finite its magnitude depending on the external toroidal field. The diagnostics measure mainly the magnetic field and the ion velocity and temperature distributions, but also the electron temperature and density. Typical initial plasma parameters are: $n_{e0} \lesssim 10^{14}$ cm^{-3}, $T_{i0} \simeq T_{e0} \sim 20$ eV, and $B \lesssim 300$ G (Ono *et al.*, 1996, 1997).

As in MRX, the reconnection dynamics is mainly determined by the field ratio B_X/B_p. For small B_X/B_p, in particular in the counterhelicity case $B_X = 0$, the merging process is fast and gives rise to strong ion heating $T_i \gg T_{i0}$, with T_i up to 200 eV, while the electron temperature is not changed substantially. With increasing B_X, reconnection rate and ion heating decrease and both become rather small for $B_X/B_p \gg 1$. (The reconnection rate is also proportional to the accelerating coil current.) The behavior can be associated with the character of the ion motions in the current layer. If the ion gyroradius ρ_i at the X-line exceeds the width δ of the current layer, the ions are nonadiabatic, i.e., no longer magnetized. The particles in the outflow region exhibit complicated three-dimensional motions and can easily be randomized, which corresponds to efficient heating. If, however, B_X becomes sufficiently large, such that $\rho_i < \delta$, the ions remain magnetized where the magnetic moment and hence $T_{i\perp}$ are conserved, which precludes thermalization of the outflow kinetic energy. The observed behavior can be related to the effect of the Hall term in Ohm's law, discussed in detail in section 6.2. This term, which catalyzes the fast reconnection in high-β plasmas, is important in the region where ions are nonmagnetized, decoupling them from the strongly magnetized electrons. However, the fluid description presented in this chapter is evidently inadequate to deal with the nonadiabatic ion heating requiring a kinetic treatment, which will be presented in chapter 7.

There is an interesting analogy with the heating rates observed in tokamaks and reversed-field pinches (RFP). In a tokamak, where the toroidal field is much larger than the poloidal field (or the helical field in a kink or tearing mode reconnection process) the electrons are hotter than the ions, $T_e > T_i$, (except in the case of strong auxiliary heating). By contrast, in an RFP with $B_t \lesssim B_p$, where B_t is only sustained by continuous reconnection processes, the ions are much hotter $T_i \gg T_e$.

7

Microscopic theory of magnetic reconnection

In chapter 6 we considered noncollisional reconnection processes in the framework of the two-fluid theory, where the term "noncollisional" implies that collisional effects are weak and not important for the reconnection dynamics but still sufficient to make the pressure tensor isotropic. Dissipation coefficients such as η, μ_i, μ_e, though very small, must remain finite to avoid singularities of the solution. The noncollisional fluid description is particularly suitable for quasi-neutral, primarily electromagnetic processes associated with the scale-lengths $c/\omega_{pi}, c/\omega_{pe}$, where electrostatic waves on the Debye scale as well as finite Larmor-radius effects can be neglected.

Coulomb collisions are negligible if the electron mean free path λ_e is long compared with the gradient scale-length of the process under investigation. The mean free path of electrons, the most rapidly colliding particle species, is

$$\lambda_e \simeq v_{te}/\nu_e \sim \lambda_D\,(n\lambda_D^3), \qquad (7.1)$$

since the collision frequency is approximately $\nu_e \sim \omega_{pe}/(n\lambda_D^3)$, where $\lambda_D = \sqrt{T_e/(4\pi ne^2)} = v_{te}/\omega_{pe}$ is the Debye length and $v_{te} = \sqrt{T_e/m_e}$ the electron thermal velocity. The parameter $\lambda_e/L_\parallel$, $L_\parallel$ = typical gradient scale-length along the magnetic field, is a measure of the plasma collisionality. (As explained in section 6.1, in the perpendicular direction the Larmor radius plays the role of the mean free path.) Table 7.1 gives this parameter for some laboratory and space plasmas. The edge region of a tokamak discharge and the ionospheric plasma are strongly collisional (in the latter, collisions with neutrals dominate), the tokamak core plasma and the solar corona are weakly collisional, whereas the magnetotail and, more generally, the interplanetary plasma are collisionless.

In a collisionless plasma dissipation occurs by wave-particle resonance effects, of which the most common is Landau damping. These effects are strongest for slow electrostatic waves, which may resonate with the

Table 7.1. *Collisionality parameter $\lambda_e/L_{\parallel}$ for typical laboratory and space plasma conditions.*

tokamak edge plasma	10^{-1}
tokamak core plasma	10
ionosphere	10^{-2}
geomagnetic tail	10^6
solar corona	10^2

main body of the ion or electron distribution function. To describe such resonance processes a kinetic approach is required. In this chapter we first give a brief introduction to microscopic plasma theory, section 7.1, where, starting from the Vlasov equation, we discuss the different resonance effects which may give rise to instability, and the main nonlinear saturation processes. Sections 7.2–7.4 deal with several microinstabilities, which generate turbulent resistivity and may thus enhance magnetic reconnection: the ion-sound instability, section 7.2; the lower-hybrid-drift instability, section 7.3; and the whistler anisotropy instability, section 7.4. We study the linear theory, give the quasi-linear estimates of the resulting turbulent transport coefficients, and discuss the nonlinear properties. Section 7.5 deals with the collisionless tearing mode, which has received much attention in the magnetospheric physics community. The tearing mode may be driven by electron and ion Landau resonance, but both processes turn out to be rather inefficient. Lastly, section 7.6 gives a brief introduction to particle simulation methods and discusses recent two- and three-dimensional simulation results on collisionless reconnection.

7.1 Vlasov theory and microinstabilities

In the absence of Coulomb collisions, particles interact only through electric and magnetic fields generated self-consistently by collective electron and ion motions, which are described by Vlasov's equation for the phase space distribution functions $f_j(\mathbf{x}, \mathbf{v}, t)$ of the particle species j,

$$\partial_t f_j + \mathbf{v} \cdot \nabla f_j + \frac{q_j}{m_j} \left[\mathbf{E}(\mathbf{x}, t) + \frac{1}{c} \mathbf{v} \times \mathbf{B}(\mathbf{x}, t) \right] \cdot \frac{\partial}{\partial \mathbf{v}} f_j = 0, \qquad (7.2)$$

$$\nabla \cdot \mathbf{E} = \sum_j 4\pi q_j \int d^3 v f_j, \qquad (7.3)$$

$$\frac{1}{c} \partial_t \mathbf{B} = -\nabla \times \mathbf{E}, \qquad (7.4)$$

$$\nabla \times \mathbf{B} - \frac{1}{c}\partial_t \mathbf{E} = \sum_j \frac{4\pi}{c} q_j \int d^3 v \mathbf{v} f_j. \tag{7.5}$$

Because of the self-consistency conditions (7.3) and (7.5), Vlasov's equation (7.2) constitutes a nonlinear integrodifferential equation for f_j. Since (7.2) can be interpreted as Liouville's equation for one particle moving under the influence of the fields $\mathbf{E}$ and $\mathbf{B}$, the solution for a stationary system is a functional of the constants of motion, in particular the particle energy

$$W_j = \tfrac{1}{2} m_j v^2 + q_j \phi, \tag{7.6}$$

and, in the case of spatial symmetry, the canonical momentum in the direction of the ignorable coordinate x_α

$$p_{j\alpha} = m_j v_\alpha + \frac{q_j}{c} A_\alpha, \tag{7.7}$$

where ϕ and $\mathbf{A}$ are the scalar and vector potentials, $\mathbf{E} = -\nabla\phi - \partial_t \mathbf{A}/c$, $\mathbf{B} = \nabla \times \mathbf{A}$. Hence the equilibrium distribution has the general form

$$f_{j0} = f_j(W_j, p_{j\alpha}). \tag{7.8}$$

Insertion into the self-consistency conditions yields integrodifferential equations for ϕ and A_α which, together with the boundary conditions, determine the equilibrium configuration. It is interesting to note that in the absence of spatial symmetry, where $f = f(W)$, the current density, i.e., the r.h.s. of (7.5) vanishes, which makes the existence of exact nonsymmetric collisionless equilibria questionable.

7.1.1 *Linear Vlasov theory*

We now linearize Vlasov's equation about a stationary state, $f_j = f_{j0} + f_j^{(1)}$:

$$\partial_t f_j^{(1)} + \mathbf{v} \cdot \nabla f_j^{(1)} + \frac{q_j}{m_j}\left(\mathbf{E}_0 + \frac{1}{c}\mathbf{v} \times \mathbf{B}_0\right) \cdot \frac{\partial}{\partial \mathbf{v}} f_j^{(1)}$$

$$= -\frac{q_j}{m_j}\left(\mathbf{E}^{(1)} + \frac{1}{c}\mathbf{v} \times \mathbf{B}^{(1)}\right) \cdot \frac{\partial}{\partial \mathbf{v}} f_{j0}, \tag{7.9}$$

which has the formal solution
$$f_j^{(1)}(\mathbf{x}, \mathbf{v}, t) = \delta f_j\big(\mathbf{x}_0(\mathbf{x}, \mathbf{v}, t), \mathbf{v}_0(\mathbf{x}, \mathbf{v}, t)\big)$$

$$-\frac{q_j}{m_j} \int_0^t dt' \left(\mathbf{E}^{(1)}(\mathbf{x}', t') + \frac{1}{c}\mathbf{v}' \times \mathbf{B}^{(1)}(\mathbf{x}', t')\right) \cdot \frac{\partial}{\partial \mathbf{v}'} f_{j0}(\mathbf{v}'). \tag{7.10}$$

Here δf_j is the disturbance at $t = 0$ depending on the initial values $\mathbf{x}_0$, $\mathbf{v}_0$ of a particle as functions of the (unperturbed) orbit, and the time integral in the last term is taken along this orbit. Inserting expression (7.10) into (7.3) and (7.5) gives a system of homogeneous linear equations for the field perturbations, which yields the dispersion relation.

Let us first consider the simplest case of a homogeneous unmagnetized plasma $f_{j0}(\mathbf{v})$, $\mathbf{E}_0 = \mathbf{B}_0 = 0$, with the further restriction to electrostatic perturbations $\mathbf{B}^{(1)} = 0$. Spatial Fourier transformation of (7.9) gives

$$\partial_t f_{j\mathbf{k}} + i\mathbf{k} \cdot \mathbf{v} f_{j\mathbf{k}} = -\frac{q_j}{m_j} \mathbf{E}_\mathbf{k} \cdot \frac{\partial}{\partial \mathbf{v}} f_{j0}. \tag{7.11}$$

The appropriate method to solve the initial value problem is the Laplace transform

$$f_\omega = \int_0^\infty dt\, e^{i\omega t} f(t), \quad \text{Im}\{\omega\} > 0, \tag{7.12}$$

$$f(t) = \int_C \frac{d\omega}{2\pi} e^{-i\omega t} f_\omega, \tag{7.13}$$

where the contour of integration C should remain above all singularities of f_ω in the complex ω-plane. Laplace transformation of (7.11) gives the solution in terms of the initial perturbation $\delta f_\mathbf{k}$:

$$f_{j\mathbf{k}\omega} = \frac{\delta f_{j\mathbf{k}}}{i(\mathbf{k} \cdot \mathbf{v} - \omega)} - \frac{q_j}{m_j} \frac{\mathbf{E}_{\mathbf{k}\omega} \cdot \partial f_{j0}/\partial \mathbf{v}}{i(\mathbf{k} \cdot \mathbf{v} - \omega)}. \tag{7.14}$$

Inserting this expression into (7.3) and using $\mathbf{E}_{\mathbf{k}\omega} = -i\mathbf{k}\phi_{\mathbf{k}\omega}$ we obtain

$$k^2 \phi_{\mathbf{k}\omega} = \frac{1}{\epsilon(\mathbf{k}, \omega)} \sum_j 4\pi q_j \int d^3v \frac{\delta f_{j\mathbf{k}}}{i(\mathbf{k} \cdot \mathbf{v} - \omega)}, \tag{7.15}$$

where

$$\epsilon(\mathbf{k}, \omega) = 1 - \sum_j \frac{4\pi q_j^2}{m_j k^2} \int d^3v \frac{\mathbf{k} \cdot \partial f_{j0}/\partial \mathbf{v}}{\mathbf{k} \cdot \mathbf{v} - \omega} \tag{7.16}$$

is called the dielectric function. The time evolution of the potential disturbance $\phi_\mathbf{k}(t)$ is determined by the singularities of $\phi_{\mathbf{k}\omega}$, which arise from two sources, the zeros of $\epsilon(\mathbf{k}, \omega)$ and the poles from the integral involving the initial disturbance $\delta f_{j\mathbf{k}}$. The latter, being proportional to q_j, gives the field oscillations generated by the free propagation of the particles contained in $\delta f_{j\mathbf{k}}$, which are therefore also called ballistic modes ($\propto e^{-i\mathbf{k}\cdot\mathbf{v}t}$). Ballistic modes depend only on the initial disturbance and are, in general, damped.[*]

The zeros of $\epsilon(\mathbf{k}, \omega)$ arising from the collective particle interactions ($\propto q_j^2$) represent the natural (electrostatic) oscillations in a collisionless

---

[*] For special initial conditions, ballistic modes give rise to an interesting nonlinear effect, *viz.*, plasma echos (Gould *et al.*, 1967).

plasma depending only on the equilibrium distribution functions f_{j0}. The equation

$$\epsilon(\mathbf{k}, \omega) = 0 \qquad (7.17)$$

is called the dispersion relation for electrostatic modes, which will be treated in more detail for particular microinstabilities in the subsequent sections. Here, we briefly discuss some general properties. To simplify the notation, we extract the homogeneous equilibrium particle density writing $f_{j0} = n_{j0} F_j$, where charge neutrality requires $\sum_j n_{j0} q_j = 0$. The dielectric function contains the resonance denominator $(\mathbf{k} \cdot \mathbf{v} - \omega)$ indicating that particles moving with the phase velocity of the wave play a special role. The relation $\mathbf{k} \cdot \mathbf{v} = \omega$ is a special case of the general resonance condition, that particles, for which $\mathbf{E}^{(1)} \cdot \mathbf{v}$ does not change sign, can efficiently exchange energy with the wave. In the cold plasma case $F_j = \delta(\mathbf{v} - \mathbf{u}_j)$ the resonance in the integral in (7.16) vanishes and the dispersion relation (7.17) is reduced to a simple algebraic equation,

$$1 - \sum_j \frac{\omega_{pj}^2}{(\mathbf{k} \cdot \mathbf{u}_j - \omega)^2} = 0, \qquad (7.18)$$

where $\omega_{pj}^2 = 4\pi n_{j0} q_j^2 / m_j$. In a hot plasma the shape of the distribution function plays a crucial role. While the dispersion relation must, in general, be solved numerically, an approximate analytical solution is possible if the instability is weak $\gamma = \omega_i \ll \omega_r$. In this case we may expand the propagator,

$$\frac{1}{\mathbf{k} \cdot \mathbf{v} - \omega} \simeq \frac{1}{\mathbf{k} \cdot \mathbf{v} - \omega_+} + \frac{i\gamma}{(\mathbf{k} \cdot \mathbf{v} - \omega_+)^2}, \qquad (7.19)$$

where $\omega_+ = \omega_r + i0$ results from the requirement on the integration path C in the inverse Laplace transform (7.13), such that the v-integration follows the Plemelj formula

$$\frac{1}{\mathbf{k} \cdot \mathbf{v} - \omega_+} = P \frac{1}{\mathbf{k} \cdot \mathbf{v} - \omega_r} + i\pi \delta(\mathbf{k} \cdot \mathbf{v} - \omega_r). \qquad (7.20)$$

Insertion into the dispersion relation and separation of real and imaginary parts gives, after integration by parts,

$$1 - \sum_j \omega_{pj}^2 \int d^3 v \frac{F_j}{(\mathbf{k} \cdot \mathbf{v} - \omega_r)^2} = 0, \qquad (7.21)$$

$$\pi \sum_j \frac{\omega_{pj}^2}{k^2} \int d^3 v \, \mathbf{k} \cdot \frac{\partial F_j}{\partial \mathbf{v}} \delta(\mathbf{k} \cdot \mathbf{v} - \omega_r) + 2\gamma \sum_j \omega_{pj}^2 \int d^3 v \frac{F_j}{(\mathbf{k} \cdot \mathbf{v} - \omega_r)^3} = 0. \qquad (7.22)$$

A simple expression for ω can be derived for modes with large phase velocity $(\omega_r/k)^2 \gg v_{te}^2 = \int dv\, v^2 F_e(v)$. Expanding the denominator in the integral in (7.21) we obtain

$$1 - \frac{\omega_{pe}^2}{\omega_r^2}\left(1 + 3\frac{k^2 v_{te}^2}{\omega_r^2}\right) = 0,$$

or

$$\omega_r^2 \simeq \omega_{pe}^2 + 3k^2 v_{te}^2 = \omega_{pe}^2(1 + 3k^2\lambda_D^2), \tag{7.23}$$

which represent (electron) plasma oscillations or Langmuir waves. Substitution in (7.22) gives

$$\gamma = \omega_{pe}\frac{\pi}{2}\frac{\omega_{pe}^2}{k^2}\left.\frac{\partial F_e}{\partial v_\parallel}\right|_{v_\parallel = \omega/k} \tag{7.24}$$

$$\simeq -\omega_{pe}\sqrt{\frac{\pi}{8}}\frac{1}{(k\lambda_D)^3}\,\mathrm{e}^{-1/2k^2\lambda_D^2},$$

where the latter expression is obtained for a Maxwellian distribution

$$F_e(\mathbf{v}) = \frac{1}{(2\pi v_{te}^2)^{3/2}}\mathrm{e}^{-v^2/2v_{te}^2}. \tag{7.25}$$

Hence there is a finite damping rate proportional to the gradient of the distribution function taken at the phase velocity of the wave, which is called Landau damping. It originates from the $\mathbf{k}\cdot\mathbf{v} - \omega_+$ resonance in the dielectric function. Particles with velocities slightly exceeding the phase velocity lose energy to the wave, they "emit waves" in analogy to the Cherenkov effect, while particles with somewhat lower velocities are accelerated by the wave, i.e., extract energy from the wave. Hence if the population of the slower particles is higher than that of the faster ones corresponding to $\partial F/\partial v|_{v=\omega/k} < 0$, the net effect is wave damping.[*] For more general non-Maxwellian distributions, however, $\partial F/\partial v|_{v=\omega/k}$ may be positive, leading to growth of the wave amplitude, a process which is called inverse Landau damping. Since such instabilities depend on the shape of the velocity distribution function, they are called microscopic instabilities or simply microinstabilities. The most important electrostatic microinstabilities giving rise to anomalous resistivity are the ion-sound instability and the lower-hybrid-drift instability, sections 7.2 and 7.3., respectively, for which we will study the linear properties and the nonlinear saturation.

---

[*] An alternative interpretation of this effect can be given as a superposition and phase mixing of van Kampen modes (van Kampen, 1955).

If the plasma is embedded in a magnetic field $\mathbf{B}_0 = B_0 \mathbf{e}_z$, the particle orbits are no longer straight lines with constant velocity but helices about the magnetic field, which considerably complicates the formalism (for a detailed analysis we refer to the classical book on plasma waves by Stix, 1962, chapter 8). The integrand in the integral over particle trajectories in (7.10) now contains the factor $\exp\{i(\omega - k_z v_z)(t - t') + ik_\perp (v_\perp/\Omega_j) \cos \Omega_j (t - t')\}$, where $\Omega_j = |q_j| B_0/m_j c$ is the cyclotron frequency. Using the Bessel function identity

$$e^{\lambda \cos \Omega t} = \sum_{-\infty}^{\infty} I_n(\lambda) e^{in\Omega t},$$

the expression for the perturbed distribution function becomes an infinite sum over harmonics of the cyclotron frequency of the form

$$f_{j\mathbf{k}\omega}(v_\perp, v_z) = \sum_{-\infty}^{\infty} \frac{A_n(\lambda_j)}{k_z v_z - n\Omega_j - \omega_+}, \quad \lambda_j = (k_\perp v_\perp/\Omega_j)^2. \tag{7.26}$$

Hence in addition to the Landau resonance $n = 0$ where the particle velocity along the field equals the phase velocity $v_z = \omega/k_z$, there is a series of further resonances involving Ω_j. The effect is easily understood for long parallel wavelength $k_z \ll k_\perp$, where the resonance frequency is just a multiple of Ω_j. In a strong magnetic field, where $\Omega_j \gg \omega$, the Landau resonance is most important, which means that the particle dynamics is essentially one-dimensional along the field, whereas for $\Omega_j \sim \omega$ the cyclotron resonance $n = 1$ dominates.

While short-wavelength oscillations in a plasma are essentially electrostatic, long-wavelength modes are predominantly electromagnetic. Since in this context we are mainly interested in modes excited in the plasma by wave–particle resonance, where phase velocities are of the order of the thermal velocities, we can neglect the displacement current in (7.5) (the latter is important primarily to describe plasma interaction with electromagnetic radiation such as radio waves or laser light). Equation (7.5) now reduces to Ampère's law, which implies quasi-neutrality, i.e., Poisson's equation reduces to the condition that the charge density, the r.h.s. of (7.3), vanishes.[*]

Genuine electromagnetic microinstabilities in a homogeneous plasma are caused, for instance, by electron temperature anisotropy, which destabilizes whistler waves (considered in some detail in section 7.4). On the other hand, there is the collisionless analog of resistive MHD instabilities,

---

[*] That quasi-neutrality is valid for wavelengths long compared to the Debye length can be seen from the following simple argument considering Poisson's equation $-\nabla^2 \phi = 4\pi e(n_i - n_e)$. The electron density contribution can be estimated from the force balance $n_e e \nabla \phi \simeq -\nabla p_e \simeq -T_e \nabla n_e$, hence $4\pi e n_e \simeq \phi/\lambda_D^2$, such that the l.h.s. of Poisson's equation, $-\nabla^2 \phi = k^2 \phi$, is negligible for $k^2 \lambda_D^2 \ll 1$.

where the free energy originates from the global magnetic configuration, but is released by wave–particle resonance instead of collisions. Here the well-known paradigm is the collisionless tearing mode, which will be discussed in section 7.5.

7.1.2 Quasi-linear theory

Since microinstabilities are driven by wave-particle resonance, which depends on the local properties of the distribution function, the growth rate may be modified by a minor change of the particle distribution. This process is described by the quasi-linear approximation (Vedenov *et al.*, 1961, 1962; Drummond & Pines, 1962). Here the equilibrium distribution function $f_{j0}(\mathbf{v})$ is allowed to evolve under the influence of finite-amplitude perturbations, while the latter still follow the linearized equation with the momentary $f_{j0}(t)$:

$$\partial_t f_{j0} = -\frac{q_j}{m_j} \int \frac{d^3k}{(2\pi)^3} i\mathbf{k}\phi_{-\mathbf{k}} \cdot \frac{\partial}{\partial \mathbf{v}} f_{j\mathbf{k}}, \tag{7.27}$$

$$\partial_t f_{j\mathbf{k}} + i\mathbf{k} \cdot \mathbf{v} f_{j\mathbf{k}} = \frac{q_j}{m_j} i\mathbf{k}\phi_{\mathbf{k}} \cdot \frac{\partial}{\partial \mathbf{v}} f_{j0}, \tag{7.28}$$

$$-k^2\phi_{\mathbf{k}} = \sum_j 4\pi q_j \int d^3v f_{j\mathbf{k}}, \tag{7.29}$$

restricting consideration at this point to the simplest case of electrostatic perturbations in a homogeneous unmagnetized plasma.

For further analytical evaluation of these equations the assumption is made that the instability is sufficiently weak, such that (7.28) can be solved in the WKB approximation,

$$f_{j\mathbf{k}}(\mathbf{v}, t) = f_{j\mathbf{k}}(\mathbf{v}) \exp\left\{-i \int^t \omega'_{\mathbf{k}} dt'\right\}, \tag{7.30}$$

where $\omega_{\mathbf{k}} = \omega_{\mathbf{k}r} + i\gamma_{\mathbf{k}}$ is the solution of the dispersion relation $\epsilon(\mathbf{k}, \omega) = 0$ with the momentary distribution function $f_{j0}(t)$ (if there are multiple solutions, the most unstable or least stable is to be chosen), and $f_{j\mathbf{k}}$ is given by (7.14)

$$f_{j\mathbf{k}} = \frac{q_j}{m_j} \phi_{\mathbf{k}} \frac{\mathbf{k} \cdot \partial f_{j0}/\partial \mathbf{v}}{\mathbf{k} \cdot \mathbf{v} - \omega_{\mathbf{k}+}}. \tag{7.31}$$

Since we are interested in the finite-amplitude behavior, we ignore the contribution from the initial perturbation δf in (7.14). Insertion into (7.27) gives the quasi-linear diffusion equation

$$\partial_t f_{j0} = \frac{\partial}{\partial \mathbf{v}} \cdot \mathscr{D} \cdot \frac{\partial f_{j0}}{\partial \mathbf{v}} \tag{7.32}$$

with

$$\mathscr{D} = \pi \frac{q_j^2}{m_j^2} \int \frac{d^3k}{(2\pi)^3} \mathbf{kk} |\phi_\mathbf{k}|^2 \delta(\mathbf{k} \cdot \mathbf{v} - \omega_\mathbf{k}), \tag{7.33}$$

while the fluctuation amplitude changes according to the momentary growth rate,

$$\partial_t |\phi_\mathbf{k}|^2 = 2\gamma_\mathbf{k}(t) |\phi_\mathbf{k}|^2. \tag{7.34}$$

In the expression for the diffusion tensor (7.33) only the resonant contribution resulting from the first term in (7.19) is considered, which gives rise to the δ-function in (7.20), since it describes the dominant irreversible change of the distribution. The non-resonant part ($\propto \gamma$) gives the reversible adiabatic response $\mathscr{D}_{ad} \propto \partial_t |\phi_\mathbf{k}|^2$, which can in general be neglected for resonantly driven instabilities.

A condition for the validity of this diffusion approximation[*] is that unstable waves are dispersive with phase velocities $v_{ph} = \omega_k/k$ extending over a sufficiently broad range,

$$k\Delta v_{ph} \gg \gamma. \tag{7.35}$$

This requirement is most stringent for a plasma embedded in a strong magnetic field $\mathbf{B}_0 = B_0 \mathbf{e}_z$ with $\Omega_e > \omega_{pe}$, where the electron dynamics is one-dimensional along the field such that the quasi-linear equation becomes

$$\partial_t f_0 = \frac{\partial}{\partial v_z} D \frac{\partial f_0}{\partial v_z}, \tag{7.36}$$

$$D(v_z) = \pi \frac{q_j^2}{m_j^2} \int \frac{dk}{2\pi} k^2 |\phi_k|^2 \delta(kv_z - \omega_k). \tag{7.37}$$

The classical paradigm is the "bump-on-the-tail" distribution, i.e., a weak diffuse electron beam ("gentle bump") propagating along the magnetic field at a speed v_b exceeding the thermal velocity of the bulk distribution, see fig. 7.1. Unstable modes have frequencies $\omega_k \simeq \omega_{pe}$ and growth rates given by (7.24). The diffusion coefficient is finite only in the unstable range, where $\partial f_0/\partial v|_{v=\omega_{pe}/k} > 0$, which leads to a flattening of the beam as indicated in the figure. In this state the instability is switched off and the final mode spectrum is obtained from the equation

$$|\phi_k(t)|^2 = |\phi_k(0)|^2 \exp\left\{2\int_0^t \gamma_k' dt'\right\}$$

---

[*] The quasi-linear approximation is often used in dynamical systems, where one can distinguish between fast short-wavelength fluctuations and a slowly evolving background, for instance in mean-field electrodynamics (section 5.1.3) or in the estimate for the saturation of the tearing mode (section 4.1.3). In microscopic plasma theory this approximation has received particular scrutiny and become the subject of longstanding discussions, see e.g., Laval & Pesme (1983).

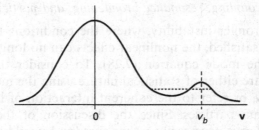

Fig. 7.1.   Bump-on-the-tail distribution. The dashed horizontal line indicates the result of the quasi-linear diffusion.

for $t \to \infty$, which depends on the initial fluctuation level. If the latter is much lower than the saturation level, the spectrum will be strongly peaked at the mode with the largest growth rate, which corresponds essentially to a single large-amplitude wave $\phi_0 \cos(kz - \omega_{pe}t)$. Only electrons with velocities sufficiently different from the phase velocity will still move on only slightly perturbed orbits, while those with $|v - v_{ph}| < \sqrt{e\phi/m_e} \equiv \Delta v_{tr}$ will be trapped in the potential wells of the wave, hence their orbits will be strongly modified (see the last paragraph in section 7.1.3). A nonlinear condition for the validity of the quasi-linear approximation is that waves should phase mix sufficiently rapidly, such that coherent trapping effects remain unimportant, which not only requires a broad range of unstable phase velocities, (7.35), but also that the corresponding waves are growing to similar levels such that

$$\Delta v_{ph}^{\text{nonlinear}} > \Delta v_{tr}. \qquad (7.38)$$

In an unmagnetized or weakly magnetized plasma, electrons resonate with the parallel velocity $v_\parallel = \mathbf{v} \cdot \mathbf{k}/k$. For any mode direction, waves try to generate a plateau in the distribution function. Since the unstable spectrum has a certain angular range, a cone about the beam direction in the case of a beam-driven instability, there is a competition of plateau formation in the different directions. As a result no strict plateau can be formed, such that the final velocity distribution has stabilizing gradients for all modes leading to the decay of the fluctuations and to more efficient thermalization than in the one-dimensional case. Examples of such behavior are discussed in the subsequent sections. It should also be noted that for a three-dimensional spectrum the condition, (7.35), regarding wave dispersion is less stringent, since even for nondispersive waves $\omega_\mathbf{k}/k = const$ the resonance extends over a broad velocity range because of the different propagation directions.

7.1.3 Mode coupling, resonance broadening, and particle trapping

In the case of stronger instability, where the conditions for quasi-linear behavior are not satisfied, the nonlinear effects can no longer be neglected as assumed in the mode equation (7.28). To consider these effects the theoretical tools are either of statistical nature using the methods of weak turbulence theory, or refer to the coherent interaction of large-amplitude waves with plasma particles. Since the discussion of these approaches given in this section is necessarily rather cursory, I would like to refer the reader interested in more details to the classical treatises by Kadomtsev (1965) and Galeev & Sagdeev (1973, 1983).

The concept of weak turbulence rests on the assumption that the individual modes $\mathbf{k}$ preserve approximately their linear dispersion properties, i.e., that the nonlinear frequency shift is small,

$$\Delta\omega_{\mathbf{k}} \equiv \omega_{\mathbf{k}} - \omega_{\mathbf{k}0} \ll \omega_{\mathbf{k}0}.$$

Waves with different phase velocities (different either in magnitude or in propagation direction) interact only weakly. One can therefore assume that in the lowest-order approximation modes are uncorrelated. This is called the random-phase approximation, which has a more solid justification for an ensemble of propagating waves than for fluid turbulence consisting of non-propagating eddies.

Instead of the linear equation (7.11) we now consider the full nonlinear equation for the perturbed distribution function

$$\partial_t f_{j\mathbf{k}} + i\mathbf{k} \cdot \mathbf{v} f_{j\mathbf{k}} = \frac{q_j}{m_j} i\phi_{\mathbf{k}} \mathbf{k} \cdot \frac{\partial}{\partial\mathbf{v}} f_{j0}$$

$$+ \frac{q_j}{m_j} \frac{\partial}{\partial\mathbf{v}} \int \frac{d^3k'}{(2\pi)^3} i\mathbf{k}' \Big(\phi_{\mathbf{k}'} f_{j\mathbf{k}-\mathbf{k}'} - \langle\phi_{\mathbf{k}'} f_{j\mathbf{k}-\mathbf{k}'}\rangle\Big). \quad (7.39)$$

Laplace transformation gives

$$f_{j\mathbf{k}\omega} = \mathbf{g}_{j\mathbf{k}\omega} \cdot \mathbf{k}\phi_{\mathbf{k}\omega} f_{j0}$$

$$+ \mathbf{g}_{j\mathbf{k}\omega} \cdot \int \frac{d^3k'd\omega'}{(2\pi)^4} \mathbf{k}' \Big(\phi_{\mathbf{k}'\omega'} f_{j\mathbf{k}-\mathbf{k}',\omega-\omega'} - \langle\phi_{\mathbf{k}'\omega'} f_{j\mathbf{k}-\mathbf{k}',\omega-\omega'}\rangle\Big), \quad (7.40)$$

where $\mathbf{g}$ is the free particle propagator

$$\mathbf{g}_{j\mathbf{k}\omega} = \frac{1}{\mathbf{k}\cdot\mathbf{v} - \omega_+} \frac{q_j}{m_j} \frac{\partial}{\partial\mathbf{v}}. \quad (7.41)$$

Iterative solution of (7.40) and application of the random-phase approximation results in the equation for the wave intensity $I_{\mathbf{k}\omega} = |\phi_{\mathbf{k}\omega}|^2$. For weak instability $\gamma_{\mathbf{k}}/\omega_{\mathbf{k}} \ll 1$ the dispersion relation

$$\epsilon(\mathbf{k}, \omega + i\gamma) = \epsilon_r + i\epsilon_i \simeq (\omega - \omega_{\mathbf{k}})\partial_\omega\epsilon_r + i\epsilon_i = 0$$

has the solution $\omega = \omega_k$, $\gamma = \gamma_k = -\epsilon_i/\partial_\omega \epsilon_r$, such that to lowest order $I_{k\omega} = I_k \delta(\omega - \omega_k)$. The slow time-dependence of $I_k(t)$ is determined by the wave kinetic equation

$$\partial_t I_k = 2\gamma_k I_k + \pi(\partial_\omega \epsilon_r)^{-2} \int \frac{d^3 k'}{(2\pi)^3} |M_{k'k''}^{(2)}|^2 I_{k'} I_{k''} \delta(\omega_k - \omega_{k'} - \omega_{k''})$$

$$+ 2I_k(\partial_\omega \epsilon_r)^{-1} \int \frac{d^3 k'}{(2\pi)^3} \text{Im} \left\{ \frac{M_{kk''}^{(2)} M_{k''k}^{(2)}}{\epsilon(k'', \omega_k - \omega_{k'})} - M_{kk'k''}^{(3)} \right\} I_{k'}, \quad (7.42)$$

where

$$M_{kk'}^{(2)} = \sum_j \frac{4\pi q_j}{k^2} \int d^3 v \; \mathbf{g}_{k\omega} \cdot \left[\mathbf{k}'(\mathbf{g}_{k''\omega''} \cdot \mathbf{k}'') + \mathbf{k}''(\mathbf{g}_{k'\omega'} \cdot \mathbf{k}') \right] f_{j0},$$

$$\mathbf{k}'' = \mathbf{k} - \mathbf{k}', \quad \omega'' = \omega_k - \omega_{k'},$$

and $M^{(3)}$ is a similar expression involving three factors of $\mathbf{g}$. In addition to the quasi-linear change of $f_{j0}(t)$, we have two nonlinear processes, *viz.*, resonant mode coupling and nonlinear wave-particle interaction. Resonant mode coupling, the second term on the r.h.s. of (7.42) corresponds to the induced decay of mode $\mathbf{k}$ into modes $\mathbf{k}'$, $\mathbf{k}''$,

$$\omega_k = \omega_{k'} + \omega_{k''}, \quad (7.43)$$

the probability being proportional to the intensities ("occupation numbers" in quantum mechanical language) of the decay products. It is easy to see that the resonance condition, (7.43), can only be satisfied, if the waves have negative dispersion $\partial \omega_k / \partial k < 0$ or if waves of different types interact, for instance two high-frequency Langmuir waves and one low-frequency ion-sound wave.

The second process, nonlinear wave–particle interaction, results from the wave-particle resonances in the expressions in brackets in the last term in (7.42),

$$\omega_k - \omega_{k'} = (\mathbf{k} - \mathbf{k}') \cdot \mathbf{v}, \quad (7.44)$$

the resonance of particle $\mathbf{v}$ with the beat wave $\mathbf{k} - \mathbf{k}'$. The process is the analog of stimulated Compton scattering, the inelastic scattering of a wave on a particle, $\hbar \omega_k - \hbar \omega_{k'} = \Delta E = \Delta \mathbf{p} \cdot \mathbf{v}$, where from momentum conservation the change of particle momentum is $\Delta \mathbf{p} = \hbar(\mathbf{k} - \mathbf{k}')$. In this way Langmuir waves, whose phase velocity ω_{pe}/k is much higher than the thermal velocity, can interact nonlinearly with the bulk of the electron distribution. Since the last term in (7.42) is proportional to I_k, it gives rise to a nonlinear change of the growth rate, while resonant mode coupling acts as a sink (or possibly also a source) of modal energy.

Weak turbulence theory is a consistent model for a weakly excited dispersive medium. It is, however, found that for many turbulent plasmas the nonlinear processes just discussed are rather insignificant compared with the quasi-linear effect. To go beyond the quasi-linear approximation, nonlinear corrections to the particle orbits are important. A mathematically consistent formalism is Kraichnan's direct interaction approximation (Kraichnan, 1959), which is, however, simply too unwieldy when applied to the Vlasov equation (Orszag & Kraichnan, 1967; Biskamp, 1968) to be of much benefit without severe ad hoc assumptions. It is therefore practicable to follow a route based more on physical reasoning. Dupree (1966) has introduced the concept of broadening of the wave-particle resonance (see also Dum & Dupree, 1970). The basic ingredient of the theory is a modification of the free-particle propagator, i.e., the $(\mathbf{k} \cdot \mathbf{v} - \omega_k)$ resonance in the quasi-linear diffusion coefficient (7.33), by taking into account the effect of just this diffusion. In particular, one replaces the expression $\exp\{-i\mathbf{k} \cdot (\mathbf{x} - \mathbf{v}t)\}$ containing the free-particle orbit by the average $\langle\exp\{-i\mathbf{k} \cdot \mathbf{x}(t)\}\rangle$ over the true orbits, which is calculated in the diffusion approximation. In a one-dimensional statistically homogeneous problem this is described by a diffusion in velocity space

$$\langle e^{-ikx(t)}\rangle = e^{-ikx+ikvt-\frac{1}{3}k^2 D_v t^3}, \tag{7.45}$$

where D_v is determined self-consistently by the nonlinear integral equation

$$D_v(v) = \frac{q_j^2}{m_j^2} \int \frac{dk}{2\pi} k^2 |\phi_k|^2 \mathrm{Re} \int_0^\infty dt\, e^{i(kv-\omega)t-\frac{1}{3}k^2 D_v(v)t^3}, \tag{7.46}$$

generalizing the quasi-linear expression, (7.37). The magnitude of the resonance broadening can easily be estimated. For $(D_v/k)^{1/3} < \Delta v$, where Δv is the range of unstable phase velocities, the effect is small, while in the opposite case $(D_v/k)^{1/3} > \Delta v$ resonance broadening dominates over free-particle propagation and the time integral in (7.46) is approximately

$$\mathrm{Re} \int_0^\infty dt\, e^{i(kv-\omega)t-\frac{1}{3}k^2 D_v t^3} \simeq 1.3(k^2 D_v)^{-1/3}. \tag{7.47}$$

Substitution in (7.46) yields D_v in terms of the fluctuation energy $\langle\widetilde{E}^2\rangle = \int(dk/2\pi)k^2|\phi_k|^2$,

$$D_v \sim (q_j/m_j)^{3/2}\langle\widetilde{E}^2\rangle^{3/4}/k_0^{1/2}, \quad k_0^{-2/3} = \langle 1/k^{2/3}\rangle \tag{7.48}$$

and the diffusion velocity

$$\delta v = (D_v/k_0)^{1/3} \sim (q_j/m_j)^{1/2}\langle\widetilde{E}^2\rangle^{1/4}/k_0^{1/2} \sim (q_j\phi/m_j)^{1/2}, \tag{7.49}$$

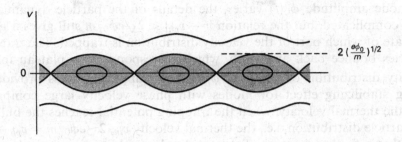

Fig. 7.2. Phase-space orbits of particles in the potential $\phi_0 \sin x$. Areas of trapped particles are shaded.

which is just the velocity range of a particle trapped by the potential ϕ of the wave and forced to move with the phase velocity of the wave, as we will discuss in a minute.

In the case of a multi-dimensional wave spectrum, resonance broadening can be caused by *spatial* particle diffusion $D_\perp$, in particular diffusion across a strong magnetic field, which is an important stabilizing effect for drift waves. The process has been studied by Dupree (1967) in the case of weak magnetic shear, where the resonance function assumes the form

$$\langle e^{i\mathbf{k}\cdot\mathbf{x}(t)} \rangle = e^{-i\mathbf{k}\cdot\mathbf{x}+ik_\parallel v_\parallel t - k_\perp^2 D_\perp t}, \tag{7.50}$$

and in the case of strong shear by Hirschman & Molvig (1979)

$$\langle e^{i\mathbf{k}\cdot\mathbf{x}(t)} \rangle = e^{-i\mathbf{k}\cdot\mathbf{x}+ik_\parallel v_\parallel t - \frac{1}{3}(k_\parallel' v_\parallel)^2 D_\perp t^3}, \tag{7.51}$$

where $k_\parallel' = k_\perp/L_s$, $L_s =$ shear length, (4.125), $L_s^{-1} = \hat{s}/Rq$ in a cylindrical column, $\hat{s} =$ shear parameter, (3.16).

For fluctuation levels, at which resonance broadening is important, the dynamics is often dominated by a few modes, such that the picture of particles interacting with a coherent large-amplitude wave is more appropriate than a statistical description of particle orbits. Assume a stationary electrostatic wave $\phi(x,t) = \phi_0 \sin k(x - v_{ph}t)$. For convenience we consider the particle orbits in the frame v_{ph} of the wave, i.e., particles moving in the static potential $\phi_0 \sin kx$. From energy conservation

$$W = \frac{m}{2}v^2 + e\phi_0 \sin kx = const$$

we see that there are two classes of particles, those with $W < \phi_0$ are trapped in the potential wells of the wave, while those with $W > \phi_0$ are passing above the potential crests in both directions with only slightly modulated orbits. The phase space diagram is illustrated in fig. 7.2. If the mode amplitude $\phi_0(t)$ varies, the details of the particle dynamics is more complicated, but the relation $|v - v_{ph}| < 2\sqrt{e\phi_0/m}$ still gives a good estimate of which part of the velocity distribution is trapped. The trapped particles bounce back and forth, which corresponds to a plateau in the velocity distribution of width $\propto \phi_0^{1/2}$. The trapping process provides a strong stabilizing effect for modes with phase velocity large compared with the thermal velocity, when the trapping potential reaches the bulk of the particle distribution, i.e., the thermal velocity v_{th}, $2\sqrt{e\phi_0/m} = v_{ph} - v_{th}$, such that a further increase of the amplitude would trap and accelerate a substantial number of particles. An example of this process is ion trapping in the current-driven ion-sound instability, see section 7.2.2.

7.1.4 Turbulent transport coefficients

Small-scale turbulent fluctuations, such as are generated by a microinstability, act as quasi-collisions and give rise to anomalous transport coefficients. Here particular interest is in transport effects in Ohm's law, i.e., in the electron equation of motion, since these can lead to enhanced magnetic reconnection rates. We consider Ohm's law, (6.5) or (6.20), splitting the dynamical variables into background and fluctuating parts,

$$\mathbf{E}_0 + \frac{1}{c}\mathbf{v}_{e0} \times \mathbf{B}_0 = -\frac{1}{n_0}\langle \tilde{n}_e\tilde{\mathbf{E}}\rangle - \frac{1}{c}\langle \tilde{\mathbf{v}}_e \times \tilde{\mathbf{B}}\rangle + \eta\mathbf{j}_0 - \eta_2\nabla^2\mathbf{j}_{e0},^{*} \tag{7.52}$$

where the inertia and pressure terms are neglected for simplicity. The terms in brackets represent the turbulent transport effects. Their relative importance is seen in two limiting cases.

(a) *Electrostatic fluctuations* $\tilde{\mathbf{B}} \simeq 0$, described by the first bracketed term in (7.52), which can be related to a turbulent resistivity η_T^{es},

$$\eta_T^{es}\mathbf{j}_0 = \frac{v_{\text{eff}}}{\omega_{pe}^2}\mathbf{j}_0 = -\frac{1}{n_0}\langle \tilde{n}_e\tilde{\mathbf{E}}\rangle. \tag{7.53}$$

Note that this expression is independent of the wavelength, valid for Debye-length turbulence $k\lambda_D \sim 1$ as in the case of ion-sound instability (section 7.2), as well as for quasi-neutral electrostatic turbulence with wave numbers $k\lambda_D \ll 1$ resulting, for instance, from lower-hybrid-drift instability (section 7.3).

In the quasi-linear approximation the $\langle \tilde{n}_e\tilde{\mathbf{E}}\rangle$ term can be written as a linear function of the modal wave energy $|E_\mathbf{k}|^2$. From (7.14), neglecting

---

* In the second term on the r.h.s. $\tilde{v}_e$ should, properly speaking, read $\tilde{n}v_e/n_0$, but in electromagnetic processes, where this term is important, density fluctuations are usually not essential.

the contribution from the initial perturbation δf, we obtain

$$n_{jk\omega} = -\chi_j(\mathbf{k}, \omega)\frac{k^2}{4\pi q_j}\phi_{\mathbf{k}\omega}, \tag{7.54}$$

where

$$\chi_j = -\frac{\omega_{pj}^2}{k^2}\int d^3v\frac{\mathbf{k}\cdot\partial F_j/\partial\mathbf{v}}{\mathbf{k}\cdot\mathbf{v}-\omega_+} = -\omega_{pj}^2\frac{\partial}{\partial\omega}\int d^3v\frac{F_j}{\mathbf{k}\cdot\mathbf{v}-\omega_+} \tag{7.55}$$

is the linear susceptibility of the particle species j, hence

$$q_j\langle\widetilde{n}_j\widetilde{\mathbf{E}}\rangle = \int\frac{d^3k}{(2\pi)^3}\frac{|E_\mathbf{k}|^2}{4\pi}\mathbf{k}\,\text{Im}\big\{\chi_j(\mathbf{k}, \omega_\mathbf{k}+i\gamma_\mathbf{k})\big\}. \tag{7.56}$$

Since $\epsilon(\mathbf{k}, \omega_\mathbf{k}+i\gamma_\mathbf{k}) \equiv 1 + \chi_i + \chi_e = 0$ using expression (7.16), one has $\text{Im}\{\chi_i\} = -\text{Im}\{\chi_e\}$, which gives the momentum conservation for electrostatic waves (with $q_e = -e$)

$$e\langle\widetilde{n}_e\widetilde{\mathbf{E}}\rangle = q_i\langle\widetilde{n}_i\widetilde{\mathbf{E}}\rangle. \tag{7.57}$$

We thus find that the anomalous resistivity vanishes for pure electron waves $\widetilde{n}_i = 0$. Using (7.53), (7.56), (7.57) the turbulent collision frequency can hence be written in the form

$$\begin{aligned}
v_{\text{eff}} &= -\frac{8\pi}{n_0 m_e u^2}\Big\langle\mathbf{k}\cdot\mathbf{u}\,\text{Im}\{\chi_i(\mathbf{k})\}\Big\rangle\frac{\widetilde{E}^2}{8\pi}\\
&= \frac{8\pi\omega_{pi}^2}{n_0 m_e u^2}\Big\langle\frac{\mathbf{k}\cdot\mathbf{u}}{k^2 v_{ti}^2}\,\text{Im}\{\zeta_i Z(\zeta_i)\}\Big\rangle\frac{\widetilde{E}^2}{8\pi},
\end{aligned} \tag{7.58}$$

where $\mathbf{j}_0 = -en_0\mathbf{u}$, the brackets $\langle\rangle$ indicate the average over the spectrum, $\widetilde{E}^2 = \int|E_\mathbf{k}|^2 d^3k/(2\pi)^3$, and the last expression is valid for a Maxwellian distribution, see section 7.2.1.

(b) *Quasi-incompressible electromagnetic fluctuations*, where the density can be assumed constant $\widetilde{n}_e \simeq 0$. Here the second bracketed term in (7.52) dominates,

$$\eta_T^{\text{em}}\mathbf{j}_0 = \frac{1}{c}\langle\widetilde{\mathbf{v}}_e\times\widetilde{\mathbf{B}}\rangle. \tag{7.59}$$

This expression is similar to the electromotive force ϵ, (5.13), in mean-field electrodynamics, where the analogy also indicates that the term may have a complicated tensorial form such as the α- and β-effects, cf. (5.17), in particular in the presence of a mean magnetic field. However, in contrast to mean-field electrodynamics, in microscopic theory dealing with short-wavelength oscillations $kc/\omega_{pi} > 1$, $\widetilde{\mathbf{v}}_e$ differs in general from the mass flow $\widetilde{\mathbf{v}} \simeq \widetilde{\mathbf{v}}_i$. Often one has $\widetilde{\mathbf{v}}_e \gg \widetilde{\mathbf{v}}_i$ such as in the whistler mode introduced in section 6.2.

This raises a point which is often ignored in the discussion of anomalous resistivity. It is usually assumed that the turbulence acts as an effective resistivity, which implies a momentum transfer to the ions. This is true for electrostatic modes, where (7.57) indicates that $\langle \tilde{n}_e \widetilde{\mathbf{E}} \rangle = 0$, if there is no coupling to the ions. For electromagnetic modes such a conclusion is not valid, since for $\tilde{\mathbf{v}}_i = 0$ the term $\langle \tilde{\mathbf{v}}_e \times \widetilde{\mathbf{B}} \rangle$ is proportional to $\langle \tilde{\mathbf{j}} \times \widetilde{\mathbf{B}} \rangle$, which does not, in general, vanish. In the whistler instability excited, for instance, by electron temperature anisotropy, ions are not coupled and hence electron momentum is conserved. Such oscillations may, however, give rise to anomalous electron viscosity, which also enhances reconnection. (The turbulent electron viscosity generated by the electron Kelvin–Helmholtz instability at spatial scales below the electron inertia length $kd_e > 1$ arises from the electron Reynolds stress tensor $m_e \langle \mathbf{v}_e^{(1)} \mathbf{v}_e^{(1)} \rangle$, which is ignored in (7.52).)

It should be mentioned that small-scale turbulence also leads to heating, where the heating rates of electrons and ions depend on the type of instability and its nonlinear properties. The quasi-linear expressions for the turbulent transport coefficients, such as (7.56), are functions of the distribution function, which is not known a priori but evolves self-consistently under the influence of the turbulent spectrum, the mean magnetic field and also the boundary conditions (open or periodic system), as discussed in the subsequent sections.

7.2 Ion-sound instability

7.2.1 Linear stability characteristics

If the electric current density in a plasma exceeds a certain threshold value, electrostatic turbulence is excited. We assume that both particle species have Maxwellian velocity distributions with temperatures $T_j = m_j v_{tj}^2$, which are, in general, not equal in a collisionless plasma, and that the electron distribution is shifted by $\mathbf{u}$ with respect to the ion distribution:

$$F_i = \frac{1}{(2\pi v_{ti}^2)^{3/2}} \exp\{-v^2/2v_{ti}^2\}, \quad F_e = \frac{1}{(2\pi v_{te}^2)^{3/2}} \exp\{-(\mathbf{v}-\mathbf{u})^2/2v_{te}^2\}.$$

$$(7.60)$$

Using (7.16) and (7.55) the dispersion relation can be written in the following form:

$$1 + \sum_j \chi_j = 1 - \frac{1}{2k^2} \sum_j \frac{\omega_{pj}^2}{v_{tj}^2} Z'(\zeta_j) = 0, \qquad (7.61)$$

where

$$\zeta_i = \omega/\sqrt{2}kv_{ti}, \quad \zeta_e = (\omega - \mathbf{k} \cdot \mathbf{u})/\sqrt{2}kv_{te}, \qquad (7.62)$$

and $Z(\zeta)$ is the plasma dispersion function (tabulated in Fried & Conte, 1961)

$$Z(\zeta) = \frac{1}{\sqrt{\pi}} \int_{-\infty}^{\infty} dt \, \frac{e^{-t^2}}{t - \zeta}, \qquad (7.63)$$

which satisfies the relation

$$Z'(\zeta) = -2[1 + \zeta Z(\zeta)]. \qquad (7.64)$$

The Z-function has the power-series expansion for small argument $|\zeta| \ll 1$

$$Z(\zeta) = i\sqrt{\pi}e^{-\zeta^2} - 2\zeta(1 - \tfrac{2}{3}\zeta^2 \pm ...) \qquad (7.65)$$

and the asymptotic expansion for large argument $|\zeta| \gg 1$

$$Z(\zeta) = i\sigma\sqrt{\pi}e^{-\zeta^2} - \frac{1}{\zeta}\left(1 + \frac{1}{2\zeta^2} + ...\right), \qquad (7.66)$$

where $\sigma = 0, 1, 2$ for $\mathrm{Im}\zeta > 0, = 0, < 0$, respectively.

The dispersion relation (7.61) describes the electrostatic (electron-ion) two-stream instability. Before discussing the general threshold conditions for instability it is convenient to consider two special cases, which can be treated analytically.

(a) $u \gg v_{te}$. Here the thermal spread of the distribution functions is negligible, and the dispersion relation reduces to the algebraic form, (7.18),

$$1 = \frac{\omega_{pi}^2}{\omega^2} + \frac{\omega_{pe}^2}{(\omega - k_\parallel u)^2}, \qquad (7.67)$$

where $k_\parallel u = \mathbf{k} \cdot \mathbf{u}$. This equation describes the two-stream instability in a cold plasma, called Buneman instability (Buneman, 1959). Figure 7.3, which shows the r.h.s. of (7.67), illustrates the properties of the solution. While for short-wavelength modes $k_\parallel u > \omega_{pe}$ (omitting corrections of order ω_{pi}/ω_{pe}) the minimum of this function is lower than unity, giving rise to four real roots, for $k_\parallel u < \omega_{pe}$ two conjugate complex roots arise, of which one corresponds to instability. Since ω is independent of $k_\perp$, $\mathrm{Im}\{\omega\} > 0$ implies a broad angular spectrum of unstable modes $\mathbf{k}$. Maximum growth rates and frequencies are of the order of ω_{pi}.

When u becomes smaller than v_{te}, ion thermal effects give rise to heavy Landau damping which, in general, switches off the two-stream instability, unless the ion temperature is sufficiently low.

(b) $u \ll v_{te}$, $T_i \ll T_e$. Sound waves or ion-sound waves, as they are called in a plasma, are normally strongly damped, since the ion-sound

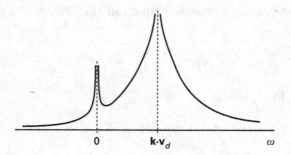

Fig. 7.3. Schematic drawing of the r.h.s. of (7.67).

velocity $c_s = \sqrt{(T_e + 3T_i)/m_i}$ [*] is only slightly larger than the ion thermal velocity, $c_s = 2v_{ti}$ for $T_i = T_e$, such that there is a large negative gradient of $F_i(v)$ at the ion-sound resonance $\omega/k \simeq c_s$. However, the damping becomes exponentially weak for $T_i \ll T_e$, where $c_s \simeq \sqrt{T_e/m_i} \gg v_{ti}$. Hence $|\zeta_i| \gg 1$, which allows to use the asymptotic expansion (7.66) for $Z(\zeta_i)$, while for $u \ll v_{te}$ one has $|\zeta_e| \ll 1$, which allows us to apply the small-argument expansion (7.65) for $Z(\zeta_e)$. With these approximations the dispersion relation (7.61) becomes

$$1 - \frac{\omega_{pi}^2}{\omega^2} + \frac{1}{k^2\lambda_D^2} + i\sqrt{\pi}\,\frac{\omega - k_\parallel u}{\sqrt{2}kv_{te}}\,\frac{1}{k^2\lambda_D^2} = 0, \tag{7.68}$$

which can easily be solved approximately, since $\gamma \ll \omega_r$:

$$\omega_r^2 = \frac{k^2c_s^2}{1 + k^2\lambda_D^2}, \tag{7.69}$$

$$\gamma = \sqrt{\frac{\pi}{8}\frac{m_e}{m_i}}\,\frac{k_\parallel u - \omega_r}{(1 + k^2\lambda_D^2)^{3/2}}. \tag{7.70}$$

Equation (7.69) shows that ion-sound waves are dispersive with phase velocity ω/k decreasing for $k\lambda_D \gtrsim 1$. Instability arises if the drift velocity exceeds the threshold $u_c = \omega/k_\parallel$; $u_c \simeq c_s$ for the most important modes with $k\lambda_D < 1$. For $c_s \ll u \ll v_{te}$ the growth rate is $\gamma \simeq 0.6\sqrt{m_e/m_i}\,k_\parallel u$.

We now briefly discuss the dispersion relation (7.61) for general temperature ratio T_e/T_i. A detailed numerical study has been performed by Fried & Gould (1961) who give, in particular, the critical drift velocity u_c

---

[*] In collisionless theory the hydrodynamic expression (3.26) $c_s = \sqrt{\gamma p/\rho}$ is generalized to $c_s = \sqrt{(\gamma_e p_e + \gamma_i p_i)/\rho}$, which here becomes $\sqrt{(T_e + 3T_i)/m_i}$, since electrons are isothermal ($\gamma_e = 1$) because of $c_s \ll v_{te}$, whereas the ions are adiabatic ($\gamma_i = 3$) because of $c_s \gg v_{ti}$. Note that in a collisionless plasma only one degree of freedom is excited, hence $\gamma = 3$.

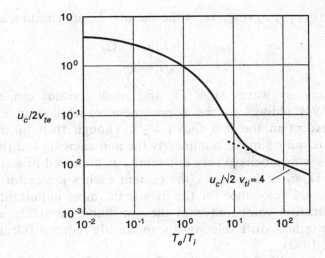

Fig. 7.4. Critical drift velocity $u_c/2v_{te}$ for the two-stream instability as a function of the temperature ratio T_e/T_i (from Fried & Gould, 1961).

as a function of the temperature ratio T_e/T_i, see fig. 7.4. For $T_e \lesssim T_i$ the drift velocity must exceed the electron thermal speed for instability. Only if $T_e/T_i > 10$ does u_c become of order c_s, as given by the approximate dispersion relation (7.70). Figure 7.4 indicates that for $T_e/T_i > 20$ there is a simple asymptotic behavior $u_c \simeq 4\sqrt{2}v_{ti}$. [*]

If the plasma is embedded in a mean magnetic field, the stability analysis becomes much more intricate, since the appearance of the cyclotron frequencies Ω_j and the orientation of the field with respect to both the current direction and the wave vector $\mathbf{k}$ make the dispersion relation rather complicated. In the context of anomalous resistivity it is useful to distinguish between the low-β case, where the current flows along the magnetic field, and the high-β case, where the current is perpendicular to the field. In the parallel-current situation with $\mathbf{B} = \mathbf{e}_z B$ the susceptibilities χ_j in (7.61) are replaced by

$$\chi_{j,\mathbf{k}\omega} = \frac{\omega_{pj}^2}{k^2 v_{tj}^2}\left[1 + \frac{\omega - k_z u_j}{\sqrt{2}|k_z|v_{tj}}e^{-\lambda_j}\sum_{n=-\infty}^{\infty} I_n(\lambda_j)Z(\zeta_j^{(n)})\right], \quad (7.71)$$

[*] This behavior is dominated by ion Landau damping, which in the large T_e/T_i-range is important for modes with $k\lambda_D \gg 1$. Since ion Landau damping is neglected in the approximate dispersion relation (7.68), the threshold $k_\parallel u = \omega$ is valid only for not too large k.

where $\lambda_j = (k_\perp \rho_j)^2$, $\rho_j = v_{tj}/\Omega_j$ is the thermal Larmor radius, and

$$\zeta_j^{(n)} = \frac{\omega - k_z u_j + n\Omega_j}{\sqrt{2}|k_z|v_{tj}}. \tag{7.72}$$

For low-frequency waves $\omega \ll \Omega_j$ and small Larmor radius $\lambda_j \ll 1$ equation (7.71) reduces to the nonmagnetic expression in (7.61) with $k \to |k_z|$ (except in the first factor k^{-2}). Though their linear stability properties are quite similar, nonlinearly the nonmagnetic and the strongly magnetized systems behave very differently, as discussed in section 7.2.2.

For $\omega \gtrsim \Omega_i$, $k_z \ll k_\perp$, $\lambda_i \sim 1$ the current excites ion-cyclotron waves, where the $n = 1$ resonance for the ions is the most important one. The main modification compared with the ion-sound instability is that for $T_e \sim T_i$ the critical drift velocity is significantly reduced (Drummond & Rosenbluth, 1962).

For many applications the high-β perpendicular-current configuration is the most interesting one. Here the electron plasma frequency ω_{pe} is much higher than the electron cyclotron frequency Ω_e, typically $\omega_{pe} \gg \Omega_e \sim \omega_{pi} \gg \Omega_i$, such that ion-sound waves $\omega \sim \omega_{pi}$ can resonate with the electron-cyclotron motion and the ions are essentially unmagnetized. A cross-field current may drive the coupling between electron Bernstein waves and (Doppler-shifted) ion-sound waves unstable, which is called electron-cyclotron-drift instability (see, e.g., Forslund *et al.*, 1970). The most interesting feature of this instability has been its insensitivity to the temperature ratio T_e/T_i, which results from the fact that (perpendicular) electron Bernstein waves are undamped and the positive slope (seen from the electron frame) of the ion distribution function drives the instability, i.e., acts as inverse Landau damping. The instability is essentially limited to the plane perpendicular to the magnetic field, since finite $k_\parallel$ introduces strong electron Landau damping. More importantly, the instability saturates at low amplitude due to nonlinear broadening of the electron-cyclotron resonance (Lampe *et al.*, 1971), such that, for efficient generation of turbulence and anomalous resistivity, the threshold conditions of the nonmagnetic ion-sound instability are restored. We shall briefly come back to this point in section 7.2.2 when discussing the nonlinear properties of the ion-sound instability in more detail.

7.2.2 Nonlinear saturation and long-time behavior

The transport effects produced by the two-stream instability cannot be derived from the linear stability characteristics, but require knowledge of the fully nonlinear behavior. While the linear properties are essentially independent of the mean magnetic field (except for the electron-cyclotron-drift instability, which is, however, not important nonlinearly), the saturation mechanism is strongly influenced by both the magnitude and the direction of the field. We therefore distinguish between the low-β parallel-current case, where the instability is very inefficient due to quasi-linear flattening of the electron distribution, and the high-β cross-field-current case, where saturation occurs by ion trapping. Moreover, the development of the turbulence depends on the boundary conditions: open or periodic system, constant applied electric field, or constant current.

Low-β parallel-current configuration Most quasi-stationary laboratory plasmas and many plasmas in space, e.g., in the solar corona, have low β, such that currents are flowing almost parallel to B_0. The electron cyclotron-frequency is high, typically $\Omega_e \gtrsim \omega_{pe}$. In such a plasma the electrons are magnetized, which makes their dynamics one-dimensional along the field. Here the strongest nonlinear effect is quasi-linear plateau formation described by (7.36) and (7.37), in the short range of unstable phase velocities $c_s \geq v_z > v_{ti}$, (7.69). This change of f_{e0} implies only very weak heating, $\Delta T_{e,\text{eff}} \sim m_e u^2 c_s^2 / v_{te}^2 = (m_e/m_i)(u/v_{te})^2 T_e$. In a periodic system with constant current the turbulence will subsequently decay, while in an open system, where electrons are continuously resupplied with a Maxwellian distribution, a low level of turbulence is sustained giving rise to weak anomalous resistance. The more interesting situation arises in the case of a constant applied electric field E_0.* Since the ion-sound instability cannot prevent electron acceleration or runaway, the system evolves into a state with $u \gtrsim v_{te}$, where violent two-stream instability sets in. Hence one may expect a balance between electron heating and acceleration, such that u/v_{te} remains of order unity. The process has been studied numerically by particle simulations (Boris *et al.*, 1970; Biskamp & Chodura, 1973a), which reveal the following features:

(a) The system evolves in a self-similar way

$$u \sim v_{te} \sim \langle k \rangle^{-1} \propto t, \quad \langle \tilde{E}^2 \rangle / T_e \sim const, \qquad (7.73)$$

and the electron distribution function remains self-similar. The ion energy increase is slow, in particular the ions are not trapped by the wave potential and hence are not heated irreversibly.

(b) Drift velocity $u = \langle v \rangle$ and electron thermal velocity v_{te}, $v_{te}^2 = \langle v^2 \rangle - \langle v \rangle^2$ remain equal, $u = v_{te}$, apart from oscillations about each other.

---

* These considerations do not apply to tokamak plasmas. Though these are nearly collisionless, residual collisions are not negligible since the applied electric field is weak, usually below the critical field for electron run-away.

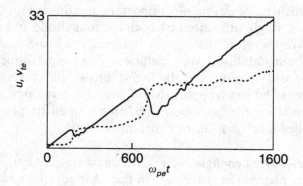

Fig. 7.5. $u(t)$ (solid line) and $v_{te}(t)$ (dashed line) in a one-dimensional system evolving under the influence of a constant electric field. The mass ratio is $m_i/m_e = 1836$ (from Biskamp & Chodura, 1973a).

This equality can be interpreted in terms of a simple extremum principle. Consider the energy balance

$$\tfrac{1}{2}u^2(t) + \tfrac{1}{2}v_{te}^2(t) = \frac{e}{m_e}E_0 \int_0^t dt'\, u(t'), \tag{7.74}$$

where the electrostatic field energy $\langle \tilde{E}^2 \rangle$, which is always small compared with the thermal energy, is neglected and so are the initial values. If the evolution is self-similar, u/v_{te} is constant and from (7.74) $u \propto v_{te} \propto t$. Let us look for the state of maximum thermal energy production by a given field E_0. Writing $u = \alpha(e/m_e)E_0 t$ and varying the thermal energy with respect to α, one obtains

$$\frac{\delta}{\delta \alpha} v_{te}^2 = \left[(e/m_e)E_0 \right]^2 t^2 (1 - 2\alpha) = 0, \tag{7.75}$$

which gives $\alpha = 1/2$, hence $u = v_{te}$, in agreement with the simulation result. This implies that the two-stream turbulence driven by a constant electric field maximizes dissipation. The time evolution of u and v_{te} is plotted in fig. 7.5.

This behavior cannot be explained by quasi-linear theory. Instead the picture of the interaction with a nonlinear coherent wave is more appropriate. The actual state of the system is quiescent for most of the time, characterized by a nearly stationary large-amplitude electrostatic wave, a Bernstein-Green-Kruskal wave of wavenumber k_0. The presence of this wave does not, however, prevent free particle acceleration, $\dot{u} \simeq (e/m_e)E_0$, see fig. 7.5, while v_{te} remains nearly constant, i.e., the width of the distribution function does not change, until the wave becomes unstable to

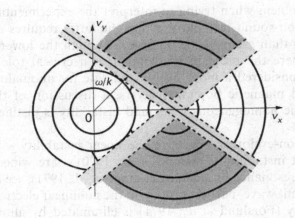

Fig. 7.6. Velocity-space region of resonant wave–electron interaction. Only the non-shaded part can be freely accelerated.

pairwise coalescence $k_0 \to k_0/2$. This is a rapid process connected with an increase of the wave amplitude which, by trapping a major fraction of the electrons, strongly decreases the drift velocity and broadens the distribution, such that $v_{te} > u$. From this state-free acceleration restarts. The time interval between coalescence bursts increases as k_0^{-1}.

If we want to describe the average behavior in terms of an anomalous resistivity η_T as a function of the plasma state, we may write

$$\eta_T = E_0/enu = E_0/env_{te} \propto T_e^{-1/2}. \qquad (7.76)$$

In the absence of a mean magnetic field where the resonance condition is $\mathbf{k \cdot v} = \omega_k$ instead of $k_z v_z = \omega_k$, resonant wave–particle interaction occurs for most of the electron distribution, as illustrated in fig. 7.6. Only in a small cone with cone-angle $\theta_0 \sim c_s/u$ can electrons be freely accelerated, such that the stabilizing distortion of the electron distribution is weak and ion trapping becomes the dominant nonlinear process as discussed in the following subsection.

High-β cross-field-current configuration Historically the primary application of current-driven ion-sound turbulence as the source of anomalous resistivity has been to collisionless magnetic shocks, especially in the resistive regime with Mach numbers $M_A \sim 2$–3, where experimental observations have revealed a high level of electrostatic fluctuations (for a review on collisionless shocks see, e.g., Biskamp, 1973). Since wavelengths $k\lambda_D \lesssim 1$ are much shorter than the gradient scale-length of the shock front, the system may be assumed homogeneous with a current perpendicular to the magnetic field, and since $\omega \sim \Omega_e \sim \omega_{pi} \gg \Omega_i$, ions are unmagnetized.

The major problem when trying to interpret the experimental findings as the result of ion-sound instability is that the latter requires a larger temperature ratio than seems to be observed. Though the lower-hybrid-drift instability, where the pressure gradient plays a crucial role (see section 7.3), is now considered a more likely candidate for anomalous resistivity in shocks and magnetic boundary layers, a discussion of the nonlinear behavior of the homogeneous ion-sound instability is justified in its own right.

Since electron-cyclotron effects in the linear instability – the electron-cyclotron-drift instability (Forslund *et al.*, 1970) – are wiped out at low amplitude by resonance broadening (Lampe *et al.*, 1971), see section 7.2.1, and the coherent wave–particle process in the nonlinear electron-cyclotron-drift instability (Forslund *et al.*, 1971) is eliminated by allowing a finite angular spread of the mode spectrum (see, e.g., Biskamp & Chodura, 1973b), the only effect of the perpendicular magnetic field is to prevent anisotropic distortion of the electron distribution.

Hence we can take the electron distribution to be isotropic in the plane perpendicular to $\mathbf{B}_0$. Since the gyrating electrons see an effectively isotropic spectrum, we determine the quasi-linear diffusion coefficient by assuming a spectrum $|\phi_{\mathbf{k}}|^2 = |\phi_k|^2$, such that $D_{ij} = D\delta_{ij}$ in (7.33) (Galeev & Sagdeev, 1973, p. 76):

$$
\begin{aligned}
D &= \pi \frac{e^2}{m_e^2} \int \frac{k \, dk \, d\theta}{(2\pi)^2} |E_k|^2 \delta(kv_\perp \cos\theta - \omega_k) \\
&= \pi \frac{e^2}{m_e^2} \int_0^\infty \frac{k \, dk}{(2\pi)^2} |E_k|^2 \frac{1}{kv_\perp \sin\theta} \Bigg|_{kv_\perp \cos\theta = \omega_k} \\
&= \pi \frac{e^2}{m_e^2} \int_0^\infty \frac{k \, dk}{(2\pi)^2} \frac{|E_k|^2}{\sqrt{k^2 v_\perp^2 - \omega_k^2}}.
\end{aligned}
\tag{7.77}
$$

We are mainly interested in the global shape of the distribution function for $v \sim v_{te} \gg \omega_k/k$, such that the ω_k^2 term in the denominator may be neglected, hence

$$
D = \frac{A}{v_\perp},
\tag{7.78}
$$

where

$$
A = 4\pi \frac{e^2}{m_e^2} \left\langle \frac{1}{k} \right\rangle W, \quad W = \frac{1}{8\pi} \int \frac{d^2 k}{(2\pi)^2} |E_k|^2 = \frac{\langle \widetilde{E}^2 \rangle}{8\pi},
$$

and the brackets indicate the average over the spectrum $|E_k|^2$. Using $f_e = f_e(v_\perp)$ and introducing the new time variable $\tau = \int^t A(t') dt'$, the diffusion equation (7.32) assumes the form

$$
\partial_\tau f_e = \frac{\partial}{v_\perp \partial v_\perp} \frac{1}{v_\perp} \frac{\partial f_e}{v_\perp \partial v_\perp}.
\tag{7.79}
$$

This equation has a self-similar solution valid asymptotically for long times, when the system becomes independent of the initial distribution,

$$f_e = \frac{c_0}{\tau^{2/5}} \exp\{-v_\perp^5/\tau\}. \tag{7.80}$$

Similar square-shaped electron distributions have been observed in numerical simulations of ion-sound turbulence (Dum *et al.*, 1974; Biskamp *et al.*, 1975) and also in the solar wind (see, e.g., Montgomery *et al.*, 1970).

Since the perpendicular magnetic field inhibits electron run-away which dominates in the parallel-field case, inverse electron Landau damping continues driving the instability, such that saturation must occur either by mode–mode coupling or by nonlinear wave–ion interaction, i.e., ion trapping, where the latter effect is found to dominate. It generates a high-energy tail in the ion distribution function in the direction of the electron drift with an effective temperature $T_{i,\text{tail}} \simeq T_e$, which resonates directly with the ion-sound waves, $v_{i,\text{tail}} \sim \omega_k/k \sim c_s$. The fraction α of ions populating this tail can be estimated from the condition that ion Landau damping should balance the inverse electron Landau damping (Vekshtein & Sagdeev, 1970; Manheimer & Flynn, 1974):

$$\gamma_i + \gamma_e = \omega_{pi} u/v_{te} - \alpha\omega_{pi} = 0 \tag{7.81}$$

for the most unstable modes $k\lambda_D \sim 1$, hence $\alpha \sim u/v_{te}$. Vekshtein & Sagdeev have determined the ratio u/v_{te} using a simple self-consistency argument first given by Sagdeev (1967). Assuming that the growth of wave energy is balanced by absorption by the ions in the tail, while the electron thermal energy increases by turbulent heating, they obtain the relation $T_{i,\text{tail}} \sim T_e$ and

$$\alpha \sim u/v_{te} \sim (m_e/m_i)^{1/4}. \tag{7.82}$$

Extensive particle simulations of the current-driven ion-sound instability and turbulence have been performed for both $B_0 = 0$ (e.g., Biskamp & Chodura, 1971) and for $\mathbf{B}_0 \perp \mathbf{j}_0$ in the plane perpendicular to the magnetic field (e.g., Dum *et al.*, 1974). In both cases, nonlinear saturation of the instability occurs when ion heating sets in, fig. 7.7. Phase space plots show that this process corresponds to ion trapping, though the fluctuating field energy $W = \langle\tilde{E}^2\rangle/8\pi$ is rather low, $W/nT_e \lesssim 10^{-2}$, whereas a rough estimate seems to indicate that ion trapping requires $W/nT_e \sim 1$. It is, however, easy to evaluate the wave energy exactly for a thermal ion $v = v_{ti}$ trapped by a sinusoidal ion wave of amplitude ϕ_0 and wavelength and phase velocity corresponding to the most unstable mode $k\lambda_D = 1/\sqrt{2}$, $\omega_k/k = \sqrt{2/3}c_s$. From the trapping condition

$$2e\phi_0/m_i = \left(\sqrt{2/3}c_s - v_{ti}\right)^2$$

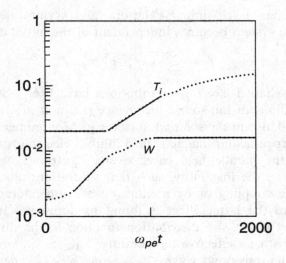

Fig. 7.7. Time evolution of the ion temperature and fluctuation field energy demonstrating saturation of the ion-sound instability by ion heating. The mass ratio in this simulation is $m_i/m_e = 400$ (from Biskamp & Chodura, 1971).

one obtains

$$W/nT_e = 0.014 \left(1 - \sqrt{3T_i/2T_e}\right)^4,$$

which is in fact small, $\lesssim 0.01$, consistent with the values observed in the numerical simulations.

Using (7.56) and (7.68) the turbulent resistivity η_T, (7.53), can be written as a function of the turbulent wave energy W,

$$\eta_T \omega_{pe} = \frac{v_{\text{eff}}}{\omega_{pe}} = a_2 \sqrt{2\pi} \left\langle \frac{\mathbf{k} \cdot \mathbf{u} - \omega_k}{ku} \frac{\mathbf{k} \cdot \mathbf{u}}{ku} \frac{1}{k\lambda_D} \right\rangle \frac{W}{nT_e} \simeq 0.7 \left\langle \frac{\cos^2 \theta}{k\lambda_D} \right\rangle \frac{W}{nT_e},$$
(7.83)

where again the average is over the spectrum and the numerical coefficient a_2 accounts for the change in the electron distribution from a Maxwellian, $a_2 = 0.285$ for the distribution (7.80).

Both the fraction α of high-energy ions, (7.82), and the expression (7.83) for the turbulent resistivity are verified by simulations (Dum et al., 1974). With W/nT_e typically 0.01 one has $v_{\text{eff}} \simeq 0.005\omega_{pe}$. Sagdeev (1967) derived a simple formula for the turbulent collision frequency by evaluating the integral over the wave spectrum in the wave kinetic equation:

$$v_{\text{eff}} \simeq 10^{-2} \omega_{pi} \frac{T_e}{T_i} \frac{u}{c_s}.$$
(7.84)

Using relation (7.82) and interpreting the ion temperature as the mean ion energy $T_{i,\text{eff}} = \alpha T_{i,\text{tail}}$, Sagdeev's formula agrees with the simulation result $v_{\text{eff}} \sim 10^{-2}\omega_{pe}$.

It is true that the particle simulations do not show a quasi-stationary turbulent state in the nonlinear evolution of the ion-sound instability. Instead, the turbulence level W/nT_e decreases after reaching its maximum. This behavior is mainly a consequence of the spatial periodicity of the system, where high-energy ions (and also run-away electrons in the unmagnetized case) are confined. Thus these tails tend to become overpopulated, tipping the balance relation (7.81) off to the negative side. In a collisionless shock the shock front passes across the turbulent system in a time $\sim 10^2\omega_{pi}^{-1}$, which is of the order of the period covered in the simulations.

7.3 Lower-hybrid-drift instability (LHDI)

We have seen in the preceding section that the ion-sound instability, when driven by a cross-field current in a high-β plasma, is much more effective in generating anomalous resistivity than in the parallel-current low-β case since, on time-scales characteristic of the nonlinear evolution, the electrons are turned around by the magnetic field, which prevents run-away and quasi-linear stabilization. The main shortcoming of the ion-sound instability when applied to high-β laboratory and space plasmas is the requirement of high temperature ratio $T_e/T_i \gg 1$ contrary to the observation of $T_e/T_i < 1$ in many systems.

At low temperature ratio, a different powerful instability may arise with long quasi-neutral wavelength $k\lambda_D \ll k\rho_e \sim 1$ and low frequency $\omega \ll \Omega_e$. Here electrons are magnetized, while the ions are still unmagnetized $k\rho_i \gg 1$, $\omega \gg \Omega_i$. In the presence of an electric field E_x the electrons perform an E×B drift $v_{Ey} = -cE_x/B$, see (6.66), assuming the magnetic field in the z-direction as usual. A density disturbance is transported along with $\mathbf{v}_E$ forming an electron drift wave. In this frequency range the plasma also exhibits the lower-hybrid resonance, which arises when electron and ion motions in the direction of the electric field $E_x e^{-i\omega t}$ are synchronized.[*] The coupling of the drift wave and the lower-hybrid wave by a finite inhomogeneity of the system gives rise to instability, the lower-hybrid-drift instability (LHDI) (Krall & Liewer, 1971). For weak inhomogeneity

---

[*] Since the ions are unmagnetized, the ion response is $v_{xi} = i(e/m_i)E_x/\omega$, while for the magnetized electrons the velocity in the x-direction is given by the polarization drift, see (6.68), $v_{xe} = i(\omega/\Omega_e)cE_x/B$, hence resonance $v_{xi} = v_{xe}$ occurs at $\omega = \omega_{LH} = (\Omega_e\Omega_i)^{1/2}$. This single-particle picture of the lower-hybrid resonance is correct at sufficiently high plasma density $\omega_{pe}^2/\Omega_e^2 = 4\pi nm_ec^2/B^2 > 1$. At lower density, collective plasma effects reduce the resonance frequency, see (7.95).

(compared with the wavelength) the equilibrium profiles can be assumed linear,

$$\mathbf{B}(x) = B(1 - x/L_B)\mathbf{e}_z, \quad n(x) = n(1 - x/L_n), \quad T_e(x) = T_e(1 - x/L_T). \quad (7.85)$$

In the quasi-local approximation the x-dependence of the profiles is only considered in the derivatives. In addition, there is a uniform electric field $E_x = E = -\partial_x \phi$. The electron equilibrium distribution is a function of the constants of motion $f(W_e, p_{ey}, p_{ez})$, see (7.8), with $W_e = \frac{1}{2}m_e v^2 + eEx$, $p_{ey} = m_e v_y - eA_y/c = m_e(v_y - \Omega_e x)$, $p_{ez} = m_e v_z$. Expanding the distribution function about a Maxwellian for weak gradients n', T', ϕ', the equilibrium distribution f_{e0} can be chosen in the following form (see Krall, 1968)

$$f_{e0} = \frac{n}{(2\pi v_{te}^2)^{3/2}} \left[1 + \left(\frac{1}{L} + \frac{1}{L_T} \frac{v_\perp^2}{2v_{te}^2} \right) \left(\frac{v_y}{\Omega_e} - x \right) \right] \exp \left[-(\tfrac{1}{2}m_e v^2 + eEx)/T_e \right],$$
$$(7.86)$$

where $L^{-1} = L_n^{-1} - L_T^{-1} - eE/T_e$. We are mainly interested in a narrow current layer $L \lesssim \rho_i$, so that the ions are unmagnetized also on the equilibrium scale. The ion distribution function can be taken as a simple Maxwellian

$$f_{i0} = \frac{n}{(2\pi v_{ti}^2)^{3/2}} \exp[-(\tfrac{1}{2}m_i v^2 - eEx)/T_i]. \quad (7.87)$$

The current density, which is carried only by the electrons, contains both the diamagnetic current (see (6.70)) and the E×B drift,

$$j_y = -enu, \quad u = v_{*e} + v_E = \frac{cT_e}{eB}\left(\frac{1}{L_n} + \frac{1}{L_T}\right) - \frac{cE}{B}. \quad (7.88)$$

Ampère's law $\nabla \times \mathbf{B} = (4\pi/c)\mathbf{j}$ leads to the relation

$$\frac{1}{L_B} = -\frac{\beta_e}{2}\left(\frac{1}{L_n} + \frac{1}{L_T} - \frac{eE}{T_e}\right), \quad \beta_e = \frac{8\pi n T_e}{B^2}. \quad (7.89)$$

To estimate the natural magnitude of the E×B drift we note that, to confine the ions, the change of the potential across the system L_n is given by $T_i \sim eE/L_n$, hence $v_E \sim v_{*e} T_i/T_e$. The electric field points in the direction of the density gradient, i.e., $\mathbf{v}_E$ is in the direction of the diamagnetic drift. In the case $T_e \ll T_i$ the diamagnetic velocity v_{*e} is small, such that $u \simeq v_E$.

For electrostatic perturbations, the dispersion relation for modes in the current direction $\mathbf{k} = k_y \mathbf{e}_y$ becomes (see Krall & Liewer, 1971)

$$1 + \chi_i + \chi_e = 0 \quad (7.90)$$

with

$$\chi_i = -\frac{1}{2k_y^2} \frac{\omega_{pi}^2}{v_{ti}^2} Z'(\zeta_i) = \frac{\omega_{pi}^2}{k_y^2 v_{ti}^2}[1 + \zeta_i Z(\zeta_i)], \quad \zeta_i = \frac{\omega}{\sqrt{2}k_y v_{ti}} \quad (7.91)$$

as in (7.61), and

$$\chi_e = \frac{\omega_{pe}^2}{\Omega_e^2} \frac{1 - I_0(\lambda_e)e^{-\lambda_e}}{\lambda_e} + \frac{\omega_{pe}^2}{k_y^2 v_{te}^2} \frac{k_y v_\Delta}{\omega - k_y v_E}, \quad \lambda_e = (k_y \rho_e)^2, ^* \qquad (7.92)$$

where v_Δ is a generalized drift velocity arising from the inhomogeneity of the configuration

$$v_\Delta(k_y) = \frac{cT_e}{eB} I_0(\lambda_e)e^{-\lambda_e} \left\{ \frac{1}{L_n} - \frac{1}{L_B}\left[1 - \lambda_e\left(1 - \frac{I_1(\lambda_e)}{I_0(\lambda_e)}\right)\right] \right.$$
$$\left. - \frac{1}{L_T}\lambda_e\left(1 - \frac{I_1(\lambda_e)}{I_0(\lambda_e)}\right) \right\}. \qquad (7.93)$$

The dispersion relation assumes a particularly simple form for long wave-length $k_y \rho_e \ll 1$ and large phase velocity $\omega/k_y \gg v_{ti}$. Expansion of the Bessel functions for small λ_e and the Z-function for large ζ_i gives

$$(\omega - k_y v_E)\left(1 - \frac{\omega_{LH}^2}{\omega^2}\right) = -\frac{1}{k_y^2 \rho_e^2} k_y v_\Delta \qquad (7.94)$$

with

$$\frac{1}{\omega_{LH}^2} = \frac{1}{\omega_{pi}^2} + \frac{1}{\Omega_i \Omega_e} \simeq \frac{1}{\Omega_i \Omega_e} \quad \text{for } \omega_{pe} > \Omega_e, \qquad (7.95)$$

which illustrates the coupling of drift wave and lower-hybrid wave by the inhomogeneity ($\propto v_\Delta$). If the inhomogeneity and the electric field are sufficiently weak such that $v_E v_\Delta \ll c_{se}^2 = T_e/m_i$, the solution of (7.94) is an unstable drift wave $\omega = k_y v_E + i\gamma$, $\gamma \ll k_y v_E$. The most effective coupling, leading to the maximum growth rate, occurs when both modes are in resonance, i.e., have the same frequency $\omega_r = k_y v_E = \omega_{LH}$, hence $k_y = \omega_{LH}/v_E$. Insertion into (7.94) gives

$$2\gamma^2 \simeq v_E v_\Delta / \rho_e^2 = \omega_{LH}^2 v_E v_\Delta / c_{se}^2. \qquad (7.96)$$

Instability requires $v_E v_\Delta > 0$. This is normally satisfied for low β_e, since by (7.89) one has $L_B^{-1} \ll L_n^{-1}, L_T^{-1}$, and hence $v_\Delta \simeq cT_e/(eBL_n) \sim v_{*e}$, such that $v_E v_\Delta > 0$ with the natural direction of the electric field along the density gradient as discussed before. The free-energy source of the instability is the E×B drift, while the plasma inhomogeneity only taps the energy from this reservoir.

For higher drift velocity $v_E v_\Delta \gtrsim c_{se}^2$ the growth rate becomes large $\gamma \sim \omega_r \sim \omega_{LH}$, which occurs at wavenumbers $k_y \rho_e \sim (v_\Delta/v_E)^{1/2}$. Hence in

---

* Contrary to chapter 6 and following the conventional notation, in this chapter on collisionless processes λ_e does not designate the electron mean free path.

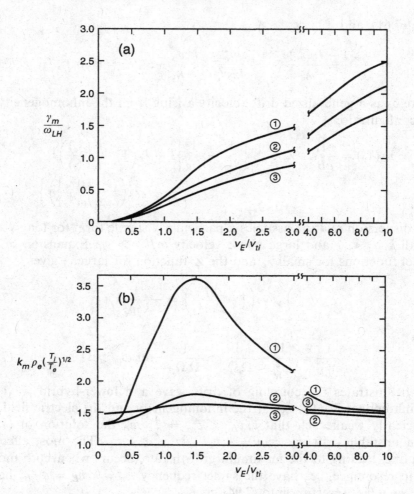

Fig. 7.8. Plots of the maximum LHDI growth rate (a) and the corresponding wavenumber (b) versus v_E/v_{ti} for ① $T_e/T_i = 2, \beta_e = 0.4$; ② $T_e/T_i = 1, \beta_e = 0.2$; and ③ $T_e/T_i = 0.2, \beta_e = 0.04$ (from Davidson & Gladd, 1975).

the usual case $v_E \sim v_A$ the approximation $\lambda_e \ll 1$ leading to the simple algebraic form, (7.94), is no longer valid. Krall & Liewer have solved numerically the full (electrostatic) dispersion relation (7.90) with (7.92) and (7.93).

Since most applications of the LHDI are for high-β plasmas, both in the laboratory and in space, the previous electrostatic treatment has been generalized to include electromagnetic perturbations (Davidson & Gladd, 1975; Gladd, 1976; Davidson *et al.*, 1977). It is found that finite-pressure effects tend to weaken the instability, such that the system becomes stable for $\beta > \beta_c$, which is, however, not very restrictive, since β_c is rather

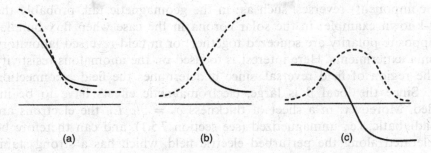

(a)   (b)   (c)

Fig. 7.9. Schematic drawing of magnetic field (full) and pressure (dashed) profiles in a magnetic shock (a); in a θ-pinch (b); in a field-reversed configuration (c).

high, $\beta_c \gtrsim 1$, increasing with v_E/v_{ti} and decreasing with increasing T_e/T_i. Figure 7.8 shows plots of maximum growth rate and the corresponding wavenumber as functions of v_E/v_{ti} for three cases differing in β_e and T_e/T_i.

The stability analysis is usually performed for flute-like modes $\mathbf{k} \cdot \mathbf{B} = 0$, where the contributions from electron Landau and cyclotron resonances (see (7.26)) vanish and hence the largest growth rates are obtained (Gladd, 1976). For $k_z \neq 0$ strong electron Landau damping occurs if $\omega/k_z \to v_{te}$. Since $\omega/k_z \sim v_E k_y/k_z$, the effect becomes important for

$$\frac{k_z}{k_y} > \left(\frac{m_e}{m_i}\right)^{1/2} \frac{v_E}{v_{ti}} \sqrt{\frac{T_i}{T_e}}, \tag{7.97}$$

quenching the instability. For smaller k_z, however, the parallel damping is weak, such that for $v_E > v_{ti}$ and $T_i > T_e$ the instability covers a substantial parallel-wavenumber range in spite of the small mass ratio factor.

The local approximation, to which the preceding analysis of the LHDI has been restricted, is suitable for configurations where the magnetic field does not change sign, such as in collisionless shocks and θ-pinches. Here the maximum values of pressure gradient, electric field and hence drift velocity are located in the region of finite magnetic field, see fig. 7.9. The anomalous resistivity generated by the instability leads to resistive broadening of the magnetic profile, $L_B \gg c/\omega_{pe}$, and to efficient turbulent heating. A nonlocal description is required only if gradients become very steep, for instance in the initial phase of a θ-pinch implosion where $L_B \sim c/\omega_{pe}$, such that $kL_B \sim 1$. The nonlocal electrostatic theory of the LHDI developed by Batchelor & Davidson (1976) and Davidson (1976) applies as long as the drift approximation for the electron orbits is valid, i.e., the total magnetic field does not become too weak. A different situation arises in configurations wherein the magnetic field (not only just one component) reverses, such as: in the geomagnetic tail, probably the best-known example; in the solar corona in the case when flux bundles of opposite polarity are squeezed together; or in field-reversed laboratory plasma experiments. Here interest is focused on the anomalous resistivity in the region of field reversal, since it determines the field reconnection rate. Since the local β is large, electromagnetic effects have to be included. Moreover, in a sheet of thickness $\delta_e = \sqrt{\rho_e L_B}$ the electrons are nonadiabatic, i.e., unmagnetized (see section 7.5.1), and can therefore be accelerated along the perturbed electric field, which has a strong stabilizing effect. Taking these effects into account, Huba *et al.* (1980) find that the eigenfunctions of the unstable modes are restricted to regions on both sides of the field reversal with practically no overlap, even for low temperature ratio $T_e/T_i \sim 0.1$. This behavior indicates that the LHDI does not, at least not directly, lead to fast collisionless reconnection.

To estimate the magnitude of the turbulent resistivity generated by the LHDI and its scaling with the relevant physical parameters, the nonlinear behavior of the instability must be understood. With the assumption that the linear characteristics of the mode are not strongly changed, an approximate expression is obtained from quasi-linear theory, (7.58), which, however, requires knowledge of the the saturation field amplitudes. (General quasi-linear expressions are found in the review paper by Davidson & Krall, 1977).

Several nonlinear saturation processes compete: plateau formation in the ion distribution function in the low-drift regime $\omega/k_y \sim u \lesssim v_{ti}$, when the ion Landau resonance is important; or ion trapping in the high-drift regime $\omega/k_y > v_{ti}$, which is similar to the main stabilization process in the cross-field current-driven ion-sound instability, section 7.2.2; or electron resonance broadening (Huba & Papadopoulos, 1978).

An estimate, or rather an upper limit, of the fluctuation amplitude is obtained by equating the wave energy to the free energy of the unstable system, the electron flow energy,

$$\frac{\partial}{\partial \omega}(\omega \epsilon)\frac{\widetilde{E}^2}{8\pi} = \frac{1}{2}n_0 m_e u^2, \tag{7.98}$$

where ϵ is the dispersion function $\epsilon = 1 + \chi_i + \chi_e$. (The wave energy contains both the field energy and the kinetic energy of the oscillating particles, see, e.g., Stix, 1962). For lower-hybrid waves of long wavelength $\lambda_e < 1$ and small drift one has simply $\epsilon = 1 - \omega_{pi}^2/\omega^2 + \omega_{pe}^2/\Omega_e^2$, hence (7.98) becomes

$$2\left(1 + \frac{\omega_{pe}^2}{\Omega_e^2}\right)\frac{\widetilde{E}^2}{8\pi} = 2\frac{\omega_{pi}^2}{\omega_{LH}^2}\frac{\widetilde{E}^2}{8\pi} = \frac{1}{2}n_0 m_e u^2. \tag{7.99}$$

From (7.58) we find the following estimate of the turbulent collision frequency using $\mathrm{Im}\{\zeta_i Z(\zeta_i)\} \lesssim 1$,

$$\nu_{\text{eff}} \lesssim 2\pi \frac{\omega_{LH}^2}{k_y v_{ti}} \frac{u}{v_{ti}}. \tag{7.100}$$

Quasi-linear estimates also indicate that ions and electrons are heated at similar rates.

Numerical simulations of the nonlinear development of the LHDI were performed by several groups in the 1970s and 1980s. The simulations used a nonradiative electromagnetic particle code in a 2D slab perpendicular to the magnetic field with inhomogeneity in x, periodic boundary conditions in y and parameters $u \gtrsim v_{ti}$ and $T_i \gg T_e$ (for a brief review of particle simulation methods see section 7.6). Attention had originally been focused on explaining the broad profiles and heating in a θ-pinch discharge, i.e., a configuration as sketched in fig. 7.9(b) (Winske & Liewer, 1978). The LHDI was found to develop in the region of strong field and pressure gradients, saturating at amplitudes close to the energy estimate, (7.99), and leading to rapid broadening of the equilibrium profiles. However, as θ-pinches became less attractive in fusion plasma physics, the emphasis of LHDI simulations shifted to field-reversed configurations, especially in the geomagnetic tail. Here, typical parameters are $T_e/T_i \sim 0.1$ and $u \sim v_{ti}$, as discussed in section 8.3, hence the LHDI appeared as the ideal candidate for anomalous resistivity to explain the fast reconnection associated with the observed magnetic activity. Particle simulations of the LHDI in a neutral sheet by Winske (1981), Tanaka & Sato (1981) and Brackbill *et al.* (1984) revealed the following:

(a) The primarily electrostatic LHDI is only active in the flanks of the current profile but is excluded from the neutral line, where the local $\beta \to \infty$, in agreement with the linear theory. Hence the LHDI cannot be the direct cause for fast magnetic reconnection in a neutral sheet.

(b) There is, however, a different primarily electromagnetic mode located at the neutral surface, which grows more slowly than the LHDI with longer wavelength but finally reaches large amplitude, thus giving rise to anomalous resistivity η_T^{em}, (7.59). This electromagnetic mode, which is called drift-kink or kinetic kink mode, has recently attracted renewed interest (Ozaki *et al.*, 1996; Pritchett & Coroniti, 1996; Zhu & Winglee, 1996; Lapenta & Brackbill, 1997, 2000; Hesse *et al.*, 1998). Here, the main motivation is to find a dynamical process with a wave vector component in the current direction, in order to destabilize the tearing mode which (in the presence of a weak normal field component) is stable for strictly two-dimensional perturbations, as discussed in section 7.5.3. A comprehensive stability investigation of the LHDI in a weakly two-dimensional current sheet has been performed by Yoon *et al.* (1994). Most of the recent stability studies use particle simulations considering a relatively thin Harris sheet equilibrium, (7.131), with $a \lesssim \rho_i$ (corresponding to $u_i \lesssim v_{ti}^*$ as easily seen from (7.129)) and $T_i > T_e$. The observed scaling of the growth rate of the drift-kink mode is

$$\gamma \sim \frac{u_i}{v_{ti}}\Omega_i \sim \Omega_i \qquad (7.101)$$

for a thin current sheet with $a \sim \rho_i$ and hence $u_i \sim v_{ti}$. This indicates that the mode is essentially a Kelvin–Helmholtz instability driven by the shear of the ion flow. Note that the perpendicular magnetic field has no stabilizing effect. The interpretation of this mode in terms of the Kelvin–Helmholtz instability has been corroborated by Hesse *et al.* (1998). However, the dependence of the growth rate on the temperature ratio T_e/T_i implies that the electrons also play an important part.

7.4 Whistler anisotropy instability

The cross-field current-driven instabilities discussed in the preceding sections tend to increase the electron energy preferentially in the plane perpendicular to the magnetic field and thus may lead to substantial temperature anisotropy $T_{e\perp} > T_{e\parallel}$. Such a system is susceptible to a different type of microinstability which, contrary to the electrostatic current-driven modes, is primarily electromagnetic. Since the classical paper by Weibel (1959), anisotropy-driven instabilities have been studied by many authors considering linear and nonlinear properties both in unmagnetized and in magnetized plasmas, see, e.g., Davidson *et al.* (1972), where also earlier references are found. Here we restrict consideration to $B_0 \neq 0$, where the electrons are magnetized. The anisotropy instability excites whistler waves, where only the electrons participate, the ions forming a static neutralizing background. Whereas in the LHDI the most unstable modes are flute-like, $\mathbf{k} \cdot \mathbf{B}_0 = 0$, the whistler anisotropy instability is strongest for parallel propagation, $\mathbf{k} \times \mathbf{B}_0 = 0$.

In order to simplify the algebra we neglect the electrostatic part of the fluctuations and consider only the most unstable case $\mathbf{k} \parallel \mathbf{B}_0$, $\mathbf{k} = k\mathbf{e}_z$, which eliminates the complicated sums over Bessel functions such as in expression (7.71). The background distribution is assumed isotropic in the plane perpendicular to $\mathbf{B}_0$, i.e., $f_{e0}(\mathbf{v}) = f_{e0}(v_\perp, v_\parallel)$. After Fourier and Laplace transformation as described in section 7.1.1, the equations (7.3), (7.4), (7.5) for the perturbed fields now become

$$\mathbf{k} \cdot \mathbf{E}_{k\omega} = 0, \qquad (7.102)$$

---

* The Harris equilibrium is usually considered in the frame where the electric potential vanishes, i.e., $v_E = 0$, while the ions have the drift velocity u_i, see section 7.5.

$$\omega \mathbf{B}_{\mathbf{k}\omega} = ck\mathbf{e}_z \times \mathbf{E}_{\mathbf{k}\omega}, \tag{7.103}$$

$$i\omega \mathbf{E}_{\mathbf{k}\omega} + ick\mathbf{e}_z \times \mathbf{B}_{\mathbf{k}\omega} = -4\pi e \int d^3 v \mathbf{v} f_{e\mathbf{k}\omega}, \tag{7.104}$$

where $f_{e\mathbf{k}\omega}$ follows (7.9),

$$i(-\omega + kv_\parallel)f_{e\mathbf{k}\omega} - \Omega_e \mathbf{v} \times \mathbf{e}_z \cdot \frac{\partial}{\partial \mathbf{v}} f_{e\mathbf{k}\omega} = \frac{e}{m_e}\left(\mathbf{E}_{\mathbf{k}\omega} + \frac{1}{c}\mathbf{v} \times \mathbf{B}_{\mathbf{k}\omega}\right) \cdot \frac{\partial}{\partial \mathbf{v}} f_{e0}, \tag{7.105}$$

As before, the electron cyclotron-frequency is defined as a positive quantity, $\Omega_e = eB_0/m_e c$. Introducing polar velocity coordinates $(v_\perp, \theta)$ in the plane perpendicular to $\mathbf{B}_0$, $v_x + iv_y = v_\perp e^{i\theta}$, the differential operator on the l.h.s. of (7.105) reduces to

$$-\mathbf{v} \times \mathbf{e}_z \cdot \frac{\partial}{\partial \mathbf{v}} = \frac{\partial}{\partial \theta}. \tag{7.106}$$

Because of (7.102) we can write

$$\mathbf{E}_{\mathbf{k}\omega} = E_{xk}\mathbf{e}_x + E_{yk}\mathbf{e}_y.$$

Since the magnetic field $\mathbf{B}_0$ induces mode amplitudes to rotate, it is useful to introduce the representation

$$E_k^\pm = E_{xk} \pm iE_{yk}. \tag{7.107}$$

Eliminating $\mathbf{B}_{\mathbf{k}\omega}$ by use of (7.103), the operator on the r.h.s. of (7.105) assumes the form

$$(\mathbf{E}_{\mathbf{k}\omega} + \frac{1}{c}\mathbf{v} \times \mathbf{B}_{\mathbf{k}\omega}) \cdot \frac{\partial}{\partial \mathbf{v}} = \frac{1}{2}(E_k^- e^{i\theta} + E_k^+ e^{-i\theta})\left[\frac{kv_\perp}{\omega}\frac{\partial}{\partial v_\parallel} + \left(1 - \frac{kv_\parallel}{\omega}\right)\frac{\partial}{\partial v_\perp}\right]$$

$$+ \frac{i}{2}(E_k^- e^{i\theta} - E_k^+ e^{-i\theta})\frac{1}{v_\perp}\left(1 - \frac{kv_\parallel}{\omega}\right)\frac{\partial}{\partial \theta}. \tag{7.108}$$

The r.h.s. of (7.105) now becomes

$$(\mathbf{E}_{\mathbf{k}\omega} + \frac{1}{c}\mathbf{v} \times \mathbf{B}_{\mathbf{k}\omega}) \cdot \frac{\partial}{\partial \mathbf{v}} f_{e0}$$

$$= \frac{1}{2}(E_k^- e^{i\theta} + E_k^+ e^{-i\theta})\left[\frac{kv_\perp}{\omega}\frac{\partial}{\partial v_\parallel} + \left(1 - \frac{kv_\parallel}{\omega}\right)\frac{\partial}{\partial v_\perp}\right]f_{e0}. \tag{7.109}$$

Writing $f_{\mathbf{k}\omega}$ in the form

$$f_{\mathbf{k}\omega} = f_k = f_k^+(v_\perp, v_\parallel)e^{i\theta} + f_k^-(v_\perp, v_\parallel)e^{-i\theta} \tag{7.110}$$

omitting the subscript ω, we can solve (7.105) directly

$$f_k^\pm = -i\frac{e}{m_e}\frac{E_k^\mp}{kv_\parallel - \omega \pm \Omega_e}\left[\frac{kv_\perp}{\omega}\frac{\partial}{\partial v_\parallel} + \left(1 - \frac{kv_\parallel}{\omega}\right)\frac{\partial}{\partial v_\perp}\right]f_{e0}. \tag{7.111}$$

The dispersion relation is obtained by inserting this expression into (7.104), using (7.107), and averaging over θ

$$(k^2c^2 - \omega^2)E_k^{\pm} + i\omega 4\pi e \int d^3v v_{\perp} f_k^{\mp} = 0, \qquad (7.112)$$

which gives

$$k^2c^2 - \omega^2 + \frac{\omega_{pe}^2}{2} \int d^3v v_{\perp} \frac{1}{kv_{\parallel} - \omega \mp \Omega_e} \left(kv_{\perp} \frac{\partial}{\partial v_{\parallel}} + (\omega - kv_{\parallel}) \frac{\partial}{\partial v_{\perp}} \right) F_e = 0, \qquad (7.113)$$

with $F_e = f_{e0}/n_0$. It can easily be checked that, for slow phase velocity, $\omega^2/k^2 \ll c^2$, and in the cold-plasma limit this equation reduces to the whistler dispersion relation, (6.35). As discussed in section 6.2, whistler waves follow the sense of rotation of the electron cyclotron motion, i.e., they are right-hand circularly polarized when propagating in the direction of $\mathbf{B}_0$, and left-hand circularly polarized when propagating in the opposite direction. For a bi-Maxwellian distribution

$$F_e = \frac{1}{(2\pi)^{3/2}} \frac{1}{v_{e\perp}^2 v_{e\parallel}} \exp\left\{ -\frac{v_{\perp}^2}{2v_{e\perp}^2} - \frac{v_{\parallel}^2}{2v_{e\parallel}^2} \right\}$$

the dispersion relation (7.113) becomes (Sudan, 1963; Kennel & Petschek, 1966; for a review of linear theory see also Gary, 1993)

$$k^2c^2 - \omega^2 - \omega_{pe}^2 \left[\left(\frac{T_{e\perp}}{T_{e\parallel}} - 1 \right) \left(1 + \zeta_e^{\pm} Z(\zeta_e^{\pm}) \right) + \frac{\omega}{\sqrt{2}kv_{e\parallel}} Z(\zeta_e^{\pm}) \right] = 0, \qquad (7.114)$$

where $\zeta_e^{\pm} = (\omega \pm \Omega_e)/\sqrt{2}kv_{e\parallel}$ and $Z(\zeta)$ is the plasma dispersion function (7.63). The dispersion relation (7.114) must, in general, be solved numerically, but the stability threshold can be obtained analytically by considering the imaginary part of the dispersion relation for marginal stability $\text{Im}\{\omega\} = 0$. Instability occurs for any anisotropy,

$$\frac{T_{e\perp}}{T_{e\parallel}} - 1 > 0, \qquad (7.115)$$

where the unstable wavenumber range is

$$k < \frac{\omega_{pe}}{c} \left(\frac{T_{e\perp}}{T_{e\parallel}} - 1 \right)^{1/2} \qquad (7.116)$$

corresponding to frequencies

$$\omega < \left(1 - \frac{T_{e\parallel}}{T_{e\perp}} \right) \Omega_e. \qquad (7.117)$$

For large anisotropy the maximum growth rate is $\gamma \lesssim \Omega_e$ for $k \lesssim \omega_{pe}/c$.

The main nonlinear effect in the evolution of the whistler anisotropy instability is the change of the distribution function, which is described by the quasi-linear equation (Kennel & Engelmann, 1966; see also Davidson, 1972)

$$\partial_t f_{e0}(v_\perp, v_\parallel) = \frac{e}{m_e} \int \frac{dk}{2\pi} \left\langle \left(\mathbf{E}_{-k} + \frac{1}{c}\mathbf{v} \times \mathbf{B}_{-k} \right) \cdot \frac{\partial}{\partial \mathbf{v}} f_k \right\rangle, \qquad (7.118)$$

where the brackets indicate the average over the gyroangle θ. Using (7.108), (7.110) and (7.111) one obtains, after some algebra,

$$\partial_t f_{e0} = -\frac{ie^2}{2m_e^2} \sum_{+,-} \int \frac{dk}{2\pi} |E_k|^2 \left[\frac{kv_\perp}{\omega_k} \frac{\partial}{\partial v_\parallel} + \left(1 - \frac{kv_\parallel}{\omega_k} \right) \frac{1}{v_\perp} \frac{\partial}{\partial v_\perp} v_\perp \right]$$

$$\times \frac{1}{kv_\parallel - \omega_k \pm \Omega_e} \left[\frac{kv_\perp}{\omega_k} \frac{\partial}{\partial v_\parallel} + \left(1 - \frac{kv_\parallel}{\omega_k} \right) \frac{\partial}{\partial v_\perp} \right] f_{e0}. \qquad (7.119)$$

Here the sum is over both propagation directions $\omega_k/k > 0$ and $\omega_k/k < 0$ and $|E_k|^2 = \mathbf{E}_{-k} \cdot \mathbf{E}_k = E_{-k}^- E_k^+ = E_{-k}^+ E_k^-$.

The tendency in the nonlinear evolution becomes clear from the two global constants of motion, which are derived by forming second-order moments of the quasi-linear equation (7.119) using the dispersion relation (7.113):

$$\frac{d}{dt} \left[W_\perp + \int \frac{dk}{2\pi} \frac{|B_k|^2}{8\pi} \left(2 + \frac{\omega_{pe}^2}{k^2 c^2} \right) \right] = 0, \qquad (7.120)$$

$$\frac{d}{dt} \left[W_\parallel - \int \frac{dk}{2\pi} \frac{|B_k|^2}{8\pi} \left(1 + \frac{\omega_{pe}^2}{k^2 c^2} \right) \right] = 0. \qquad (7.121)$$

Here we have assumed low frequency $\omega^2 \ll k^2 c^2$ [which corresponds to neglecting the displacement current, the first term on the l.h.s. in (7.104)] and substituted $|E_k|^2 = (\omega/kc)^2 |B_k|^2$ from Faraday's law, (7.104). $W_\perp = \frac{1}{2} m_e \int d^3 v v_\perp^2 f_{e0}$ and, analogously, $W_\parallel$ are the perpendicular and parallel kinetic energies. Summation of (7.120) and (7.121) gives the total energy conservation; note that in the low-frequency limit $|E_k|^2 \ll |B_k|^2$.

Equations (7.120), (7.121) indicate that $W_\perp$ decreases and $W_\parallel$ increases as the magnetic mode energy grows, thus reducing the temperature anisotropy. As the latter decreases, high-k modes become stable according to (7.116). Particle simulations (Ossakow et al., 1972) and solutions of the quasi-linear equation (Hamasaki & Krall, 1973), which are in good agreement, show that for large initial anisotropy $T_{e\perp}/T_{e\parallel} > 2$ this ratio decreases rapidly, until the wave energy saturates. The system does, however, not relax to a fully isotropic state, which is not too surprising, since the quasi-linear diffusion leads to a non-Maxwellian shape of the distribution function, which may have a higher stability threshold than (7.115) for the bi-Maxwellian.

Whistler turbulence, whether driven by temperature anisotropy or by electron Kelvin–Helmholtz instability discussed in section 6.2.4, involves only electron dynamics and hence cannot act as anomalous resistivity. It does, however, give rise to anomalous electron viscosity $\mu_{e,\text{eff}}$ or hyperresistivity $\eta_2 = (c/\omega_{pe})^2 \mu_{e,\text{eff}}$, see (6.22), which can also cause collisionless reconnection. A simple estimate gives

$$\mu_{e,\text{eff}} \sim \langle \gamma/k^2 \rangle|_{\max} \lesssim \Omega_e(c/\omega_{pe})^2. \tag{7.122}$$

7.5 The collisionless tearing mode

Contrary to the small-scale microscopic processes excited by the instabilities discussed up to now in this chapter, which occur in a homogeneous or almost homogeneous environment, the collisionless tearing mode is a macroinstability driven by the magnetic energy of a necessarily inhomogeneous system. As discussed in section 4.1, the system is divided into an ideal outer region which determines the free energy, and a narrow dissipative layer which allows reconnection to occur. In a collisionless plasma dissipation is due to collective resonance processes, primarily Landau damping. Contrary to the broad variety of equilibrium configurations which have been investigated with respect to resistive tearing modes, the literature on the collisionless mode is almost exclusively focused on the plane sheet pinch as a model of the geomagnetic tail and its most important magnetic activity, the substorm phenomenon. While the latter will be discussed in more detail in chapter 8, here consideration is restricted to the theoretical aspects, the linear and nonlinear properties of the collisionless tearing mode in a (nearly) one-dimensional neutral sheet configuration.

Let me first point out a peculiarity, namely the choice of the coordinate system, which might cause some confusion to the unbiased reader when browsing through the literature on the subject covering the last 30 or so years. The classical paper on the collisionless tearing mode by Laval *et al.* (1966) uses the coordinates we have adopted in the treatment of the resistive mode in section 4.1, indicated in fig. 4.1, x across the sheet, i.e., in the direction of the (main) inhomogeneity, y in the direction of (or opposite to) the reversing magnetic field, and z along the current, which is the symmetry direction of the system. (By the way, Furth *et al.* (1963) in their classical article on resistive modes chose a different system, where x and y are interchanged.) In other studies, that appeared in the later-

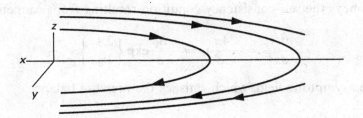

Fig. 7.10. Coordinate system now generally used in the theory of the collisionless tearing mode. Sketched is a weakly two-dimensional configuration.

1960s, e.g., by Hoh (1966) and by Schindler & Soop (1968), y and z are interchanged compared with fig. 4.1. Since the mid-1970s, the coordinate system commonly used in magnetospheric physics, i.e., x along the sheet, y along the current, and z, the vertical axis, across the sheet, see fig. 7.10, has become the prevailing choice in tearing mode theory. It is therefore convenient to adopt this coordinate system in this section, wherein we treat the basic theory of the collisionless tearing mode, in order also to conform with the notation in the subsequent chapter on magnetospheric processes.

Most studies of the collisionless tearing mode start from the Harris equilibrium (Harris, 1962), which is very convenient in analytical theory. As indicated in (7.8) the distribution function depends on the particle energies $W_j = \frac{1}{2}m_jv^2 + q_j\phi$ and the canonical momenta $p_{jy} = m_jv_y + (q_j/c)A_y$ in the direction of the ignorable coordinate. In the Harris equilibrium the distribution is

$$f_{0j} = C\exp\{(u_jp_{jy} - W_j)/T_j\}, \tag{7.123}$$

C = normalizing constant, which can be written in the form of a shifted Maxwellian

$$f_{0j} = \frac{n(z)}{(2\pi v_{tj}^2)^{3/2}}\exp\left\{-m_j\frac{v_x^2 + (v_y - u_j)^2 + v_z^2}{2T_j}\right\}. \tag{7.124}$$

Choosing the frame such that

$$\frac{u_i}{T_i} = -\frac{u_e}{T_e}, \tag{7.125}$$

the electrostatic potential vanishes. The density is

$$n_e = n_i = n(z) = n_0\exp\{2A_y/(B_0a)\}, \tag{7.126}$$

and A_y obeys the self-consistency condition resulting from Ampère's law

$$\frac{d^2 A_y}{dz^2} = -\frac{4\pi}{c} j_y = -\frac{B_0}{a} \exp\left\{\frac{2A_y}{B_0 a}\right\}. \qquad (7.127)$$

B_0 is the asymptotic field, which satisfies the pressure balance

$$\frac{B_0^2}{8\pi} = n_0 (T_e + T_i), \qquad (7.128)$$

and the equilibrium scale-length a is obtained from the relation

$$\frac{2}{B_0 a} = \frac{e}{c} \frac{u_i - u_e}{T_i + T_e}, \qquad (7.129)$$

since $u_i / T_i = (u_i - u_e)/(T_i + T_e)$ because of (7.125). Equation (7.127) has the solution

$$A_y = -B_0 a \ln[\cosh(z/a)], \qquad (7.130)$$

which corresponds to the well-known Harris sheet

$$B_x = B_0 \tanh(z/a), \qquad \frac{4\pi}{c} j_y = \frac{B_0}{a} \text{sech}^2(z/a). \qquad (7.131)$$

It is interesting to remember that (7.127) and its solution, (7.131), have also been discussed in MHD theory as the most probable current profile, see section 3.2.1. Note that here we use the vector potential A_y instead of the flux function ψ, where $\psi = -A_y$. In this section we restrict consideration to two-dimensional perturbations described by the y-component of the vector potential, such that the perturbed current is only in the y-direction and field perturbations B_y are neglected. We may therefore omit the subscript writing A and j instead of A_y and j_y.

7.5.1 Linear stability theory

A perturbation $f_j^{(1)}$ of the equilibrium evolves according to the linearized Vlasov equation (7.9). We extract the adiabatic change from the perturbed distribution function

$$f_j^{(1)} = \frac{\partial f_{0j}}{\partial A_0} A_1 + \tilde{f}_j, \qquad (7.132)$$

where, from (7.9), the nonadiabatic part $\tilde{f}_j$ obeys the equation

$$\frac{d\tilde{f}_j}{dt} \equiv \partial_t \tilde{f}_j + \mathbf{v} \cdot \nabla \tilde{f}_j + \frac{q_j}{m_j}\left(\mathbf{E}_0 + \frac{1}{c}\mathbf{v} \times \mathbf{B}_0\right) \cdot \frac{\partial \tilde{f}_j}{\partial \mathbf{v}} = -(\mathbf{E}_1 \cdot \mathbf{v})\frac{\partial f_{0j}}{\partial \phi_0}, \qquad (7.133)$$

and from (7.123)

$$\frac{\partial f_{j0}}{\partial \phi_0} = q_j \frac{\partial f_{j0}}{\partial W_j} = -\frac{q_j}{T_j} f_{j0}.$$

The equation for the perturbed current density $\nabla^2 A_1 = -(4\pi/c)j_1$ becomes, neglecting the displacement current and making the Fourier ansatz $A_1 \propto e^{ikx}$,

$$\frac{d^2 A_1}{dz^2} - \left(k^2 - \frac{4\pi}{c}\frac{\partial j_0}{\partial A_0}\right)A_1 = -\frac{4\pi}{c}\sum_j q_j \int v_y \tilde{f}_j d^3v = -\frac{4\pi}{c}\tilde{j}, \qquad (7.134)$$

where, for the Harris equilibrium, (7.127) and (7.131) give

$$\frac{4\pi}{c}\frac{\partial j_0}{\partial A} = \frac{2}{a^2}\text{sech}^2(z/a). \qquad (7.135)$$

As in the case of the resistive instability the analytical treatment is limited to a sufficiently broad current sheet $a \gg \rho_i$, where the nonideal or nonadiabatic region at the neutral surface can be considered in simplified geometry. The theory of the collisionless tearing mode, which has been developed by Hoh (1966), Laval *et al.* (1966), and Dobrowolny (1968), is algebraically rather involved. Here, we are content with a qualitative derivation of the growth rate. It is useful to distinguish between adiabatic particles, which follow drift orbits, and nonadiabatic particles. The term refers to the adiabatic invariance of the magnetic moment $\mu_{Bj} = m_j v_\perp^2/2B$, which is valid if the variation of the magnetic field across the gyroradius is sufficiently weak, $(\rho_j/B)dB/dz \lesssim 0.1$. A more quantitative picture of the particle trajectories is obtained from the conservation of the canonical momentum p_{jy} (7.7) which, in the vicinity of the field reversal, can be written in the form

$$v_y \mp \Omega_j \frac{(z^2 - z_0^2)}{2a} = 0, \qquad (7.136)$$

using $A_0 \simeq -(B_0/2a)z^2$ and $\Omega_j = eB_0/m_jc$, and $\mp$ stands for ions and electrons, respectively. Here $z = z_0$ is the point at which $v_y = 0$. For $z_0 \gg \sqrt{\rho_j a}$, $\rho_j = \rho_j(|z| \to \infty) = v_y/\Omega_j$, where the adiabaticity condition holds, i.e., far away from the field reversal, the excursion Δz is small, since v_y is limited by energy conservation. With $\frac{1}{2}(z^2 - z_0^2) \simeq z_0\Delta z$ we find for adiabatic particles

$$\Delta z \simeq \frac{v_y}{\Omega_j}\frac{a}{z_0} = \rho_j(z_0). \qquad (7.137)$$

The orbits are illustrated in fig. 7.11(a), the guiding center moving with

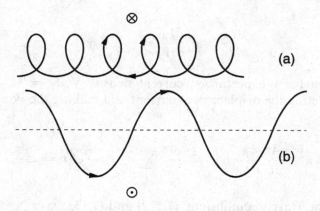

Fig. 7.11.   Electron orbits in the sheet pinch. (a) Adiabatic, not crossing the neutral surface; (b) nonadiabatic, crossing the neutral surface. Ions move in the opposite directions, respectively.

the ∇B-drift[*]

$$\mathbf{v}_{Bj} = \mu_{Bj} \frac{c}{q_j} \frac{\mathbf{b} \times \nabla B}{B}.$$ (7.141)

In the field-reversal layer the adiabaticity parameter $(\rho_j/B)dB/dz$ diverges, hence the magnetic moment is no longer conserved. For $z_0 \simeq 0$ we see that v_y does not change sign. The particles cross the neutral line

---

[*] As the reader may check, these drifts are retrograde, i.e., opposite to the directions of the respective contributions to the current density, for instance $v_{Bi,y} < 0$, while $j_y = (c/4\pi)dB_x/dz > 0$. It may therefore be interesting to see how guiding center and fluid theory are reconciled. In the guiding center approach the current density consists of two contributions $\mathbf{j} = \mathbf{j}_d + \mathbf{j}_M$, where $\mathbf{j}_d = \sum_j q_j n \langle \mathbf{v}_{Bj} \rangle$ is the guiding center drift motion and $\mathbf{j}_M$ the magnetization current,

$$\mathbf{j}_M = \frac{c}{4\pi} \nabla \times \mathbf{M}.$$ (7.138)

The magnetization

$$\mathbf{M} = -4\pi \sum_j n \langle \mu_{Bj} \rangle \mathbf{b}$$ (7.139)

is negative, which causes the plasma to be diamagnetic. Substitution of μ_{Bj} gives

$$j_{My} = \frac{c}{4\pi} \frac{dM_x}{dz} = -\sum_j cT_j \frac{n}{B} \left(\frac{1}{n} \frac{dn}{dz} - \frac{1}{B} \frac{dB}{dz} \right).$$ (7.140)

It can now easily be seen that the second term on the r.h.s. of (7.140) cancels the drift contribution j_{dy} leaving us with the net current density, called the diamagnetic current, consistent with the force balance $\mathbf{j} \times \mathbf{B} = c\nabla(p_i + p_e)$.

meandering along the layer of width δ_j, fig. 7.11(b),

$$\Delta z \simeq \delta_j = \sqrt{\frac{m_j v_{tj} c}{e B'(0)}} = \sqrt{\rho_j a}, \tag{7.142}$$

ions in positive (= current) direction, electrons in negative.

It is mainly the nonadiabatic particles in the resonant layer δ_j which contribute to the r.h.s. of (7.134), since these can be accelerated by the reconnecting induction field $E_y = -\partial_t A_1/c = -\gamma A_1/c$, where γ is the growth rate. In the layer the solution of (7.133) becomes

$$\tilde{f}_j = \gamma \frac{q_j}{cT_j} f_{j0} \int_{-\infty}^{t} v_y(t') A_1(t') dt' \tag{7.143}$$

and

$$\int_{-\infty}^{t} v'_y A'_1 dt' \simeq \frac{v_y A_1}{i(kv_x - i\gamma)} \simeq \pi v_y A_1 \delta(kv_x), \tag{7.144}$$

while the solution outside the layer is nonresonant and hence negligible. Insertion into (7.134) and integration over the resonant layer gives

$$\Delta' A_1(0) = -\frac{4\pi}{c} \int dz\, \tilde{j} \simeq -\gamma \sqrt{\frac{\pi}{2}} \sum_j \frac{\omega_{pj}^2}{c^2} \frac{\delta_j}{kv_{tj}} A_1(0). \tag{7.145}$$

Δ' is the jump of the derivative of A_1 at the resonant surface, cf. (4.10). This corresponds to the asymptotic matching of the layer solution to the ideal external solution described in section 4.1, where Δ' is determined by the boundary conditions imposed on the latter. As appropriate for the tearing mode, we have assumed that $A_1 \simeq A_1(0)$ is constant across the layer. The contribution from the electrons is larger than that from the ions by a factor $(T_i m_i/T_e m_e)^{1/4}$, hence for $T_i \gtrsim T_e$ the former dominates. Using (7.128) and the expression (4.23) for Δ' we obtain the growth rate for the collisionless tearing mode

$$\gamma \simeq \frac{v_{te}}{a} \left(\frac{\rho_e}{a}\right)^{3/2} \left(1 + \frac{T_i}{T_e}\right) (1 - k^2 a^2), \tag{7.146}$$

which agrees with the exact expression up to a numerical factor of order unity. The mode is also called the electron tearing mode, since it is driven by electron inertia, $\gamma \propto m_e^{1/4}$.

7.5.2 Nonlinear saturation

As we have seen, the instability is driven by the resonant acceleration of nonadiabatic electrons. Now, the growth of the mode generates a normal field component $B_n = B_z = kA_1$ in the resonant layer, which affects the motion of these particles. If the field is strong enough to make the electrons adiabatic, the instability should be switched off. Inserting the total field $B = \sqrt{k^2 A_1^2 + (z/a)^2 B_0^2}$ into the adiabaticity criterion

$$(\rho_e/B)dB/dz \lesssim 0.1, \tag{7.147}$$

the maximum of the l.h.s. occurs at $z/a \sim kA_1/B_0$, hence the criterion is satisfied, i.e., all electrons are adiabatic, when the island width w_I (cf. (4.5)) exceeds the layer width

$$w_I = 4\sqrt{\frac{A_1 a}{B_0}} > \delta_e, \tag{7.148}$$

where we assume $ka \sim 1$.

However, this estimate of the saturation amplitude gives only an upper limit of the island width, since it ignores the spatial variation of the normal field along the sheet $kA_1 \propto e^{ikx}$. Taking this effect into account the actual saturation level is even lower. The energy principle for two-dimensional collisionless configurations has been derived by Laval *et al.* (1966). Multiplying (7.134) by $\partial_t A_1$ and using (7.133) one obtains the energy conservation law associated with the perturbation

$$\frac{dW}{dt} = \frac{d}{dt} \int d^2x \left\{ (\nabla A_1)^2 - \frac{4\pi}{c} \frac{\partial j_0}{\partial A_0}(A_1)^2 + 4\pi \sum_j \int \frac{(\tilde{f}_j)^2}{\partial f_{j0}/\partial W_j} d^3v \right\} = 0. \tag{7.149}$$

After minimization, the energy integral W can be written in the following form (Biskamp & Schindler, 1971; Schindler *et al.*, 1973)

$$W_{\min} = \int d^2x \left\{ (\nabla A_1)^2 - \frac{4\pi}{c} \frac{\partial j_0}{\partial A_0}(A_1)^2 + 4\pi \overline{(\langle\chi\rangle - \overline{\langle\chi\rangle})^2} \left| \frac{\partial \rho_0}{\partial \phi_0} \right| \right\}. \tag{7.150}$$

The mode is unstable if, and only if, $W_{\min} < 0$. Here $\chi = A_1 v_y/c - \phi_1$, $\langle\,\rangle$ is the average over the particle orbits passing through a point x, z with energy W_j and momentum p_{jy}, the overbar indicates the average over W_j, p_{jy} and the particle species j with the weight function $q_j \partial f_{j0}/\partial W_j$, and $\rho_0 = \sum_j q_j \int d^3v f_{0j}, \partial \rho_0/\partial \phi_0 = -e^2 n_0 (T_i^{-1} + T_e^{-1})$. In the one-dimensional sheet pinch we have $\langle\chi\rangle = 0$ because of the e^{ikx} dependence of the perturbations, such that $W = -\Delta'|A_1(0)|^2$, see also (4.30).

In general, however, $\langle\chi\rangle$ is finite and the last term in (7.150) exerts a strong stabilizing influence. Let us consider a weakly two-dimensional configuration corresponding to a chain of narrow islands of width $\delta_e < w_I \ll a$. The resonant particles in the region around the O-points have very complicated orbits, but here it suffices to note that these orbits are not symmetric with respect to $v_y \rightarrow -v_y$, see, e.g., the behavior in the sheet pinch as illustrated in fig. 7.11. A crude estimate gives[*]

$$(\langle \chi \rangle - \overline{\langle \chi \rangle})^2 \sim \langle \chi \rangle^2 \sim \frac{v_{te}^2}{c^2}|A_1|^2,$$

valid in a region of order δ_e^2 around an O-point (the region is in fact larger for the elongated islands assumed, but this estimate is sufficient for the argument). Hence the energy integral, (7.150), taken over one wavelength $2\pi/k$ in x, becomes approximately

$$W_{min} \simeq \left(-\frac{2\pi}{k}\Delta' + \frac{\omega_{pe}^2}{c^2}\delta_e^2 \right)|A_1(0)|^2. \tag{7.151}$$

The second term exceeds the first by a large factor $(\omega_{pe}^2/c^2)\delta_e^2/\Delta'a \sim a/\rho_e \gg 1$ (since $\rho_e \sim c/\omega_{pe}$). One can therefore conclude that $W_{min} = 0$, i.e., saturation of the tearing mode, must actually occur at much smaller island size $w_I \ll \delta_e$. This tendency is supported by a quasi-linear analysis (Biskamp *et al.*, 1970) yielding the saturation width

$$w_I \sim \delta_e \frac{\rho_e}{a}. \tag{7.152}$$

The physics behind the stabilizing term in the energy integral can readily be understood. The nonadiabatic electrons in the vicinity of the O-point experience a uniform acceleration by the induction field $E_y = -\partial_t A_1/c$, instead of the oscillating response, (7.144), in the linear phase of the instability. This generates a current-density perturbation $\tilde{j} \simeq -(e^2 n_0/m_e c)A_1$, which is independent of the growth rate. The sign is such as to oppose the change of the magnetic field according to Lenz' law. Insertion of $\tilde{j}$ into (7.145) gives (7.151).

We thus find that the collisionless tearing mode saturates at very low amplitude,[†] contrary to the resistive mode which continues to grow, though at reduced growth rate, to macroscopic amplitude as discussed in section 4.1.2. Residual collisions, or effective collisions due to small-scale fluctuations (anomalous resistivity), however, allow also further growth of the collisonless tearing mode.

---

[*] Because of the quasi-neutrality condition $n_e = n_i$ a finite electrostatic potential is set up $\phi_1 \sim A_1 v_{te}/c$. This does, however, not invalidate the estimate, since exact cancellation in χ does not occur.

[†] In a low-β plasma embedded in a strong magnetic field B_{0y} all particles are adiabatic. Here the collisionless tearing mode is driven by the parallel response of the electrons in the layer l_e around the resonant surface $B_{0x} = 0$, where l_e is determined by the condition $k_\parallel v_{te} \simeq \gamma$. Nonlinear saturation occurs at island size $w_I \sim l_e$ (Drake & Lee, 1977).

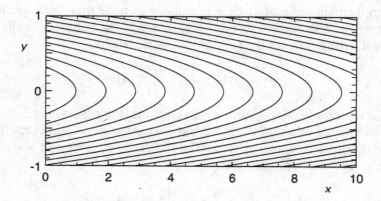

Fig. 7.12. Configuration (7.153) for $B_n/B_0 = 0.05$.

It is worth noting that for strictly collisionless conditions the stabilizing contribution by the electrons close to the O-points also dominates over the possibly destabilizing effect due to nonadiabatic ions, i.e., the ion tearing mode (see the following subsection) is stable in a one-dimensional neutral sheet configuration.

7.5.3 *The ion tearing mode*

The strictly one-dimensional neutral sheet is too idealized a model of the magnetotail. Since the latter is generated by stretching the dipole field of the Earth, its natural shape is that sketched in fig. 7.10 (at least for not too large distance from the Earth). A quasi-local model of such an elongated configuration is provided by assuming a weak constant normal field component superimposed on the Harris sheet, see fig. 7.12:

$$\mathbf{B} = B_0 \tanh(z/a)\mathbf{e}_x + B_n\mathbf{e}_z, \quad B_n \ll B_0. \tag{7.153}$$

To achieve equilibrium, a weak pressure gradient $\partial_x p = jB_n/c$ along the sheet is required but, since the gradient scale L_x is very long, $L_x/a \sim B_0/B_n$, we may regard the system as quasi-homogeneous in the x-direction, i.e., consider the x-dependence in the WKB approximation. In the vicinity of the field-reversal surface $z = 0$ the configuration (7.153) is $\mathbf{B} = B_0(z/a)\mathbf{e}_x + B_n\mathbf{e}_z$, which is called the parabolic model, since a field line $A_y = -B_0(z^2/2a) + B_nx = const$ has a parabolic shape. An approximate self-consistent 2D equilibrium has been developed by Lembège & Pellat (1982) generalizing the 1D Harris sheet solution (7.130)

$$A(x,z) = -aB_0 \ln \left[e^{\epsilon x/a} \cosh \left(\frac{z}{a} e^{-\epsilon x/a} \right) \right], \tag{7.154}$$

$\epsilon \ll 1$. Force balance is valid to first order in ϵ in the z-direction and to second order in the x-direction. As outlined in section 7.5.2, a normal field component magnetizes the electrons making their motion adiabatic, if

$$\frac{B_n}{B_0} \gtrsim \frac{\delta_e}{a} = \sqrt{\frac{\rho_e}{a}}, \tag{7.155}$$

obtained from the adiabaticity condition, (7.147). Equation (7.155) implies that the field line curvature radius $(B_n/B_0)a$ at the resonant surface exceeds the local gyroradius δ_e. However, this condition gives only an upper limit of the critical normal field, since the electron orbits in the resonant layer are already modified at much lower value. The electron tearing mode is essentially quenched when the gyroradius in the resonant plane $z = 0$, i.e., in the normal field B_n, becomes smaller than the wavelength, which implies

$$B_n/B_0 > k\rho_e \sim \rho_e/a. \tag{7.156}$$

An even more stringent criterion would be that the corresponding gyrofrequency $eB_n/m_e c$ exceeds the growth rate, i.e., by use of (7.146),

$$B_n/B_0 > (\rho_e/a)^{5/2}, \tag{7.157}$$

which means that the free-streaming electron Landau resonance, (7.144), is no longer valid. However, both analytical and numerical studies indicate that the milder condition, (7.156), is the relevant one.

Since there are no neutral points in this configuration, the large stabilizing contributions from the induced electron current, (7.151), do not appear. Schindler (1974) and Galeev & Zelenyi (1976) have therefore argued that if, B_n is sufficiently large, the nonadiabatic electron contribution in (7.145) is negligible and the tearing mode is driven by the nonadiabatic ions. Hence the growth rate becomes

$$\gamma \simeq \frac{v_{ti}}{a} \left(\frac{\rho_i}{a} \right)^{3/2}, \tag{7.158}$$

which for $T_i \gg T_e$ is larger than the growth rate of the electron tearing mode, (7.146), by a factor $(m_i T_i / m_e T_e)^{1/4}$. Also for the ion tearing mode there is an upper limit of the normal magnetic field given by the suitably modified criterion, $B_n/B_0 > k\rho_i$. Since the thinning of the current sheet observed prior to a substorm in the geomagnetic tail (see chapter 8) reduces the relative magnitude of the normal field, B_n/B_0, thus destabilizing ion tearing, this mode seems to be a natural candidate for substorm onset (Schindler, 1974).

The question of the existence of the ion tearing mode has provoked a longstanding discussion in the literature. Following the papers by Schindler

(1974) and Galeev & Zelenyi (1976), which seemed to firmly establish the presence of the mode under the conditions given above, the idea was refuted by Lembège & Pellat (1982), who showed that also in the absence of neutral points the electrons provide a strongly stabilizing effect. The effect stems from the electron part of the last term in the general energy integral, (7.149). The inductive electric field generates an electron density perturbation, which sets up an electrostatic potential in order to maintain charge neutrality. Since this compressibility effect is independent of the electron temperature, it is also present for $T_e \ll T_i$. Subsequently it was claimed in several studies that by relaxing the requirement of electron adiabaticity imposed by Lembège & Pellat, the stabilizing effect would be eliminated. In particular, Büchner & Zelenyi (1987) considered the case of a sufficiently small normal field component $B_n/B_0 < \delta_e/a$, which makes the electrons nonadiabatic and their orbits chaotic. However, Pellat *et al.* (1991) demonstrated that the result by Lembège and Pellat does not depend on the assumption of adiabatic electrons, but requires only conservation of the canonical momentum p_{ey}, i.e., two-dimensionality, which implies that the perturbed number of electrons in a flux tube remains constant. The result was later confirmed by Brittnacher *et al.* (1994), and seems to remain valid even for very thin current sheets $a \sim \rho_i$ as found numerically by Pritchett (1994). Quest *et al.* (1996) proved that the result is valid also in the presence of a "guide field" $B_{0y} \neq 0$.

Only cross-field spatial diffusion (in contrast to pitch-angle scattering in velocity space) can overcome the stabilizing effect produced by electron compressibility. Introducing such diffusion formally by adding $-k^2 D t$ in the resonance function as indicated in (7.50), the energy integral yields the following dispersion relation (Pellat *et al.*, 1991)

$$-\Delta' a + \frac{\beta_e \pi k a}{2} \frac{B_0}{B_n} \frac{\gamma}{\gamma + k^2 D} + \frac{\delta_i}{a} \left(\frac{a}{\rho_i} \right)^3 \frac{\beta_i \sqrt{\pi \gamma}}{k a \Omega_i} = 0, \qquad (7.159)$$

where $\beta_{i,e} = 8\pi n_0 T_{i,e}/B_0^2$. The first term is the free magnetic energy, the second the electron compressibility contribution (note that for finite B_n the electron Landau resonance, (7.144), is switched off), and the last comes from the ion Landau resonance, (7.145). For $D = 0$ the free energy is compensated by the electron compressibility effect. Substituting Δ' gives the stability condition

$$-\frac{2}{ka}(1 - k^2 a^2) + \frac{\beta_e \pi k a}{2} \frac{B_0}{B_n} > 0, \qquad (7.160)$$

hence the tearing instability is suppressed except for extremely long wavelength $(ka)^2 < B_n/B_0$. If D is finite, we find an electron tearing mode

$$\gamma \simeq k^2 D \frac{4}{\pi \beta_e} \frac{B_n}{B_0} \frac{1 - k^2 a^2}{k^2 a^2}. \qquad (7.161)$$

Only for very high diffusion rate

$$D > \rho_i^2 \Omega_i \frac{B_0}{B_n} \frac{\delta_i}{a} \qquad (7.162)$$

would the ion tearing mode be destabilized. Such rapid diffusion is, however, improbable, since $\rho_i^2 \Omega_i = c T_i / e B_0$ is the Bohm diffusion coefficient (cf. the footnote in section 6.1), which is considered an upper limit of cross-field diffusion due to small-scale turbulence. Moreover, the factor $B_0 \delta_i / B_n a$ must exceed unity in order to allow free, i.e., non-adiabatic ion motion, see (7.155).

7.6 Particle simulation of collisionless reconnection

In the preceding section it has been shown that the tearing mode in a collisionless neutral sheet plasma is stable. The analysis does, however, require a number of rather restrictive conditions:

(a) two-dimensional geometry;

(b) broad current profile $a \gg \rho_i$;

(c) current perturbations only in the direction of the equilibrium current;

(d) infinitesimal perturbation amplitude, in particular there is no X-line in the system.

In order to apply the theory to real-world processes, such as the substorm phenomenon, these conditions must be relaxed, which in general requires a numerical treatment. We therefore start this section with a brief introduction to the numerical method of particle simulation, before discussing the results of such computations. As will be seen, fast collisionless reconnection may occur under rather general conditions.

7.6.1 Particle simulation methods

Nonlinear collisionless plasma processes, especially in more than one spatial dimension, can, in general, only be studied numerically, and the most convenient, in fact the only viable, method is particle simulation. Since the early work, in particular by Buneman (1959) and Dawson (1962),

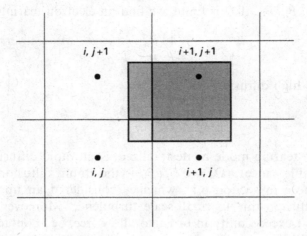

Fig. 7.13.   Illustration of the method of area weighting.

this method has been developed to a high degree of sophistication and complexity and is being applied to a broad range of phenomena (for a practical introduction see, e.g., Birdsall & Langdon, 1985, and Hockney & Eastwood, 1988). The basic idea is quite simple. Since in a collisionless plasma particles interact only through collective electric and magnetic fields, the point character of the particles can be ignored by considering only spatial scales exceeding the minimum wavelength of the collective modes. Fields and their sources, i.e., charge and current densities, are discretized on a grid of mesh size Δx and the sources are computed from the positions and velocities of the particles in a grid cell, whence the traditional acronym PIC (= particle-in-cell). Since a simulation particle represents many plasma particles, the method is also called macro-particle simulation. The distribution of the particles in phase space is always random, even if the simulation is started from a regularly distributed state, such that one is dealing with a Monte Carlo-type of approach whose inherently high noise level must be reduced in order to obtain useful results for other than the most violent processes. This is achieved by:

(a) interpolating the fields, i.e., the forces, to the actual position of a particle within a cell;

(b) smoothing the particles by giving them a finite size. In practice this is the size of a basic grid cell, such that a particle contributes to 2^D neighboring cells (D = spatial dimension) with weight factors corresponding to the fraction of the particle reaching into these cells. This process, called area weighting, is illustrated in fig. 7.13;

(c) using a sufficiently large average number of particles per cell, typically 10^2, such that the total number of particles largely exceeds that of the cells.

Particle simulation codes were originally developed to treat microscopic plasma processes involving the Debye length, the smallest collective plasma scale, hence $\Delta x \sim \lambda_D$. In electrostatic problems $\mathbf{E}$ is determined by Poisson's equation

$$\nabla \cdot \mathbf{E} = -\nabla^2 \phi = 4\pi e(\bar{n}_i - \bar{n}_e), \tag{7.163}$$

where $\bar{n}_{i,e}$ are the numbers of weighted particles in a cell. A typical electrostatic process is the ion-sound turbulence treated in section 7.2.2.

Fully electromagnetic codes have been developed mainly to simulate collective processes in laser–plasma interaction. Here, Maxwell's equations are solved directly

$$\partial_t \mathbf{B} = -c\nabla \times \mathbf{E} \tag{7.164}$$

$$\partial_t \mathbf{E} = c\nabla \times \mathbf{B} - 4\pi(\bar{\mathbf{j}}_i - \bar{\mathbf{j}}_e). \tag{7.165}$$

For slow quasi-magnetostatic processes $\omega^2 \ll k^2 c^2$ electromagnetic wave propagation can be eliminated by adopting the Darwin approximation, which consists essentially of neglecting the displacement current, the l.h.s. in (7.165), see, e.g., Busnardo-Neto *et al.* (1977). Such non-radiative codes have, for instance, been used to study the nonlinear LHDI dynamics, see section 7.3.

In contrast to these microscopic processes, direct λ_D-scale numerical simulations of magnetic reconnection, in particular of the nonlinear behavior of the collisionless tearing mode, are not possible without severe rescaling of the different time- and length-scales, which in a $\beta \sim 1$ plasma obey the ordering

$$a \gtrsim c/\omega_{pi} \sim \rho_i \gg c/\omega_{pe} \sim \rho_e \gg \lambda_D, \tag{7.166}$$

$$\omega \lesssim \Omega_i \ll \Omega_e \ll \omega_{pe}. \tag{7.167}$$

Hence, in practical simulations of macro-scale phenomena, λ_D and ω_{pe} effects should be eliminated by assuming quasi-neutrality. There are several approaches for this purpose:

(a) Hybrid codes, where one particle species, usually the electrons, is treated on a fluid level. In this case, quasi-neutrality can simply be accounted for by equating the fluid density and the particle density.

(b) Implicit schemes. Here both plasma species are represented by simulation particles, but the implicit treatment allows us to take large time-steps $\Delta t \gg \omega_{pe}^{-1}$ and grid spacings $\Delta x \gg \lambda_D$. Physically this corresponds to a strong damping of the unresolved modes, while well-resolved modes with $k\Delta x < 1$ are not affected. Such codes are very flexible, since by a suitable choice of the numerical parameters any process of interest can be investigated. The penalty to be paid is a more complex code structure and a higher noise level than in hybrid schemes. Various different implicit codes have been developed since the early 1980s, which rest on basically two different approaches, the moment-implicit method and the direct-implicit method. The moment-implicit method was introduced by Mason (1981). To solve (7.164), (7.165) (supplemented by Poisson's equation, since the continuity equation is not satisfied exactly) implicitly, the source terms $n_{i,e}$, $\mathbf{j}_{i,e}$ must be known at the new time. Instead of treating the particle equations implicitly, which would, in general, be prohibitively expensive, the source terms are computed from the implicit moment equations. This is the basic concept of the widely used code "VENUS" (Brackbill & Forslund, 1982). The direct-implicit method, which was developed by several authors, e.g., Hewett & Langdon (1987), has subsequently been refined by Tanaka (1988, 1993). Here the particles are formally advanced to the new positions and velocities in terms of the (still unknown) new fields at the new positions. One can now form the new source terms which, when inserted into the field equations, gives a linear matrix problem for the new fields. This equation, which still involves summations over particles, is then suitably approximated to yield a tractable matrix equation over grid points.

The simulations which we will discuss in this section concern mainly the two-dimensional reconnection dynamics of a neutral sheet with anti-parallel magnetic field, either the simple tearing mode or the more rapidly evolving double tearing mode, or an externally driven system.* We consider first the electron dynamics, which is responsible for collisionless reconnection to occur, and subsequently discuss the ion dynamics, which determines the macroscopic behavior of the configuration. Finally, some recent results of three-dimensional particle simulations are presented, which combine the characteristics of the tearing mode with the interchange-type dynamics of the generalized lower-hybrid-drift instability.

7.6.2 Collisionless electron dynamics

The reconnection characteristics in a high-β weakly collisional plasma have been studied in section 6.2, where ions and electrons are treated in the framework of two-fluid theory. On scales shorter than the ion inertia length c/ω_{pi}, the electrons decouple from the ions, and the dynamics can be described in the electron magnetohydrodynamic (EMHD) approximation, where the characteristic mode is the whistler. The electrons set up an X-point configuration with a central micro-current sheet of dimensions c/ω_{pe},

---

* For such processes, where reconnection occurs at a neutral line in an essentially laminar way, the term 'collisionless reconnection' is now usually reserved, while for systems exhibiting some small-scale three-dimensional turbulence, reconnection is said to be due to anomalous resistivity.

which adjusts to the inflow speed, such that the latter is not determined by the reconnection process but by the macro-configuration $L > c/\omega_{pi}$. Actually, reconnection takes place in a viscous sublayer $\delta_\mu < c/\omega_{pe}$ and thus relies on the presence of residual collisions, though the reconnection rate does not depend on the value of viscosity coefficient μ_e or the resistivity η.

Hence in a truly collisionless plasma $\eta, \mu_e \to 0$, the incompressible EMHD model is no longer appropriate. The main reason for its breakdown is the isotropic pressure approximation. It is clear from Ohm's law for quasi-stationary conditions,

$$\mathbf{E} = -\frac{1}{c}\mathbf{v}_e \times \mathbf{B} - \frac{1}{ne}\nabla \cdot \mathscr{P}_e - \frac{m_e}{e}\mathbf{v}_e \cdot \nabla\mathbf{v}_e, \tag{7.168}$$

that at the X-point, where $\mathbf{B}_\perp$ and $\mathbf{v}_{e\perp}$ vanish, only thermal effects comprised in the pressure tensor $\mathscr{P}_e$ can sustain a finite reconnection electric field. With the usual approximation of isotropic pressure $\mathscr{P}_{e,jk} = p_e\delta_{jk}$ the pressure contribution cancels on substitution of $\mathbf{E}$ in Faraday's law (apart from the small dynamo term $\propto \nabla n \times \nabla p_e$, see footnote in section 6.2.1). Hence collisionless reconnection requires a more general behavior of $\mathscr{P}_e$, in particular the pressure tensor should not be assumed gyrotropic, i.e., isotropic in the plane perpendicular to $\mathbf{B}$. Gyrotropy requires that the gyroperiod Ω_e^{-1} is shorter than the dynamic time-scale τ, which is no longer true in the reconnection layer, where (in the absence of a guide field) the magnetic field becomes very small and τ, the time a fluid element spends in this region, is short.

The fluid model of the electrons should therefore be generalized. The equation for the pressure tensor can be written in the following form (see, e.g., Hesse & Winske, 1994)

$$\partial_t\mathscr{P}_e = -\mathscr{D} - \mathscr{C} - \nabla \cdot \mathscr{Q}, \tag{7.169}$$

where

$$\mathscr{D} = \mathbf{v}_e \cdot \nabla\mathscr{P}_e + \mathscr{P}_e\nabla \cdot \mathbf{v}_e + \mathscr{P}_e \cdot \nabla\mathbf{v}_e + \{\mathscr{P}_e \cdot \nabla\mathbf{v}_e\}^T \tag{7.170}$$

describes convection and compression of the electron fluid,

$$\mathscr{C} = \Omega_e[\mathscr{P}_e \times \mathbf{b} + \{\mathscr{P}_e \times \mathbf{b}\}^T] \tag{7.171}$$

describes the gyrodynamics, and $\mathscr{Q}$ is the electron heat flux. $\{\ \}^T$ denotes the transposed matrix. The only approximation to be made concerns the choice of a simple expression for the heat flux. Electron particle simulations suggest that the most important influence of $\mathscr{Q}_e$ consists in isotropizing the diagonal elements of the pressure tensor (Hesse & Winske, 1998), hence one assumes

$$\nabla \cdot \mathscr{Q}_e = \frac{1}{\tau_{\text{iso}}}(\mathscr{P}_e^{\text{diag}} - p_{e0}\mathscr{I}), \tag{7.172}$$

where τ_{iso} is the isotropization time-scale, the choice of which is not crucial as long as it exceeds the electron gyroperiod, the tensor $\mathscr{P}_e^{\text{diag}}$ contains only the diagonal elements of $\mathscr{P}_e$, and p_{e0} is the average over these elements. Comparison of hybrid simulations using the fluid equations (7.168)–(7.172) for the electrons with full ion–electron particle simulations shows excellent agreement not only of the global properties, such as the reconnection rate, but even in the details of the current distribution in the diffusion region (Kuznetsova *et al.*, 2000). These studies reveal that, different from the two-fluid (EMHD) reconnection dynamics discussed in section 6.2 (where electron inertia was the main reconnection mechanism giving rise to a diffusion region scale c/ω_{pe}), for finite electron temperature and in the absence of a mean axial magnetic field, thermal effects due to nonadiabatic electrons are more important than the electron bulk flow, which leads to a width of the diffusion region of the order of the electron excursions $\delta_e = \sqrt{\rho_e L} > c/\omega_{pe}$, as also noted by Horiuchi & Sato (1997). The work of Hesse *et al.* (2000) and Kuznetsova *et al.* (2000) shows also that the reconnection rate is independent of the electron physics, both of the mass ratio m_e/m_i and of the upstream temperature, which is similar to the EMHD result in section 6.2.2. Hence for global simulations the use of a strongly simplified electron fluid model is justified. Kuznetsova *et al.* (2000) derived a phenomenological theory of the kinetic effects in Ohm's law. They give an expression for the electric field in the diffusion region

$$E \sim \partial_x B \delta_e^2 \partial_x v_{ex}. \qquad (7.173)$$

It is interesting to note that, both from this form and from the exact form obtained from (7.170), the dominant nonideal term in Ohm's law is a functional of the electron stream function ϕ_e containing second-order derivatives $\partial_s v_{ex} = -\partial_{xy} \phi_e$, which regularize ϕ_e in a similar manner to' the $\rho_s^2 \mathbf{B} \cdot \nabla \nabla^2 \phi$ term in (6.124) in the low-β case. This indicates that an X-point configuration can be sustained, thus allowing fast reconnection independent of the small parameter δ_e, characteristic of the diffusion region.

7.6.3 Collisionless ion dynamics

We have just seen that the electron dynamics and dissipation allow the reconnection region to adjust to any inflow velocity, such that the latter and hence the reconnection rate depend only on the ion dynamics. When focusing on the behavior of the ions, it is therefore numerically convenient to treat the electrons as a fluid. Collisionless reconnection on the macroscopic scale $L \gg c/\omega_{pi}$ has been studied by Shay & Drake (1998) and Shay *et al.* (1999) using a hybrid model. Since the system contains three disparate scales, i.e., c/ω_{pe}, c/ω_{pi} and the scale-length L of the global system, such simulations are rather demanding even in 2D, since high spatial resolution and a correspondingly large number of simulation particles are needed. To soften these requirements a mass ratio $m_i/m_e = 25$ is chosen, which is still sufficient to clearly separate ion and electron scales.

In the first paper cited, the authors investigate a configuration consisting of two coalescing magnetic flux bundles. With a grid spacing of $\Delta x = \Delta z = 0.2c/\omega_{pe}$ and 10^3 grid points in each direction the macro-scale – the radius of a flux bundle – is $L \leq 20c/\omega_{pi}$, which is comparable to that in the two-fluid simulations discussed in section 6.2.5. As in the latter, the ion-layer width is found to be of the order of c/ω_{pi}, independent of the ion temperature in the upstream state. The reconnection rate is fast, $u \sim 0.1v_A$, but the system is still somewhat small to reveal a definite scaling behavior. One obtains a qualitative picture of the ion velocity distribution, which in the outflow cone develops multiple beams penetrating each other, finally separating spatially as the beams move further away from the reconnection region. A similar behavior has been found in earlier hybrid simulation studies, e.g., by Krauss-Varban & Omidi (1995) and by Hoshino *et al.* (1998), who compare these features to observations in the geomagnetic tail (see section 8.4).

As mentioned above, a system size of $L \sim 20c/\omega_{pi}$ is too small to reach a conclusion about the asymptotic behavior of the reconnection rate for large $L/(c/\omega_{pi})$. The crucial question is whether the length of the ion layer becomes macroscopic $\Delta_i \sim L$, implying an asymptotically slow reconnection process $u/v_A \sim (c/\omega_{pi})/L$, or remains microscopic $\Delta_i \sim c/\omega_{pi}$, which would allow reconnection to be Alfvénic $u \sim v_A$. In a second paper Shay *et al.* (1999) study the scaling behavior using a larger system size. The configuration is now that of the tearing mode rather than that of two flux bundles, which is numerically more economic since the reconnection dynamics involves a larger fraction of total area actually computed. Moreover, because of the insensitivity to the details of the electron physics, the resolution requirements can be relaxed, allowing a larger grid spacing of $\Delta x = 0.5c/\omega_{pe}$ without affecting the global dynamics. Choosing again a mass ratio of 25, the largest system size used was $L_x = 200c/\omega_{pi}$ along the sheet and $L_z = 50c/\omega_{pi}$ across. Simulations now show a clear scaling law, $u \sim 0.1v_A$ and $\Delta_i \sim 10c/\omega_{pi}$, i.e., the reconnection rate and the layer length remain independent of the global size L.

These results can be explained by, or are at least consistent with, the properties of the subregion $c/\omega_{pe} < x, z < c/\omega_{pi}$ around the X-point, where the electrons are frozen in to the magnetic field, while the ions are unmagnetized. This region can be interpreted as a large-amplitude whistler disturbance. The magnetic field advected into the layer is rotated

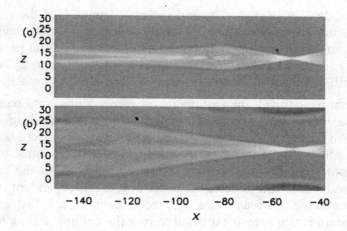

Fig. 7.14.   Electron current j_{ey} at two different times from hybrid simulations of the tearing mode. They demonstrate the invariance of the angle of the downstream cone (to the left of the X-point at $x = -50$). Scales are in units of c/ω_{pi} (from Shay *et al.*, 1999).

out of the plane, hence $B_y \sim B_x$, and from Faraday's law the out-of-plane electric field, the reconnection field, is rotated into the plane, $E_x \sim E_y$. In addition we assume that $B_z \sim B_x$, i.e., the angle formed by the separatrix in the electron outflow region is finite, which has been shown in section 6.2.2 to be so in the framework of EMHD and is corroborated for the full electron–ion system by hybrid simulations, see fig. 7.14. [By a general argument this geometric invariance holds because, in the range between c/ω_{pe} and c/ω_{pi}, there is no further intrinsic scale-length, which implies a scaling behavior, in particular the similarity solution (6.50)]. Faraday's law gives

$$\partial_t B_z \sim B_z/\tau_W \sim c\partial_y E_x \sim cE_x/\Delta_i,$$

where $\tau_W = \Delta_i^2/(\Omega_e c^2/\omega_{pe}^2)$ is the corresponding whistler time, (6.40). Hence

$$E_x \sim B_z^2/(4\pi n e \Delta_i).$$

In this region the ions are freely accelerated by the electric field E_x in the outflow direction until, for $x \sim \Delta_i$, they, too, become magnetized, which is defined by the point where the ion gyroradius $\rho_i = v_x/\Omega_i$ becomes of the order of Δ_i. With $v_x^2 = (eE_x/m_i)\Delta_i$ and the expression for E_x one obtains immediately

$$\Delta_i \sim c/\omega_{pi}, \tag{7.174}$$

while from the continuity equation it follows that $u \sim v_A$, i.e., there is Alfvénic scaling of the reconnection rate. Here we emphasize again, that the details of the reconnection process in the electron layer do not enter. It is only required that this layer remains microscopic.

Let us now inspect the outflow region more closely. While the electron current density j_{ey} is concentrated at the separatrix, the ion current density j_{iy}, which dominates outside the reconnection region, forms a sheet along the mid-plane of the outflow cone, well separated form the separatrix. Hence the magnetic field lines bend over sharply at the mid-plane. The ion distribution is highly anisotropic, representing essentially ion beams following complicated orbits. The ion current layer persists even at a distance of more than $50c/\omega_{pi}$ from the reconnection center, and there is no indication of the typical MHD features arising in an Alfvénic reconnection process, in particular the slow-mode shocks characteristic of the Petschek configuration.

Since the electrons, while providing the reconnection mechanism, do not determine the reconnection rate, their modeling may be further simplified. One often assumes a massless resistive fluid, where the induction equation has the simple form

$$\partial_t \mathbf{B} = \nabla \times (\mathbf{v}_i \times \mathbf{B}) - \frac{1}{en}\nabla \times (\mathbf{j} \times \mathbf{B}) - c\nabla \times \eta\mathbf{j}. \tag{7.175}$$

The advantage is that by neglecting c/ω_{pe} effects, the smallest scale to be resolved is c/ω_{pi}, which allows to consider the ion dynamics in a larger system. Numerical simulations using such a reduced hybrid model have been performed by several groups, in particular by Lottermoser *et al.* (1998) in which references to related work can be found. These authors study the collisionless ion dynamics arising in the externally driven reconnection in a Harris sheet of length $\sim 500c/\omega_{pi}$ using a resistivity localized at the X-point. The gross features observed are similar to those reported by Shay *et al.* (1999), in particular the formation of the extended narrow ion current sheet along the mid-plane of the outflow cone. A new feature showing up at the longer system size is a kink mode-type instability of this current sheet visible at distances from the X-point exceeding $150c/\omega_{pi}$, see fig. 7.15, which appears to be driven by the anisotropy of the ion distribution, leads to filamentation of the out-of-plane ion current and effectively thermalizes the ion distribution.

Only at a distance of several hundred c/ω_{pi}, when the ion distribution has become essentially isotropic, does a current density located mainly along the separatrix become recognizable, which can be interpreted as Petschek's slow-mode shock. Hence one finds that in order to recover MHD structures scales $L > 10^2 c/\omega_{pi}$ are required, not just $L > c/\omega_{pi}$. At smaller distances, $L < 10^2 c/\omega_{pi}$, kinetic ion effects are important.

The two-dimensional numerical studies of reconnection in a neutral sheet configuration seem to be at variance with the nonlinear theory of

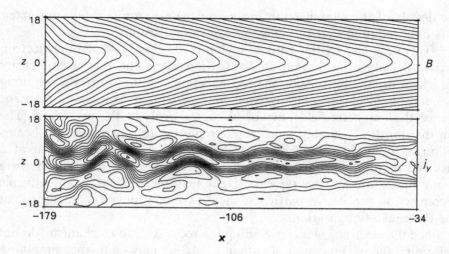

Fig. 7.15. Kink mode-type instability in the outflow region of a reconnecting neutral sheet (from Lottermoser *et al.*, 1998). Shown are magnetic field lines and the current density j_y. Spatial scales are in units of c/ω_{pi}.

the tearing mode discussed in section 7.5.2, which predicts saturation at very small island width, (7.152). It should, however, be noted that the simulations start from a perturbed equilibrium exhibiting a well-defined X-point, which corresponds to a tearing mode amplitude far exceeding the saturation level, (7.152). Moreover, the simulations include B_y fluctuations and hence the whistler, while the tearing mode theory is restricted to B_x, B_z. This effect, which is expected to be destabilizing, has not yet been studied analytically.

7.6.4 GEM Magnetic Reconnection Challenge

The project called Geospace Environmental Modeling (GEM) Magnetic Reconnection Challenge, which involves a number of groups, was initiated to compare the results from different numerical approaches to the problem of collisionless magnetic reconnection. The outcome of this endeavor is published jointly in a series of papers (see Birn *et al.*, 2000, and following articles). The goal was to identify the essential physics which is required to model reconnection in a collisionless high-β plasma.

Since the most prominent application is the magnetic activity in the geomagnetic tail, the initial equilibrium, chosen identically in these computations, is the Harris sheet, $B_x = B_0 \tanh(z/a)$, together with the plasma density profile $n(z) = n_0 \text{sech}^2(z/a) + n_\infty$, where n_∞ is the background density, and uniform temperatures T_e and T_i. In particular, the param-

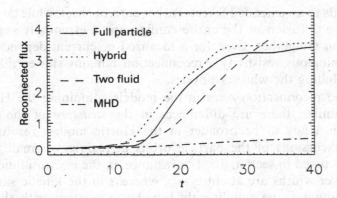

Fig. 7.16. Time evolution of the reconnected magnetic flux in the break-up of a Harris sheet configuration for different simulation models (from Birn *et al.*, 2000).

eters (in units of v_A, c/ω_{pi}, and Ω_i^{-1}) are: $B_0 = n_0 = 1$; $T_e/T_i = 0.2$; $m_i/m_e = 25$, and the dimensions of the two-dimensional computational system are $L_x = 25.6$ along the sheet with periodic boundary conditions and $L_z = 12.8$ across the sheet with ideally conducting boundary conditions. The initial perturbation described by the flux function $\widetilde{\psi}(x, z) = 0.1 \cos(2\pi x/L_x) \cos(\pi z/L_z)$ produces a magnetic island of width comparable to the equilibrium sheet width. The motivation for such a large initial perturbation is to start the reconnection process at a well-defined finite-angle X-point in order to avoid the problem arising from the properties of the small-amplitude behavior of the tearing mode, discussed in section 7.5.

The computations were performed using a variety of codes: full particle simulations; hybrid simulations (ion particles and electron fluid) including off-diagonal terms of the electron pressure tensor; two-fluid simulations with and without electron inertia (the latter model is called Hall MHD); and resistive MHD simulations. The main result is displayed in fig. 7.16, which shows the evolution of the reconnected flux $\psi(t)$ at the X-point obtained from different simulation models. As the most conspicuous feature one finds that, except for the MHD case, the reconnection rate $E = \partial_t \psi$, the slope of these curves, is essentially independent of the model, i.e., all models which include the Hall term give the same rate. Hence the rate does not depend on the actual mechanism of flux-conservation breaking, electron thermal motion, electron inertia, or resistivity. However this is only true if the value of η is sufficiently small, since excessive diffusion tends to weaken the Hall effect, i.e., the whistler, on scales below c/ω_{pi}. By contrast, in resistive MHD the reconnection rate remains much smaller than in the kinetic or two-fluid models for all values of the resistivity η.

Small η leads to an extended Sweet–Parker current sheet, while large values cause strong diffusion of the entire configuration, essentially suppressing reconnection altogether. Even for a localized or current-dependent resistivity ("anomalous resistivity") reconnection remains slower than in the models including the whistler physics.

While the reconnection *rates* in the models containing the Hall effect are very similar, there are differences in the *structure* of the diffusion region. This tends to be broader in the kinetic models resulting from the large excursions of the particle orbits in regions of small magnetic field, as discussed in section 7.5. For instance, in the electron fluid models, electron layer widths are about c/ω_{pe}, whereas in the kinetic simulations these are found to be significantly broader, consistent with the scaling $\delta_e \sim \sqrt{ac/\omega_{pe}} \sim \sqrt{a\rho_e}$, (7.142). A similar difference is observed when comparing fluid and kinetic ion models.

The simulations performed in the GEM project confirm the results, discussed in sections 6.2.2 and 7.6.2, that collisionless reconnection in a high-β plasma is independent of the electron physics. Hence a Hall MHD model, which is very convenient for large-scale simulations, seems to be sufficient to describe the essential magnetic dynamics. Such large-scale computations show that reconnection rates are also independent of the ion inertia scale c/ω_{pi} and are hence Alfvénic, i.e., scale with the Alfvén speed.

7.6.5 Three-dimensional simulations

The two-dimensional approximation in the (x, z)-plane, to which consideration was restricted up to this point, is of course an oversimplification of the true dynamics arising in a neutral sheet configuration. We have already encountered an important class of modes which are outside the scope of the 2D tearing mode geometry: the interchange-type modes $k_x = 0$, $k_y \neq 0$ driven by the LHDI; and the drift-kink instability, discussed in the last paragraph of section 7.3. These modes, which by themselves preserve magnetic topology, may have an effect on the tearing mode, acting as an anomalous resistivity. In addition, there may also be oblique tearing-like modes with $k_x \neq 0$, $k_y \neq 0$. Though linear theory suggests that these are less unstable than the $k_y = 0$ mode, they could play an important role in the nonlinear dynamics.

Three-dimensional particle simulations are now being performed by several groups (e.g., Pritchett & Coroniti, 1996; Lapenta & Brackbill, 1997; Zhu & Winglee, 1996; Hesse et al., 1998), though their relevance is still somewhat limited by the currently available computer resources which restrict both the computed system size and the achievable physical time-period. As a result the mass ratio chosen is usually small, $1 \leq m_i/m_e \leq 25$. Only by using a fully implicit algorithm, which lowers the numerical restrictions, does a higher mass ratio become feasible, as applied by Lapenta & Brackbill (1997). We restrict discussion to simulations of a Harris sheet equilibrium with periodicity both in x, i.e., along the magnetic field, and in y, i.e., along the current, deferring the effect of a normal field component to section 8.4.3, where the dynamics in the geomagnetic tail is considered. Though the details of the numerical observations and their interpretation still vary considerably, a common feature seen in these simulations is the dominance of the interchange modes, corroborating the previous 2D simulations in (y, z) geometry. The early growth of the off-center LHDI modes ($\gamma \sim \omega_{LH} \simeq \sqrt{\Omega_i \Omega_e}$) is followed by the slower drift-kink mode ($\gamma \sim \Omega_i$). On a longer time-scale (at least for sufficiently high mass ratio), these modes seem to enhance growth of reconnecting modes, both $k_y = 0$ and $k_y \neq 0$ tearing modes, of which the latter dominate nonlinearly as noticed by Lapenta & Brackbill. However, runs of higher spatial resolution and larger particle numbers are needed to put these preliminary findings on a firm ground.

8
Magnetospheric substorms

The magnetosphere is the cosmic plasma laboratory nearest to the Earth, which is therefore accessible to detailed ground and *in-situ* observations. It is, loosely speaking, a magnetic cavity generated by the interaction of the solar wind with the Earth's dipole field, which shields the Earth from direct bombardment by high-energy particles. This shield is, however, rather leaky, allowing solar-wind plasma to penetrate into the magnetosphere, which gives rise to a variety of different phenomena, the most spectacular being the aurora. The leakiness is mainly due to large-scale reconnection processes occurring at the front and in the tail of the magnetosphere. These processes form the main topic of this chapter.

The magnetosphere has a complex onion-like structure consisting of various plasma layers of distinctly different properties separated by rather sharp boundary surfaces. In section 8.1 we give a brief overview of the main features and outline the mechanisms leading to this layered structure. For a more detailed introduction to magnetospheric physics see, e.g., Baumjohann & Treumann (1996).

Reconnection is believed to be the main mechanism responsible for the magnetic processes observed in the magnetosphere, commonly called geomagnetic activity. The basic model of magnetospheric reconnection and plasma convection has been proposed by Dungey (1961) and this is considered in section 8.2. Reconnection of the dipole field (which is essentially oriented northward) with a southward component of the interplanetary field opens the magnetic cavity. The reconnected field lines are swept along by the solar wind to the nightside, until the increasing magnetic tension leads to a second reconnection process in the tail, reclosing the dipole field lines, which then contract back toward the Earth. Since the plasma is carried along with the magnetic field lines, the double reconnection process drives large-scale convection in the magnetosphere.

These reconnection processes are not stationary, but proceed in a dy-

Table 8.1. *Typical parameter values in the solar wind at the Earth's orbit.* M_A *is the Alfvén Mach number.*

	n_e(cm^{-3})	T_e(eV)	T_i(eV)	B(nT)	M_A
solar wind	5–10	10	10	10	5–10
magnetosheath	30	10	40	10–40	< 1

namic manner. The time variability is partly due to the turbulent character of the solar wind, in particular the rapid irregular changes of the magnetic field in the magnetosheath, which leads to a rather bursty reconnection behavior at the magnetopause, discussed in section 8.3. More importantly, even for periods when magnetopause reconnection is quasi-stationary, tail reconnection does not follow in the same continuous manner, but proceeds mostly in the form of sudden large-scale relaxation events. These events, which are associated with magnetospheric substorms – the basic element of the geomagnetic activity –, and their secondary effects in the ionosphere are considered in section 8.4.

8.1 The structure of the magnetosphere

The magnetic configuration and the layered structure of the plasma in the magnetosphere are caused by the combined effects of (i) the interaction between the solar wind and the dipole field and (ii) the presence of the corotating atmosphere of the Earth. In this section we will "peel the magnetospheric onion", discussing the properties of the individual layers and their mutual relationship.

8.1.1 The solar wind

The Sun emits a continuous plasma flow, called the solar wind, consisting of fully ionized hydrogen with a small admixture of helium, which reaches the Earth at supersonic speed. Typical parameters of the solar-wind plasma are summarized in table 8.1. The simplest model of the solar wind assumes a stationary radial hydrodynamic flow moving only under the effects of pressure and gravity, i.e., one neglects, in particular, the solar magnetic field. Hence the plasma properties of the flow are ignored and the flow is treated as a perfect gas with thermodynamics characterized by a polytropic coefficient γ. The basic equations for flow velocity v, mass density ρ, and pressure p are

$$\frac{d}{dR}(R^2 \rho v) = 0, \tag{8.1}$$

$$v\frac{dv}{dR} = -\frac{GM_\odot}{R^2} - \frac{1}{\rho}\frac{dp}{dR} = \frac{d}{dR}\left(\frac{GM_\odot}{R} - \frac{C\gamma}{\gamma - 1}\rho^{\gamma-1}\right), \qquad (8.2)$$

where $p = C\rho^\gamma$, $C = p_0/\rho_0^\gamma$, and $GM_\odot/R$ is the gravitational potential of the Sun. (In the isothermal case $\gamma = 1$, the last term is replaced by $C \ln \rho$.) These equations can be integrated immediately:

$$R^2\rho v = \Gamma, \qquad (8.3)$$

$$\frac{1}{2}v^2 - \frac{GM_\odot}{R} + \frac{C\gamma}{\gamma - 1}\rho^{\gamma-1} = W.^* \qquad (8.4)$$

Substituting ρ from (8.3) we obtain the Bernoulli equation for $v(R)$

$$\frac{1}{2}v^2 - \frac{GM_\odot}{R} + \frac{C\gamma}{\gamma - 1}\left(\frac{\Gamma}{R^2v}\right)^{\gamma-1} = W. \qquad (8.5)$$

Let us introduce the dimensionless variables $r = R/R_\odot$, $R_\odot$ = radius of the Sun, $u = v/c_0$, $\phi_\odot = GM_\odot/R_\odot c_0^2$, $\mu = \Gamma/R_\odot^2\rho_0 c_0$, $\epsilon = W/c_0^2$. The sound speed c_s is defined by $c_s^2(r) = \gamma p/\rho = c_0^2(\rho/\rho_0)^{\gamma-1}$, and p_0, ρ_0, c_0 are the respective values at the solar surface $r = 1$. Equation (8.5) now reads in non-dimensional form

$$\frac{1}{2}u^2 - \frac{\phi_\odot}{r} + \frac{1}{\gamma - 1}\left(\frac{\mu}{r^2u}\right)^{\gamma-1} = \epsilon. \qquad (8.6)$$

Observations indicate that the solution $u(r)$ should increase uniformly, $u' > 0$, being subsonic at small r and supersonic at large r. Analysis of (8.6) gives the small-argument behavior

$$u \sim r^{\frac{3-2\gamma}{\gamma-1}}, \quad M \sim r, \qquad (8.7)$$

and the asymptotic behavior for $r \to \infty$

$$u \to \sqrt{2\epsilon}, \quad M \sim r^{\gamma-1}, \qquad (8.8)$$

where $M = u/c_s \propto u/\rho^{(\gamma-1)/2}$ is the Mach number. Since M increases monotonically, the flow must cross the sonic point $M = 1$ at some finite radius r_c. The global behavior of the solution is usually discussed in the r, u plane (or in the r, M plane as illustrated in fig. 8.1). Consider the differential form of (8.6),

$$\frac{du}{dr}\left[u^2 - \left(\frac{\mu}{r^2u}\right)^{\gamma-1}\right] = \frac{2u}{r}\left[\left(\frac{\mu}{r^2u}\right)^{\gamma-1} - \frac{\phi_\odot}{r^2}\right]. \qquad (8.9)$$

---

*It should be mentioned that in the hydrostatic solution $v = 0$, $\rho = [1 - (\gamma - 1)\phi_\odot(1 - r^{-1})]^{1/(\gamma-1)}$, written in the units introduced below, the large gravitational potential $\phi_\odot \sim 10$ would lead to a density cutoff $\rho = 0$ at relatively small radius $r \sim 1$, which is inconsistent with observations of the interplanetary density.

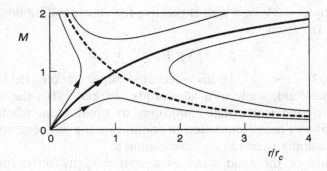

Fig. 8.1. The manifold of solutions of (8.9) in the r, M plane.

There is a critical point or saddle point determined by the conditions $du/dr = dr/du = 0$, i.e., by the vanishing of both bracketed expressions in (8.9), which yields

$$u_c^2 = \left(\frac{\mu}{r_c^2 u_c}\right)^{\gamma-1} = \left(\frac{\rho_c}{\rho_0}\right)^{\gamma-1} = c_s^2, \tag{8.10}$$

$$\frac{\phi_\odot}{2r_c} = \left(\frac{\mu}{r_c^2 u_c}\right)^{\gamma-1} = u_c^2. \tag{8.11}$$

Hence the critical point is also a sonic point $M = 1$. There exists only one solution with the required properties, the heavy line in fig. 8.1 passing through the critical point. Substitution of u_c, r_c in (8.6) determines $\epsilon = \epsilon(\mu, \phi_\odot, \gamma)$. Relation (8.11) indicates that, at the sonic point, the flow energy is only one-fourth of the potential energy, hence continued action by the pressure force is required to allow the solar wind to escape. We should mention that the gravitational field acts in a way analogous to a Laval nozzle in producing a supersonic flow.

It is interesting to note from (8.7) that the condition $u \to 0$ for $r \to 0$ entails the restriction of the polytropic coefficient to the range

$$1 < \gamma < 3/2.^* \tag{8.12}$$

By the upper inequality the polytropic coefficient must also be smaller than the adiabatic value $5/3$, which implies the presence of coronal heating, as can be understood by the following simple argument. Consider the pressure equation with a heat source $Q > 0$,

$$\partial_t p + \mathbf{v} \cdot \nabla p + \tfrac{5}{3} p \nabla \cdot \mathbf{v} = Q. \tag{8.13}$$

---

* The lower inequality is due to the polytrop ansatz in (8.2). Isothermal conditions $\gamma = 1$ are also allowed, but require a slightly different treatment as mentioned above.

Defining α by $Q = \alpha \nabla \cdot \mathbf{v}$, which is positive for an expanding flow $\nabla \cdot \mathbf{v} > 0$, (8.13) takes the form

$$\partial_t p + \mathbf{v} \cdot \nabla p + \gamma' p \nabla \cdot \mathbf{v} = 0$$

with $\gamma' = 5/3 - \alpha < 5/3$. In his celebrated paper Parker (1958) assumed an isothermal fluid, while in a later study (Parker, 1960) the model was refined, restricting isothermal conditions to small radii, where the low flow speed allows heat conduction to equilibrate the temperature, while at larger radii adiabatic conditions are assumed.

The picture of the solar wind as a stationary hydrodynamic flow is oversimplified in several respects. Since Coulomb collisions are rare outside the corona and hence the electron–ion energy exchange becomes very weak, see (6.12), only the ions should become adiabatic, while the electrons, due to their much higher heat conductivity, should be close to isothermal, which would give rise to a strong disparity of the temperatures, $T_e \gg T_i$. However, observed temperatures do not differ strongly, $T_e \sim T_i$. In fact, electron heat conduction is strongly reduced by small-scale electrostatic fluctuations in the solar wind, which leads to a reduction of T_e, while T_i is increased above the adiabatic level due to heating by Landau and cyclotron damping of magnetohydrodynamic waves. Hence a one-fluid model is actually more realistic than a two-fluid model based on classical transport effects.

Since the solar corona is a low-β plasma, the magnetic field plays an important role restricting the flow to regions of open field lines, the so-called coronal holes. Taking the Lorentz force into account in (8.2) leads to a somewhat more complicated solution manifold than that depicted in fig. 8.1 for the simple hydrodynamic flow. The asymptotic behavior is, however, qualitatively similar. As the flow expands, the pressure decreases, but the magnetic pressure decreases even more rapidly, since $Br^2 = const$ because of flux conservation, hence β rises above unity making the solar wind also super-Alfvénic. The magnetic field is therefore swept along with the flow but, since it remains frozen in the photosphere where its footpoints rotate with the Sun, the field lines are bent forming an Archimedian spiral which is divided into sectors of opposite polarity with rather sharp boundaries (tangential discontinuities) as illustrated in fig. 8.2. At the Earth's orbit the average angle with respect to the Sun–Earth line is about 45°. Though the magnetic field has therefore little influence on the gross features of the flow, its inclination with respect to the ecliptic plane has a decisive effect on the geomagnetic activity, as we will see in the subsequent sections.

The solar wind is also far from stationary, exhibiting a strongly turbulent behavior, wherein many different types of modes can be identified. For $k\rho_i < 1$ magnetic fluctuations are nonlinear Alfvén waves, which constitute an interesting example of high-Reynolds number MHD turbulence,

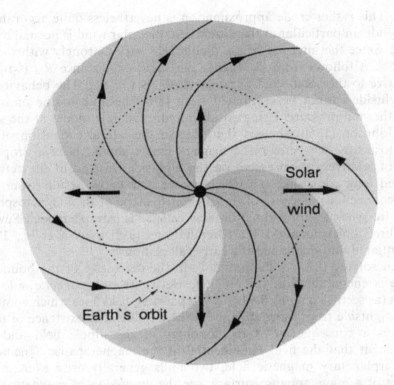

Fig. 8.2. Schematic drawing of the sector structure of the interplanetary magnetic field.

actually the only one accessible to *in-situ* measurements (see, e.g., Tu & Marsch, 1995).

8.1.2 Magnetopause and bow shock

When the solar wind interacts with the Earth's magnetic field, a tangential discontinuity, the *magnetopause*, is formed, encapsulating the field in a magnetic cavity. In the simplest model of the shape of the magnetopause the solar wind is considered as a nonmagnetic flow of charged particles, ignoring the fluid properties of the flow, in particular its supersonic character. The position of the magnetopause is determined by the balance between the magnetic pressure and the dynamic pressure of the flow

$$\rho(\mathbf{n} \cdot \mathbf{v})^2 = (\mathbf{n} \times \mathbf{B})^2, \tag{8.14}$$

where $\mathbf{n} = \nabla S / |\nabla S|$ and the surface $S(\mathbf{x}) = 0$ defines the magnetopause. This is the theory Chapman & Ferraro developed in a series of papers (e.g.,

1931).* This rather crude approximation is nevertheless quite accurate on the dayside, in particular at the nose where the solar wind is normal to the surface. Since the intensity of the dipole field varies strongly with radius, $B \sim R^{-3}$, it follows from (8.14) that the stand-off distance R_{so} is rather insensitive to the solar-wind parameters, $R_{so} \sim (\rho v^2)^{1/6}$. The behavior on the nightside cannot be calculated from (8.14) because of the omission of the thermal pressure. Using an ideal hydrodynamic model of the solar wind, Johnson (1960) showed that, due to the lateral expansion of the magnetosheath (see below), the magnetosphere should have a drop-like shape of relatively short extent on the nightside (instead of the actually observed long comet-like tail which, as we now know, results from tail-side reconnection). Because of this predicted shape of the magnetosphere, the hydrodynamic approximation was called "teardrop" model. Several fully three-dimensional MHD computations (e.g., by Wu *et al.*, 1981) substantiated and quantified the early calculations.

When solving (8.14) a discontinuity in the derivative of the boundary surface is encountered at the polar cusps, where two *magnetic nulls* are formed (as seen also in fig. 8.5 below), separating field lines which continue to the nightside from those staying on the dayside. The existence of these points is a consequence of the topology of the dipole field and the requirement that the field is tangential at the magnetopause. (Including the interplanetary magnetic field, two nulls generally arise even in the absence of a magnetopause surface, see the discussion of magnetic nulls in section 2.3.) Note that the field does not vanish on the field lines connecting the nulls with the magnetic poles.

The magnetopause constitutes a collisionless current sheet of complex internal structure. Two-fluid theory predicts a thickness of the order of the electron skin depth $\delta \sim c/\omega_{pe}$, while observations indicate that the sheet is much broader, exceeding the ion Larmor radius $\delta \gtrsim \rho_i \gg c/\omega_{pe}$. The broadening seems to be caused by microturbulence, for instance the LHDI (section 7.3). Moreover, the magnetopause may be subject to macroscopic Kelvin–Helmholtz modes. An interesting consequence of these effects is that magnetic field lines in a certain layer inside the magnetopause are coupled to the solar wind by an effective viscosity and are dragged along to the nightside, until magnetic tension drives them back to the dayside. Since the magnetospheric plasma is tightly coupled to the field lines, this process leads to plasma convection in the magnetosphere, which was first suggested by Axford & Hines (1961). However, the viscous drag is now

---

* The authors actually referred to the interaction of the Earth's field with the corpuscular radiation emitted from the Sun during a major flare, which is the origin of a geomagnetic storm. The solar wind, the continuous plasma flow from the Sun, was predicted only in the 1950s by Biermann (1951) and Parker (1958) and observed in the 1960s. For the history of these developments see, e.g., Kennel (1995).

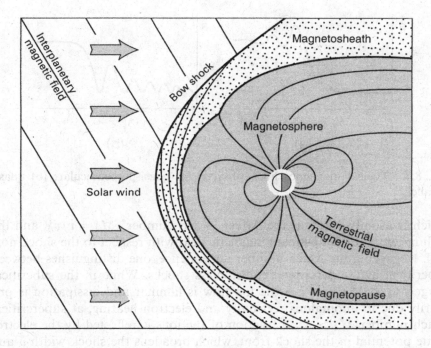

Fig. 8.3. Schematic view of the meridional cross-section of the magnetosphere, showing the bow shock and the magnetopause.

believed to be, in general, weak compared with the convection induced by reconnection with the interplanetary field, which we will consider in section 8.2.

In the previous discussion the supersonic character of the solar wind has been ignored. If a supersonic stream encounters a blunt body, a *bow shock* is formed, which slows down the flow speed and increases density and temperature, thus reducing the Mach number below unity. The position of this stand-off shock indicated in fig. 8.3 and the parameters of the downstream state are (essentially) determined by the Rankine–Hugoniot conditions resulting from mass, momentum and energy conservation in an MHD flow. The internal shock structure, however, depends on the dissipation processes. Since the solar-wind plasma is collisionless, these processes must have a collective character. Collisionless shocks have been a central topic in both laboratory and space plasma physics and, in the latter, research still continues; for a review of these ongoing activities, see Russell (1995).

The structure of the bow shock exhibits a bewildering complexity, for which theory can give only a very idealized and simplified account. The shock structure depends primarily on two parameters, the Mach number,

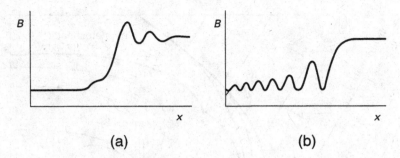

Fig. 8.4. Typical magnetic shock profiles: (a) quasi-perpendicular; (b) quasi-parallel.

which is usually taken as the Alfvén Mach number[*] $M_A = v/v_A$, and the inclination θ of the upstream magnetic field with respect to the shock normal. Regarding the Mach number dependence, one distinguishes between subcritical and supercritical collisionless shocks. While in the subcritical range $M_A < M_c(\theta) \sim 2\text{--}3$ the ion flow is laminar and dissipation is primarily due to anomalous resistivity and electron heating, at supercritical Mach number $M_A > M_c$ a fraction of the ions is reflected by the electrostatic potential in the shock front, which broadens the shock width δ and leads to strong ion heating, whence dissipation is attributed to anomalous viscosity. The bow shock is typically supercritical, $M_A \sim 10$, see table 8.1.

The inclination θ has an even stronger influence on the shock structure, in particular the shock width. For a nearly perpendicular field, reflected ions are immediately turned around back into the downstream plasma, which makes the width roughly an ion Larmor radius $\delta \sim \rho_i$. Such shocks, called quasi-perpendicular, are characterized by a trailing magnetosonic wave-train (in general nonstationary), see fig. 8.4(a). However, if the magnetic field has a substantial parallel component, the reflected ions may propagate upstream along the field, where they excite whistler waves. Note that whistlers propagate fast, with a group velocity increasing with k, see (6.35), hence there is a leading wave-train, a precursor. Wave–particle resonance gradually reduces the upstream velocity of the reflected ions by pitch-angle scattering, until these are finally swept along with the bulk flow into the downstream region. This behavior leads to a broad turbulent magnetic structure which is the characteristic feature of a parallel shock, fig. 8.4(b). The bow shock has more often than not the signature of a quasi-parallel shock.

The region between the bow shock and the magnetopause, called the *magnetosheath*, carries the compressed and heated downstream plasma,

---

[*] More appropriate would be the magnetosonic Mach number $M_{As} = v/\sqrt{v_A^2 + c_s^2}$, (3.24).

Table 8.2. *Typical parameter values of the different plasma layers in the magnetosphere.*

Layer	$n_e(\text{cm}^{-3})$	$T_e(\text{eV})$	$T_i(\text{eV})$	$B(\text{nT})$	β
ionosphere	10^5	0.1	0.1	10^4	10^{-4}
plasmasphere	10^3	1	1	10^3	10^{-3}
radiation belt	1	5×10^3	10^4–10^5	10^2–10^3	10^{-2}
plasma sheet	1	5×10^2	5×10^3	30	1
lobe plasma	10^{-2}	30	10^2	30	10^{-4}

flowing around the magnetosphere. When the positions of the bow shock and the magnetopause are given, the macroscopic behavior of the magnetosheath plasma can be computed in the MHD framework assuming isotropic pressure and a suitable polytropic coefficient. In spite of the collisionless nature, the downstream plasma is observed to be fairly well thermalized, which is caused by the interaction with the turbulent waves excited at the shock front. Ions and electrons are, however, not heated at the same rate; the ion temperature is significantly higher than the electron temperature, typically $T_i \sim 4T_e$, in contrast to the solar wind, where $T_i \lesssim T_e$. Since ion–electron energy exchange is slow, even in the presence of collective effects, T_i and T_e are essentially independent quantities. Hence in a collisionless shock the Rankine–Hugoniot conditions are not sufficient to determine the downstream state completely, which depends also on the character of the dissipation process.

8.1.3 The internal structure of the magnetosphere

The preceding discussion could evoke the impression that the magnetosphere, the magnetic cavity inside the magnetopause, is void of plasma. This is, however, only partially true. Major regions of the magnetosphere are actually filled with plasma of considerable density and pressure originating from mainly two sources: the leaky magnetopause which, because of reconnection, allows solar-wind plasma to penetrate into the magnetosphere; and evaporation from the *ionosphere*, the ionized upper part of the Earth's atmosphere. Here we give a brief overview of the properties of the different plasma layers, deferring discussion of reconnection processes to the following sections. Figure 8.5 illustrates the structure of the magnetospheric plasma, while table 8.2 summarizes typical parameters of the different plasma layers.

At this point it is useful to clarify a misconception which may arise when considering naively the role of the Earth's rotation. Magnetic field lines are frozen in to the conducting interior of the Earth, see section 5.3.3, and into

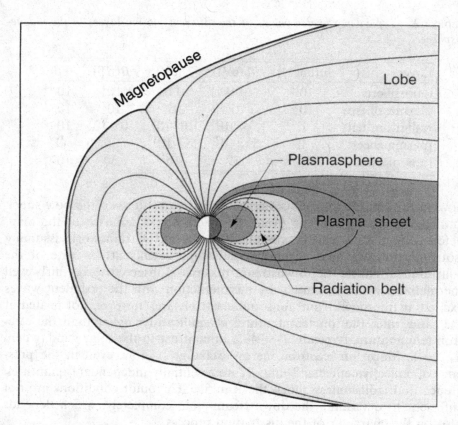

Fig. 8.5. Meridional structure of the magnetosphere (after Baumjohann & Treumann, 1996).

the magnetospheric plasma, but they may easily be cut in the intermediate poorly conducting regions, especially the neutral atmosphere. Hence the dipole field determines only the magnetic field in the magnetosphere, not the individual field lines. There is, however, a rotation effect resulting from the friction between the corotating atmosphere and the ionosphere, the latter being coupled magnetically to the non-rotating magnetosphere, which generates an electric field, as discussed further down in this section.

The ionosphere is the upper part of the Earth's atmosphere, which is partially ionized, mainly by solar ultraviolet radiation. Because of the diurnal variations of the latter the degree of ionization, determined by the balance between ionization and recombination, varies strongly (by a factor 10–100) between day and night. The electron density exhibits local maxima in the E-layer (at height 100–150 km) and the F-layer (at height 200–400 km) with the degree of ionization increasing with altitude. In the ionosphere, collisions of electrons or ions with neutrals are more frequent than collisions between charged particles, $v_{en} > v_{ei}$, $v_{in} > v_{ii}$. Due to the high collisionality the scalar resistivity in Ohm's law is replaced by a more general nonisotropic behavior, which is described by the conductivity tensor $\boldsymbol{\sigma}$,

$$\mathbf{j} = \boldsymbol{\sigma} \cdot \mathbf{E} = \sigma_\parallel \mathbf{E}_\parallel + \sigma_P \mathbf{E}_\perp - \sigma_H \mathbf{E} \times \mathbf{B}/B,$$

where σ_P is called the Pedersen conductivity and σ_H the Hall conductivity. The parallel conductivity $\sigma_\parallel \propto v_{en}^{-1}$ is the largest of these coefficients, but since $E_\parallel$ is usually much smaller than $E_\perp$, parallel and cross-field currents are of the same order, a consequence of the continuity equation $\nabla \cdot \mathbf{j} = 0$. Since in the E-layer the electron collision frequency is smaller than the electron cyclotron-frequency, $v_{en} < \Omega_e$, while for the ions the reverse is true, $v_{in} > \Omega_i$, the electrons are tied to the magnetic field, while the ions move with the neutral gas. In this regime the Hall currents are most important, carried by the electron $\mathbf{E} \times \mathbf{B}$ motion

$$\mathbf{j}_\perp \simeq -en_e c \frac{\mathbf{E} \times \mathbf{B}}{B^2}, \quad i.e., \quad \sigma_H = \frac{en_e c}{B}.$$

Tidal motions of the atmosphere due to diurnal heating and the effects of corotation give rise to ionospheric electric fields and currents, in particular the equatorial electrojet, which is measured on the ground as a decrease of the magnetic field intensity.

In addition to photo-ionization, which dominates at low latitude, ionization also occurs by the precipitation of energetic electrons from the plasma sheet (see below), which is connected magnetically to the ionosphere at higher latitudes. These lead to an increase of the conductivity and hence to strong currents and also give rise to light emission, the polar light or *aurora* localized in a belt around the magnetic poles at latitudes roughly between 65° and 75°, called the *auroral oval*. Magnetic activity in the tail, in particular substorms, produces enhanced auroral activity, which is discussed in more detail in section 8.3.

The ionosphere continues outward into a fully-ionized plasma, the *plasmasphere*. Whereas the ionosphere, being dominated by interaction with the neutral atmosphere, covers the globe uniformly, the plasmasphere is controlled by the combined effect of magnetic configuration and corotation, which gives it a belt-like shape around the equator, as shown in fig. 8.5. At low latitude the field lines remain at relatively low altitude before returning to the Earth, hence the plasma is well confined and the drag from the ionosphere forces the plasma to corotate. This implies the presence of an electric field in the non-rotating reference system,

$$\mathbf{E}_{co} = -\frac{1}{c} \mathbf{v} \times \mathbf{B} = -\frac{1}{c} (\boldsymbol{\Omega}_E \times \mathbf{R}) \times \mathbf{B},$$

Ω_E = Earth's rotation vector, $\mathbf{R}$ = radius vector. The electric field decreases with distance from the rotation axis, $E \sim R^{-2}$, since the decrease of the dipole field $B \sim R^{-3}$ outweighs the increase of the velocity $v \sim R$. At $R \sim 4R_E$, R_E = radius of the Earth, the electric field reaches the order of magnitude of the dawn-to-dusk magnetospheric electric field associated with the global magnetospheric convection (see section 8.2). This gives rise to a separatrix in the flow pattern, limiting the range of corotation and plasma confinement around the Earth. Since, outside the separatrix, plasma can freely escape into the magnetosphere, there is a sharp drop of the particle density, called the *plasmapause*. At higher latitude, where field lines reach out far into the magnetosphere, the average effect of E_{co} is too weak to enforce corotation, which limits the latitudinal extent of the plasmasphere.

Beyond the plasmapause the particle density is much lower than in the plasmasphere, but the plasma has a high-energy component with ion energies reaching more than 10^5 eV, giving rise to a finite plasma pressure $\beta \sim 10^{-2}$. This ring-shaped region is called the *radiation belt*. Particles are trapped in the region around the equator, where the magnetic field has a relative minimum, bouncing back and forth along field lines between their mirror points in the regions of higher field intensity. The basic stationary population of high-energy particles results from collisions of cosmic-ray particles in the atmosphere, the secondary particles becoming trapped. The population may, however, temporarily be strongly enhanced by high-energy particles generated in a major solar eruption, which enter deep into the magnetosphere when the latter is opened up during a geomagnetic storm. Such particles are confined in the radiation belt for many days, until they are finally lost by pitch-angle scattering and charge exchange in the ionosphere. High-energy particles perform fast drifts across field lines,

$$\mathbf{v}_{dj} = \frac{cm_j}{e_j}\left(v_{\parallel}^2 + \tfrac{1}{2}v_{\perp}^2\right)\frac{\mathbf{B} \times \boldsymbol{\kappa}}{B^2} \simeq \frac{cT_j}{e_jB}\frac{1}{R},$$

where $\boldsymbol{\kappa} = \mathbf{b} \cdot \nabla\mathbf{b}$ is the field-line curvature. These drifts generate an azimuthal current, the *ring current*, which weakens the magnetic field on the near-Earth side and thus produces the field depression observed during a magnetic storm.

Further out on the nightside, high-β plasma extends far into the magnetotail concentrated along the neutral sheet with sharp boundaries. This plasma layer is called the *plasma sheet*. Particles are supplied directly from the solar wind due to magnetic reconnection in the tail. It therefore has essentially the temperature and density characteristics of the magnetosheath plasma. The dawn-to-dusk electric field makes the particles drift slowly toward the neutral sheet and, subsequently, more rapidly along the sheet toward the Earth by reconnection processes in the tail. The precipitation

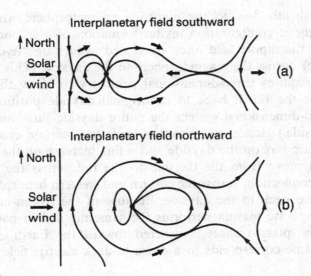

Fig. 8.6. Dungey's model of reconnection at the magnetopause in the case of (a) a southward IMF component, (b) a northward IMF component (after Dungey, 1961).

of electrons from this region into the magnetically connected high-latitude ionosphere is the origin of the aurora.

The region outside the plasma sheet, called the *lobe*, carries only a very thin plasma which does not give rise to spectacular phenomena.

8.2 Magnetospheric convection

In the hydrodynamic approximation of the solar wind–magnetosphere interaction, the effect of the relatively weak interplanetary magnetic field (IMF) is neglected, which is assumed to be just swept around the magnetosphere. Actually, however, there is a finite probability of IMF reconnection with the dipole field, which has dramatic consequences for the dynamics of the magnetospheric plasma.

Dungey's theory (Dungey, 1961) gives a qualitative account of these reconnection processes, which are illustrated in a two-dimensional schematic drawing in fig. 8.6(a). When an IMF field line (or flux tube) with a southward vertical component is pushed against the magnetopause, an X-line,* called the dayside or front-side X-line, is formed, at which reconnection with a dipole field line opens up the magnetopause. The resulting field lines, while remaining attached to the ionosphere, are dragged along with

---

* In this chapter we use the terminology neutral line or X-line instead of neutral point or X-point to underline the basic three-dimensional character of the configuration.

the magnetosheath flow to the nightside magnetosphere where, in the tail, the magnetic configuration invites formation of a second, nightside X-line. Here the dipole field lines are closed again (are "re-connected") snapping back across the magnetosphere to the dayside. (This return process clearly requires three-dimensional geometry to allow the field line to move past the Earth back to its original dayside position, since in a strictly two-dimensional system the entire dayside flux would end up on the nightside.) Hence, reconnection leads to a certain erosion of the magnetospheric flux on the dayside and a flux increase on the nightside.

Field lines move from the dayside to the tail across the polar caps and, after reconnection, across the eastern and western hemispheres of the magnetosphere back to the dayside. Because of the frozen-in condition, the magnetospheric plasma performs the same motion, in particular the plasma in the plasma sheet is ushered toward the Earth, which in a stationary frame corresponds to a dawn-to-dusk electric field $\mathbf{E} = -\mathbf{v} \times \mathbf{B}/c$.[*]

If the IMF has a northward component, reconnection on the front-side is much less efficient. Instead, IMF field lines are stretched along the polar caps generating a configuration whereby reconnection with the nightside dipole field becomes possible, at least topologically, as sketched in fig. 8.6(b). Hence a northward IMF component will give rise to an increase of the magnetospheric flux on the dayside and a flux erosion on the nightside. It is, however, intuitively clear that this process is weaker than in the case of a southward IMF component. While in the latter case reconnection on the front-side is driven by the strong dynamic pressure of the incoming solar-wind flow and is enforced in the tail because of the line-tying in the ionosphere, for a northward component the IMF flux accumulated on the front-side can easily be swept around the magnetosphere thus avoiding tailside reconnection.

The dawn-to-dusk electric field associated with the global magnetospheric convection is projected onto the high-latitude ionosphere (at low latitude, the electric field is dominated by the corotating plasmasphere). Figure 8.7 shows the equipotential lines $\phi = const$, assuming stationarity, $\mathbf{E} = -\nabla\phi$. Because of the high cross-field conductivity in the auroral oval, the electric field drives intense currents, primarily Hall currents (see

---

[*] The magnetosphere is considered in a frame stationary with respect to the Sun–Earth line, the noon–midnight line. North is in the direction of the Earth's rotation vector, i.e., the Earth rotates counterclockwise, when looking at the Earth from above (southward). Dusk is on the left, where the Sun sets for a terrestrial observer, and dawn on the right, as indicated in fig. 8.7. The points, or directions, dusk and dawn are also called west and east, respectively, such that in the dusk–dawn meridional plane, seen from the nightside, we have the usual assignment of directions north–west–south–east (counted counterclockwise).

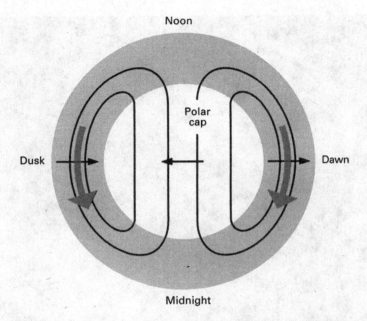

Fig. 8.7. Schematic drawing of the equipotentials of the ionospheric electric field in the auroral oval, the shaded area. Light arrows indicate the field direction, heavy arrows the plasma convection.

section 8.1.3), called the auroral electrojet. Figure 8.8 gives a view of the auroral oval.

At the time Dungey proposed his theory, the concept was rather speculative and hypothetical, since no direct evidence of the proposed reconnection processes was available. In fact, the viscous entrainment process by Axford & Hines (1961) was for some years regarded as the more important mechanism for magnetospheric convection. Over the past two decades, however, a host of observational results has been accumulated confirming the basic ingredients of the reconnection concept, in particular the reconnection characteristics on the front-side, the existence of a distant neutral X-line in the tail at a distance of 100–200 R_E, and direct measurements of the magnetospheric electric field. Observations clearly indicate that dayside reconnection starts when the vertical component of the IMF turns southward, while for northward IMF, where dayside reconnection is not possible, the magnetosphere is quiet and convection ceases. As expected, the reconnection rate depends on the magnitude of the southward field component as well as on the solar-wind velocity.

But Dungey's model can only be a conceptional framework, which must be filled with more physics to allow a semi-quantitative description of the

Fig. 8.8.   Satellite view of the auroral oval around the north pole. Also seen is the local brightening marking the occurrence of a substorm (courtesy of L. A. Frank).

different aspects of the geomagnetic activity. A crucial point is the microscopic reconnection mechanism. Dungey was aware of the collisionless nature of the magnetospheric plasma. Instead of resistivity he assumed electron inertia to be responsible for field-line merging. The issue is far from being of only academic interest. Since the IMF can, in principle, be carried around the magnetopause without reconnection, the fraction of flux actually reconnected depends on the efficiency of this microscopic reconnection process. In addition, the dynamics is far from stationary. Even for stationary dayside conditions, reconnection in the tail often occurs in the form of a rapid large-scale relaxation, the magnetospheric substorm. These topics will be considered in the following sections.

8.3 Magnetopause reconnection

Magnetic processes at the dayside magnetopause constitute a classical paradigm of driven reconnection, the solar wind pushing the IMF against the encapsulated dipole field. In principle, field-line merging may occur at any finite angle between the northward dipole field and the IMF, but

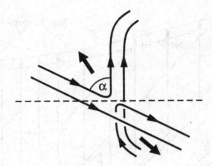

Fig. 8.9. Schematic illustration of the magnetic field behavior in magnetopause reconnection.

the process is much more efficient if the IMF has a southward compo-nent, since in this case the angle α formed by the reconnected field line is smaller than 45°, such that the magnetic tension, the slingshot effect, can rapidly transport the plasma away from the reconnection region, see fig. 8.9. Convincing observational evidence has been accumulated for a strong correlation between a southward IMF[*] and the occurrence of geomagnetic activity, characterized in particular by the number of sub-storms. Direct experimental evidence of magnetic reconnection is difficult to obtain, since the theoretically most obvious feature, the normal field component, cannot be measured with sufficient precision because of local magnetopause motions and tilting. Almost 20 years had to pass since Dungey's prediction, before a clear verification of reconnection could be observed, the high-speed tangential plasma flow just inside the magne-topause, whose magnitude and direction are consistent with the change of the magnetic field (Paschmann *et al.*, 1979). Additionally, measurements of the electric field $\mathbf{E} = -\mathbf{v} \times \mathbf{B}/c$ in the ionosphere and in the magne-tosphere above the polar cap (Baumjohann & Paschmann, 1987) show the presence of convection, a direct consequence of dayside reconnection, which is also strongly correlated with a southward IMF. Comparison of the magnetospheric electric field with the solar-wind electric field gives a direct measure of the amount of reconnected flux, which is typically 10% of the impinging IMF flux.

What do observations tell us about the actual reconnection dynamics? Though there are also phases of quasi-stationary reconnection (see, e.g.,

---

[*] It should be noted that the case of a purely southward IMF is very rare. The IMF is mainly in the equatorial plane, with only a small vertical component. Hence "southward IMF" usually means only a slight southward inclination from the horizontal direc-tion. Reconnection is facilitated if the dipole field itself obtains a sizeable equatorial component due to the tilt of the dipole axis, which is strongest at the equinoxes. This semi-annual increase of the geomagnetic activity is statistically well established.

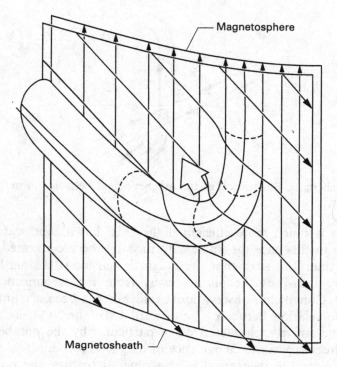

Fig. 8.10.   The Russell & Elphic model of flux transfer events (from Russell & Elphic, 1978).

Sonnerup *et al.*, 1981), the reconnection process, more often than not, occurs in a patchy and intermittent way. One source of time variation is the turbulent nature of the magnetosheath plasma, which introduces certain fluctuations of magnetic field intensity and flow velocity. However, the main cause of its bursty appearance seems to be the intrinsically dynamic character of the reconnection process. A direct indication thereof is obtained by the observation of so-called *flux transfer events* (FTEs), first reported by Russel & Elphic (1978), which are essentially bipolar pulses of magnetic field normal to the magnetopause. These are not simple surface-wave pulses, but flux bundles actually crossing the magnetopause, containing both magnetospheric and magnetosheath plasma, which can be distinguished because of their different energy spectra. Russell & Elphic visualized their observations as elbow-shaped flux tubes, sketched in fig. 8.10.

In spite of the rich collection of observational data of FTEs now available, it is difficult to extract a simple picture of the physical mechanism of this phenomenon, though there seems to be general agreement to relate the phenomenon to the reconnection process. The standard theoretical approach is based, in one way or the other, on the tearing mode in the current sheet of the magnetopause. As discussed in section 4.7, an extended current sheet is prone to break up into one or several plasmoids, see fig. 4.28. In the FTE models the emphasis lies either on the formation of multiple reconnection X-lines (e.g., Lee & Fu, 1985), or on a bursty reconnection behavior at a single X-line (e.g., Scholer, 1988). While in these 2D models the flux bundles thus generated are confined to the magnetopause, inclusion of three-dimensional effects, in particular a finite extent of the current sheet along the current, allows these flux bundles to start in the magnetosheath and end up in the magnetosphere.

However, the FTE modeling discussed up to now ignores the problem of the microscopic reconnection mechanism by assuming some rather arbitrary anomalous resistivity. Observations seem to suggest a threshold character of the reconnection process but it is unclear whether this is caused, for instance, by a finite-amplitude threshold of the collisionless tearing mode or by the excitation of some microinstability. The most promising tool for investigating these questions are 3D particle simulations discussed in section 7.6. We will come back to this point in the following section.

8.4 Magnetospheric substorms

Dayside reconnection occurs when the IMF has a southward inclination. There is, however, no reason why reconnection in the distant magnetotail should follow at the same rate. Actually, since reconnection in the tail is inherently less strongly enforced than on the front side, it tends to lag behind the front-side rate, leading to flux accumulation in the nightside magnetosphere, which is ultimately released in an eruptive way. Hence the natural behavior is not a stationary process but a relaxation oscillation analogous to the sawtooth oscillation in a tokamak plasma treated in section 6.5. The basic picture of the global substorm dynamics is illustrated in fig. 8.11. Nightside flux accumulation results in a stretching of the tail and an increase of the lobe magnetic field. This phase is called the substorm growth phase, fig. 8.11(a). In the subsequent expansion phase, fig. 8.11(b), the stretched configuration becomes unstable to tearing. A new X-line, called the near-Earth X-line, is formed, whereat the excess flux is reconnected, making the near-Earth field more dipole-like. The reconnection process, in particular the associated earthward plasma flow, gives rise to the auroral substorm, the substorm-related auroral activity. The excess tail plasma, enclosed in a region of closed field lines, a plasmoid (see section 4.7.2), is ejected along the tail away from the Earth. This process, called the recovery phase, also pulls the near-Earth X-line tailward, which finally replaces the old distant X-line, fig. 8.11(c). If conditions of southward IMF

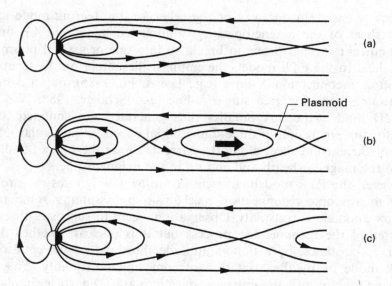

Fig. 8.11.   Qualitative picture of the changes of the magnetic configuration during a substorm: (a) growth phase; (b) expansion phase; and (c) recovery phase.

component continue, the recovery phase goes over into the growth phase of the next substorm.

This qualitative picture, which was first proposed by Schindler (1974) and developed by Hones (1979), seems to be accepted by a majority in the magnetospheric community as the basic model of the magnetic processes in the tail. However, in order to explain the various observational details obtained from satellite and ground measurements, in particular the connection with the auroral phenomena in the ionosphere, a more quantitative theory is required. (The situation can be compared with the theory of the sawtooth oscillation discussed in section 6.5. Though the basic picture, i.e., Kadomtsev's model of the resistive kink mode, was proposed more than two decades ago, a generally accepted semi-quantitative theory consistent with the different observational aspects is still to be developed.) A review of the observations on substorms and a discussion of the different theoretical models has been given by McPherron (1991).

8.4.1 *Substorm-related observations*

The global substorm phenomenon illustrated schematically in fig. 8.11 is not directly observable. Instead, there is a host of satellite observations, i.e., local measurements of particle distributions and magnetic and electric fields taken at certain somewhat accidental positions in space, each of which contribute a small piece to a global puzzle. The most complete data set comes from ground observations of ionospheric phenomena which are, however, only indirectly related to the magnetic reconnection process.

The term magnetospheric substorm insinuates that it is a part of the larger phenomenon called a geomagnetic storm. A storm arises in the wake of a major solar flare or a prominence eruption, by which a large "blob" of plasma is ejected from the Sun and strongly perturbs the solar wind and subsequently the Earth's magnetosphere, resulting in violent geomagnetic activity. The most characteristic feature of a storm is the enhancement of the ring current caused by an increase of the energetic particle population in the radiation belt, which is observed as a depression of the Earth's surface field near the equator (a quantitative measure thereof is the *Dst* index, see, e.g., Baumjohann & Treumann, 1996). While a storm lasts several days, the substorm time-scale is hours. A storm consists of a series of strong, more or less well-defined substorms, but the latter may also occur under rather quiet solar-wind conditions, requiring only a southward IMF. Hence the term *substorm* is slightly misleading.

Let us consider further the observations characterizing the different phases of a substorm. In the growth phase, fig. 8.11(a), which lasts typically 30–60 minutes, reconnection erodes the flux in the dayside magnetosphere, moving the magnetopause closer to the Earth. A direct consequence of dayside reconnection is an increase of the electric field in the polar regions. The magnetic flux convected to the nightside is mainly stored in the tail lobes, where the field intensity increases, while the plasma sheet is distinctly thinning. During this phase a certain rate of quasi-continuous ("driven") tail-side reconnection seems to occur, which brings solar-wind plasma into the magnetosphere and causes convection of the plasma-sheet plasma toward the Earth. This also increases the electric field and the current in the ionosphere, resulting in a certain increase of the *AE* index, which measures the change of the magnetic field in the auroral zone (see Baumjohann & Treumann, 1996). During the growth phase auroral activity starts but appears still in the form of individual quiet arcs,[*] which are slowly drifting equatorward.

The substorm expansion phase, fig. 8.11(b), is characterized by rather dramatic processes, which justify the term sub*storm*. The onset is abrupt, on the time-scale of a minute. The most spectacular phenomenon is the sudden brightening of the most equatorward auroral arc, which rapidly expands poleward and westward – called the westward-travelling surge – developing into a host of turbulent luminous arc-like structures, particu-

---

[*] An auroral arc has a curtain-like shape with a broad azimuthal extent $\sim 10^3$ km but very narrow width < 0.1 km. It might be tempting to associate this narrow scale with a localized process in the tail, but the arc structure seems to be mainly due to internal ionospheric effects, where the details are not yet fully understood.

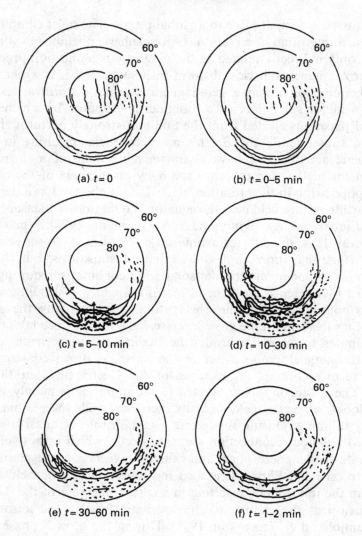

Fig. 8.12.   Schematic illustration of the development of the auroral substorm: (a) quiet aurora during substorm growth phase; (b) onset; (c),(d) expansion phase; (e),(f) recovery phase (from Akasofu, 1964).

larly in the duskward edge. In fact, it is not the bright structures that are actually moving, but more and more structures appear at the duskside, finally covering the entire nightside of the auroral oval. About 30 minutes after substorm onset the turbulent auroral activity settles into a system of quiet arcs, which slowly fade and retreat to higher latitudes. Figure 8.12 gives a schematic diagram illustrating the development of the aurora during a substorm.

Not only does the aurora suddenly flare up at substorm onset, but also the slow increase of the AE signal in the growth phase accelerates abruptly indicating the build-up of intense ionospheric currents, the auroral electrojet. A further characteristic feature at onset is a pulse of magnetic fluctuations with frequency $\sim 10^{-1}$ Hz (called Pi2 oscillations, see Baumjohann & Treumann, 1996), which corresponds to the Alfvén-wave signal from a sudden change of the magnetic configuration in the tail. In fact, *in-situ* observations show that on the near-Earth side of the tail at a distance of $10\text{--}15R_E$ the magnetic field becomes more dipole-like, a process called dipolarization, which is accompanied by rapid earthward plasma flows, while in the distant tail $> 30R_E$ plasma motions are directed away from the Earth and the magnetic field exhibits a reversed southward component.

While these observations are consistent with the general picture of a near-Earth X-line and a plasmoid ejected along the tail, as sketched in fig. 8.11, there are, however, certain features requiring a more elaborate model, for instance the location of the new X-line. Connecting field lines from the first bright arc to the tail gives a distance of less than $10R_E$, while direct satellite measurements indicate a larger distance $\simeq 20R_E$. Also, one needs to explain the mechanism by which the parallel currents are generated driving the auroral current system, the problem of the so-called substorm current wedge. Such issues will be considered in the following subsections.

8.4.2 MHD modeling of substorms

The observations outlined in the previous section support the qualitative picture of the substorm phenomenon given in fig. 8.11. However, whether and under what conditions this process actually occurs must be studied by a dynamic model of the interaction of the solar wind with the magnetosphere. Since the typical scale of the global system is $100\text{--}200R_E \sim 10^6$ km, it is obvious that small-scale physics, associated for instance with the ion and electron Larmor radii (100 km and 1 km, respectively), cannot be resolved in a global simulation, so that only the MHD approach is appropriate. This excludes also a description of the complex physics in the ionosphere, which can only be modeled in a very coarse way. In the last two decades numerical studies of the solar wind–magnetosphere system have been performed by several groups. It is clear from the geometry of the configuration that a semi-quantitative model must be fully three-dimensional. Here we may distinguish between two different lines of approach. The first deals with the entire system of the magnetosphere embedded in the solar wind, where dayside and tail-side reconnection is included and the magnetospheric configuration is set up dynamically. The second kind of modeling concentrates on the most important tail-side processes excited by the instability of a static magnetotail equilibrium and investigates, in particular, the current system driven by the plasmoid dynamics and its coupling to the ionosphere.

Modeling of the solar wind–magnetosphere interaction The issue of fundamental interest is the influence of the direction of the IMF on the behavior of the magnetosphere, in particular on the reconnection dynamics in the tail. As an example of three-dimensional simulations we discuss in some detail the work by Kageyama *et al.*, (1992) and Usadi *et al.*, (1993). These authors solve the full MHD equations in a Cartesian coordinate system. Restricting the IMF to a vertical component only (southward or northward) allows us to assume symmetry both about the equatorial and the Sun–Earth meridional plane, such that the actually computed region can be reduced to one quadrant. The size of this region is $30R_E > x > -95R_E$* and $0 \leq y, z < 50R_E$.

In the magnetosphere, reconnection can only be caused by collisionless processes, which are not yet completely understood and whose modeling in an MHD approach is rather arbitrary. Therefore the authors do not use resistivity explicitly, but rely on the intrinsic numerical magnetic field diffusion, which is equivalent to some kind of anomalous resistivity acting preferentially in regions of strong field gradients. Only in a spherical shell $5R_E < r < 8R_E$ around the Earth is finite resistivity introduced explicitly to model the cross-field resistance of the ionosphere.

A reference magnetosphere–magnetotail configuration is generated dynamically by the solar wind streaming against the dipole field. In this set-up phase the IMF is zero. After about one hour the configuration settles in a quasi-stationary state with an extended tail and a bow shock on the front side. Subsequently, a finite IMF value is introduced in the solar wind. A southward field gives rise to front-side reconnection (and hence flux erosion) in the dayside magnetosphere and an earthward shift of the magnetopause by $1–2R_E$. The reconnected flux adds to the magnetic flux in the lobes, which leads to compression, thinning and heating of the plasma sheet. About 20 minutes after the southward IMF started to interact with the magnetopause, a near-Earth X-line appears at about $20R_E$ on the tail side, where reconnection generates a plasmoid which is pushed tailward. The tail configuration recovers about 60 minutes later. The different phases of the magnetic process are illustrated in fig. 8.13.

In the case of a northward IMF, reconnection with the dayside magnetopause is not possible. Instead, field lines are pulled around the magnetosphere, tending to drape the tail, where they come into contact with the oppositely directed field beyond the cusp regions, thus inviting recon-

---

* Traditionally the positive *x*-axis is in the sunward direction with $x = 0$ at the Earth.

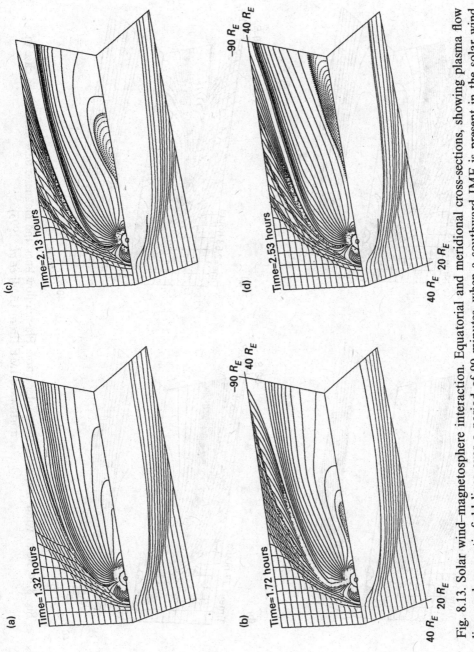

Fig. 8.13. Solar wind–magnetosphere interaction. Equatorial and meridional cross-sections, showing plasma flow lines and magnetic field lines over a period of 90 minutes, when a southward IMF is present in the solar wind (from Usadi *et al.*, 1993).

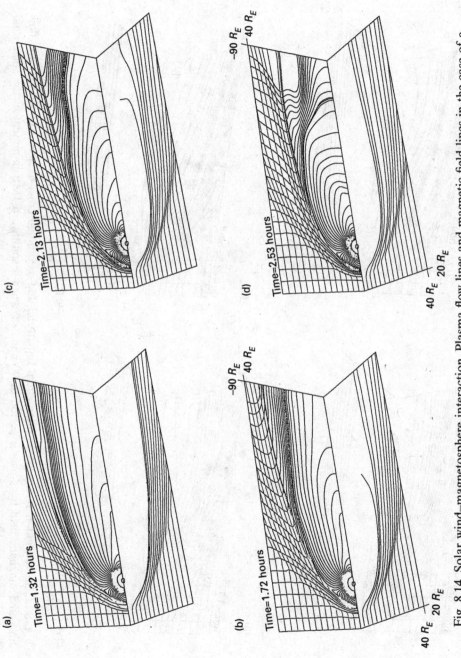

Fig. 8.14. Solar wind–magnetosphere interaction. Plasma flow lines and magnetic field lines in the case of a northward IMF (from Usadi et al., 1993).

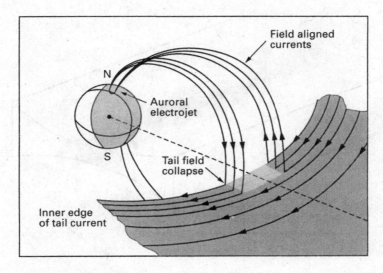

Fig. 8.15.   Substorm current wedge model relating the near-tail current disruption to the auroral electrojet (from McPherron *et al.*, 1973).

nection. This process intensifies the flux in the dayside magnetosphere, pushing the magnetopause further away from the Earth and shifting the cusps to the nightside, while the partner field lines in the reconnection process, which are now cut loose from the Earth, are blown away with the solar wind. The magnetotail assumes a more dipole-like shape which is deeply stable against tearing, inhibiting the occurence of substorms, see fig. 8.14. One thus finds that the MHD simulations corroborate the qualitative 2D picture of the global substorm dynamics shown in fig. 8.11.

Tail-side processes: the substorm current wedge   Numerical simulations of the entire solar wind–magnetosphere system are too coarse to study details of the substorm dynamics in the tail, in particular the coupling to the ionosphere. How can a magnetic event far out in the tail drive the strong ionospheric currents in the auroral electrojet? The basic concept, which dates back to the early 1970s, is the substorm current wedge model illustrated in the classical diagram by McPherron *et al.* (1973), see fig. 8.15. A crucial element in this mechanism is the azimuthal (along the *y*-direction in local coordinates) structure of the magnetotail, a feature that cannot be captured in the usual 2D picture of the plasmoid formation. The qualitative behavior of the cross-tail current sheet (and the plasma sheet) is shown in fig. 8.16. Reconnection takes place preferentially in the central high-j_y part. During this process the earthside field lines snap back, making the field more dipole-like, which reduces the current density

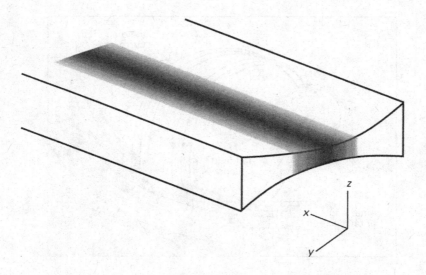

Fig. 8.16.   Schematic cross-tail view of the current sheet.

in this region. In the simple current wedge model, this current is diverted along field lines into the ionosphere, where large cross-field currents can flow because of the high Hall conductivity.

Though this picture is quite suggestive, there is the question of the dynamical relevance of the process, since the current circuit could also be closed by distributed cross-field currents in the tail itself. Is the field-aligned current large enough to account for the auroral current measured in the *AE* index? What is the actual physical mechanism generating this current? To answer these questions, a dynamical treatment including realistic three-dimensional geometry is required, whose results must be compared with satellite measurements during substorms. (For a recent review of substorm observations and their interpretation, see Baker *et al.*, 1996).

MHD simulations studying these issues consider only the magnetotail, where suitable boundary conditions account for the effect of the solar wind at the north–south and the east–west boundaries and the coupling to the ionosphere at the earthside boundary. Since it is not possible to include a realistic model of the ionosphere, it is assumed or tacitly implied that the substorm dynamics is largely independent of the details of the earthside boundary conditions, i.e., that the ionosphere responds mainly passively to the magnetic processes in the tail. Let us consider more closely the MHD studies by Birn & Hesse (1996) and Birn *et al.* (1998). The system size in these simulations is $-5R_E \geq x \geq -65R_E$, $-10R_E \leq y,z \leq 10R_E$ (again only one quadrant is actually computed), hence the Earth's dipole is located $5R_E$ outside the left-hand boundary.

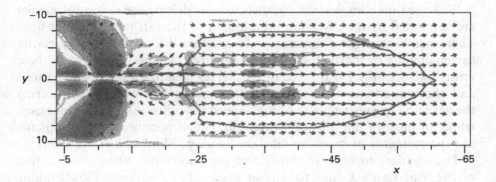

Fig. 8.17. 3D MHD substorm simulation. Regions of height-integrated $\nabla_{\parallel} j_{\parallel}$ (shaded) and flow vectors in the equatorial plane. The continuous line indicates the neutral line, where $B = 0$. The leftmost edge at $x \simeq -23R_E$ corresponds to the X-line in a 2D model and the rightmost edge at $x \simeq -60R_E$ to the O-line in the center of the plasmoid. Left of the "X-line" the plasma moves earthward, right of this line it moves tailward (from Birn *et al.*, 1998).

The initial state is a self-consistent nonresistive tail configuration with a thin current sheet in the near tail as observed shortly before substorm onset. By switching on finite resistivity, an X-line forms at $x \simeq -23R_E$, generating a plasmoid moving tailward and fast earthward flows in the near-tail region, features also expected from a 2D model. The important three-dimensional effect is the relatively narrow azimuthal extent of the earthward flow. Whereas in 2D this flow would be diverted northward and southward by the increasing field intensity, in the 3D configuration the flow is also diverted horizontally, which results in a predominantly diagonal flow direction. Because of the frozen-in property this motion also distorts the dipole field azimuthally inducing a field-line twist, since the field-line footpoints in the ionosphere do not move. This implies parallel currents in the earthward direction on the dawn side and in the tailward direction on the dusk side (the signature of so-called region 1-type parallel currents), as predicted by the simple current wedge model in fig. 8.15. The mechanism is illustrated in fig. 8.17. The simulations also give a magnitude of the total current flowing into the ionosphere of $I \simeq 2$ MA consistent with the observations. (Considered in more detail the field-line twist induced by the diverted flow at lower latitude corresponds to a current out of the ionosphere on the dawn side, called region 2 type current, but its contribution is smaller. The simple current wedge model can be generalized to include these and still other current circuits, as proposed by Birn *et al.* (1998), but such "wire" models are necessarily rather arbitrary since, in a real system, currents are distributed with parallel and perpendicular components everywhere in space.)

We have thus seen that the parallel currents driving the auroral electrojet are generated by the diversion ("braking") of the earthward plasma flow. There has been some discussion as to whether the currents are primarily driven directly by inertia forces or by pressure gradients, i.e., which of these terms balances the $\mathbf{j} \times \mathbf{B}$ force in the MHD equation. MHD simulations and observations now seem to agree that inertia is only important during the very first impulsive phase of high-speed flows just after substorm onset, while for most of the substorm expansion phase pressure forces dominate, i.e., the system is in quasi-equilibrium (see, e.g., Shiokawa *et al.*, 1998).

The simulations seem to clarify also the uncertainty about the location of the near-Earth X-line, namely at about $20R_E$, while the flow braking generating the parallel currents, which are responsible for the auroral substorm, takes place further earthward at distances of 10–$15R_E$.

8.4.3 Collisionless reconnection in the magnetotail

The three-dimensional MHD modeling provides a rather convincing picture of the global substorm dynamics including, to a certain extent, the coupling to the ionosphere. By assuming the presence of some anomalous resistivity resulting in a high effective value of η at the X-line, the time-scale is no longer dominated by the reconnection process itself but by the Alfvén time-scale of the magnetic configuration. Thus this approach ignores the physics of the collisionless reconnection mechanism, which has occupied the minds of many theorists for more than three decades. What is the collisionless process that allows the rapid local decoupling of plasma and magnetic field?

The basic theoretical aspects have been outlined in chapter 7. Here, we shall apply these ideas to the reconnection process which is assumed to occur in the substorm phenomenon. Normally the current sheet in the magnetotail is rather broad, of width $a \gtrsim R_E$, and carries a sizable normal field component B_n which makes the tearing mode strongly stable. It is, however, observed that, immediately prior to substorm onset, the current sheet thins considerably, reaching a width not much larger than c/ω_{pi}, hence the ion drift velocity becomes of the order of the thermal velocity $u_i \sim v_{ti}$. This process implies a sheet-flattening which reduces the stabilizing effect, as can be seen from the stability criterion, (7.156), such that the system approaches the tearing mode stability boundary.

There has been a longstanding debate about the existence of the (ion) tearing mode in the magnetotail configuration, see section 7.5.3. The problem seems to be decided by now, in the sense that the magnetotail is tearing-mode stable. On the other hand, the two-dimensional particle simulations discussed in section 7.6.3 indicate that, at a preformed X-line, fast quasi-Alfvénic reconnection may indeed occur. Hence there must be a further mechanism generating an X-line with a small- but finite-angle separatrix to overcome the linear stability of the tearing mode. The characteristics of the sudden substorm onset are consistent with such a threshold behavior. A possible candidate is the drift-kink mode (sections 7.3 and 7.6.4), which develops in the current direction, $k_x = 0, k_y \neq 0$. This instability is rather insensitive to the presence of a small normal field component (Pritchett, 1997). It is mainly driven by the high ion drift $u_i \sim v_{ti}$ in the narrow current sheet and has a growth rate $\gamma \sim \Omega_i$, i.e., develops on a sub-minute time-scale. There is some evidence that this process really takes place, in particular a burst of cyclotron oscillations is observed in the plasma sheet at substorm onset. This drift-kink mode turbulence can give rise to an effective resistivity, (7.59), destabilizing the tearing mode. Although three-dimensional simulations do not yet provide a converged picture of how this actually happens, the present results do hint to a destabilizing influence. Lapenta & Brackbill (1999) find that oblique tearing modes $k_x \neq 0, k_y \neq 0$ dominate, which give rise to plasmoids with a helical field structure, for which there is also some observational evidence.

8.4.4 Nonreconnective substorm model

The preceding discussion may evoke the impression that the near-Earth X-line reconnection model of the substorm is the only logical explanation of the observations. There is, however, an audible group of researchers advocating a nonreconnective substorm model associated with an instability in the magnetosphere closer to the Earth. This concept is based on data collected by satellites sweeping the region between $-8R_E$ and $-10R_E$, such as the GEOS2 and AMPTE/CCE spacecrafts. The criticism of the near-Earth X-line model is centered on the fact that field lines of the most equatorward arc, at which the substorm aurora starts, connect to the tail-side magnetosphere at about $-10R_E$, while numerical simulations and (mostly indirect) observations suggest a position beyond $-20R_E$ of the newly formed X-line. The mechanism of flow-braking and parallel-current generation indicated by the MHD simulations and described in section 8.4.2, which gives rise to ionospheric currents at latitudes lower than those corresponding to the position of the neutral line, is not fully convincing. Satellite observations at $\simeq -8R_E$ indicate that, simultaneously with the onset of an auroral substorm, a so-called current-disruption event occurs associated with the sudden appearance of strong magnetic turbulence. Though it is conceivable that this process is caused or triggered by the reconnection event at $-20R_E$, there is some observational evidence to the contrary, in that no major disturbance is propagating earthward from beyond $-10R_E$ just prior to onset (see, e.g., Roux et al., 1991; Lui, 1996). Could the auroral substorm after all not be directly related with the re-

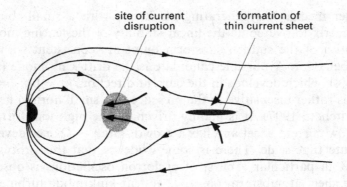

Fig. 8.18.   Plasma flow generated by a near-Earth current disruption (schematic drawing).

connection event in the more distant tail as so obviously suggested by the MHD simulations, but caused by some process nearer to the Earth, which communicates directly along field lines to the auroral latitude?

Such a process should allow snap-back (dipolarization) of the field lines which have been pulled out into the tail during the substorm growth phase due to an increase of the plasma pressure. If the latter exceeds a certain threshold, the configuration becomes unstable to the ballooning mode. The instability would lead to magnetic turbulence, in accordance with the observations, and thus generate an anomalous resistivity which would disrupt the current in the current sheet. As explained in the current-wedge model, this implies dipolarization of the field and parallel currents, which close in the ionosphere. Associated with the snapping-back of the field lines is an earthward plasma flow. Continuity requires that the earthward flow extends further into the tail, where the plasma, together with the magnetic field, is sucked in, as sketched in fig. 8.18. This leads to a thinning of the current sheet and plasmoid formation as described before, and a position of the X-line at $x \sim -20R_E$ could be expected. (A position nearer to the Earth, e.g., at $x \sim -10R_E$ is unlikely because of the large normal field component B_n.) In this picture, the ballooning instability would trigger the reconnection process which occurs further out in the tail, while the auroral substorm in the ionosphere would be mainly determined by processes nearer to the Earth. Some auroral substorms could even occur without giving rise to a major reconnection event in the tail.

As discussed in section 4.5.2, the ballooning instability tends to exhibit an explosive behavior, which would be consistent with the sudden onset characteristics of the substorm. A detailed theory of the ballooning insta-bility in the magnetosphere is, however, rather difficult, since the threshold depends on the pressure profile along the field line, while satellite measurements give only local properties, such that the configuration prior to onset must be reconstructed using scarce spatial data. In addition, kinetic effects may have a stabilizing influence. While the ideal ballooning mode requires $\beta \gtrsim 1$, kinetic processes, especially those due to finite ion Larmor radii, may raise the critical β much above unity (Cheng & Lui, 1998). It is, however, quite possible that such β-values may be reached locally. The discussion about the nature of the near-Earth current disruption is still going on, and both specific observations and more realistic theoretical modeling are required. The situation is thus not much different from that of other complex nonlinear plasma phenomena, for instance the sawtooth oscillation in tokamak plasmas (section 6.5), where, in spite of a suggestive model, the discussion about the dominant mechanism is still open.

8.4.5 Particle acceleration by magnetotail reconnection

The presence of strongly non-Maxwellian particle distributions with long superthermal tails, or even humps, is a quasi-ubiquitous property of dilute weakly collisional plasmas in space. Such distributions result from a mixing of particle beams (accelerated somewhere in space) with the local plasma. In the absence of collisions, beam propagation is affected only by collective processes, *viz.* microinstabilities, which tend to broaden such beams but do not wipe them out completely. In addition, the beam direction is easily changed by magnetic gyration, while the beam energy is conserved, leading to shell-like velocity distributions which are much less unstable than unidirectional beams. The particle energy spectrum is rather robust over long distances, such that spectral analysis may serve as a probe of some remote dynamic process and allow us to differentiate between plasmas of different origin, for instance between magnetospheric and magnetosheath plasma.

In general, one may distinguish between (i) acceleration of the entire plasma (bulk acceleration); (ii) resonant acceleration by waves, which affects only a small part of the distribution; (iii) and diffusive acceleration of individual particles. While bulk acceleration is caused by a perpendicular electric field, $\mathbf{v} = c\mathbf{E} \times \mathbf{B}/B^2$, selective particle acceleration, both resonant and diffusive, is mainly due to parallel electric fields.* In tail-side reconnec-

---

* The term "diffusive acceleration" is normally reserved for Fermi acceleration, when particles bounce back and forth between approaching magnetic clouds. A very efficient system is a quasi-parallel collisionless shock carrying strong magnetic perturbations both on the upstream and the downstream sides, which can pitch-angle scatter and reflect particles. Thus on each sweep across the shock front, the particle energy increases by a certain fraction, such that after many bounces particles may reach very high energies. This process is regarded as the most probable mechanism for cosmic-ray acceleration. In the magnetotail, however, Fermi acceleration is not important.

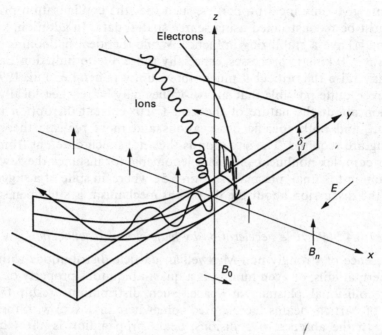

Fig. 8.19. Schematic representation of particle trajectories in a neutral sheet with a weak normal field component B_n (from Speiser, 1965).

tion, both $E_\perp$ and $E_\parallel$ acceleration processes are active. The reconnection electric field is inductive, $E_y = -\partial_t A_y/c$. Away from the neutral line the magnetic field is sufficiently strong, such that ions are adiabatic. The electric field pushes the plasma toward the Earth along the neutral sheet, the outflow cone of the reconnection region, with velocity $v_E = cE_y/B_n$, which in a fast reconnection event reaches the Alfvén velocity in the lobe, the upstream reconnection region. A typical value $v_A \sim 5 \times 10^7$ cm/s corresponds to a plasma ion energy of a few keV. The energetic plasma flow is a fundamental feature of magnetic reconnection.

Much higher energies can be reached by parallel acceleration of non-adiabatic particles in the vicinity of the neutral line. The basic mechanism was first studied by Speiser (1965). It refers to a magnetotail-like configuration in the presence of an electric field E_y, which is relevant both for a quiescent tail with a weak electric field and hence a broad current sheet, and also for substorm conditions with a strong electric field and a narrow current sheet at the near-Earth X-line. Consider a neutral sheet configuration with a weak normal field component B_n. For sufficiently small B_n, $B_n/B < \delta_j/a = \sqrt{\rho_j/a}$, particles drifting into the neutral sheet from above and below become nonadiabatic inside the layer of width δ_j,

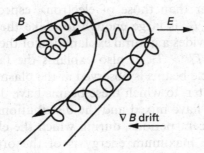

Fig. 8.20. Sketch of an ion orbit exhibiting parallel motion and mirroring, acceleration along E_y, and ∇B drifts (from Birn *et al.*, 1997).

where they meander along the sheet in the y-direction, as illustrated in fig. 7.11(b), and are accelerated by the parallel field E_y. In the absence of a normal field component such particles would be trapped in the layer until they are ejected at the current-sheet edges. The finite B_n makes this meandering motion turn slowly into the x-direction, corresponding to a gyro-orbit in the weak B_n field, until the particles escape earthward along the magnetic field. Particle trajectories are sketched in fig. 8.19. Hence the energy gain is $\Delta W_j \sim e E_y \rho_{nj}$ corresponding to a velocity increment $\Delta v \sim c E_y/B_n$, which is just the E×B velocity in the layer. This energy gain during one excursion into the neutral layer is limited, but most particles return after being reflected along the field line in the higher-field region, such that the acceleration process is repeated.

If the particle would stay on the same field line during its parallel bounce motion, its position would jump by $\Delta y \sim \rho_{nj}$ on each return to the sheet, hence the particle would finally experience the full cross-tail potential drop $\phi = E_y L$. In reality there are, however, finite cross-field drifts. The E×B drift is essentially vertical, moving the particle toward the sheet, hence successive returns to the nonadiabatic layer occur at positions shifted toward the Earth. Moreover, there is a finite ∇B drift $\mathbf{v}_B = \mu_j c(\mathbf{b} \times \nabla B)/(q_j B)$ which, in the magnetic configuration of the tail, leads to a horizontal drift in the y-direction opposite to the nonadiabatic motion in the layer, such that the particle may return to about the same y-position, as illustrated in fig. 8.20. In this case, particle orbits migrate mainly toward the Earth and the maximum energy is reached when the particles become completely adiabatic in the high near-Earth magnetic field. Test particle calculations in self-consistent MHD fields fully corroborate this picture of diffusive particle acceleration (see, e.g., Birn *et al.*, 1997).

The acceleration mechanism just outlined can account for the main features of the particle energy distributions in the plasma sheet. Ion energy gains are much larger than those of electrons, especially under normal tail conditions where B_n is large enough to make the electrons adiabatic $B_n/B > \delta_e$, which provides a natural explanation of the observed difference of the temperatures $T_i \gg T_e$. It also explains the fact that the clearest evidence of fast particle beams is observed at the plasma-sheet boundaries, while in the sheet center, to which these beams have drifted after multiple bounces, they should have mixed and the distributions should be roughly thermalized. In quiescent periods, during which the electric field is small, $E_y \sim 10^{-4}$ V/m, the maximum energy is of the order of $eE_yL_y \sim 10$ keV using a typical cross-tail dimension of $L_y \sim 30R_E$. During substorm expansion, however, the electric field in the nonadiabatic region around the near-Earth X-line is much larger, $E_y > 10^{-3}$ V/m, such that ions may reach energies exceeding 10^2 keV, which are typically observed during geomagnetic activity. Due to scattering by collisions in the ionosphere such high-energy particles may become trapped in the radiation belt.

Epilogue

It thus appears that a long-standing riddle has now been solved. Fast quasi-Alfvénic magnetic reconnection may occur under rather general conditions with a rate rather independent of the particular reconnection physics, both in high- and low-β plasmas. Ironically, the case of stationary resistive MHD, which has been regarded as the most natural framework of reconnection theory, does *not* allow fast merging. The pecularities of resistive reconnection have been the origin of the long controversy dividing the community into two camps, the adherents of the Petschek model and those of the Sweet-Parker model emphasizing current sheets. Actually, physical conditions for a stationary high-Lundquist number MHD model to apply are rarely satisfied, neither in nature nor in laboratory plasmas. Either the plasma is strongly resistive, which often implies relatively low S or, at large S-value, collisionless effects are more important than resistivity. In addition, the plasma behavior is usually highly nonstationary, sometimes fully turbulent, and such a system allows fast reconnection even in resistive MHD. It is true that the usual quasi-stationary 2D models for collisionless reconnection are also highly idealized, far from real plasma conditions. Real plasmas which tend to exhibit a whole maze of fluctuations, but these do not seem to control the reconnection *rate*.

A concept which had also polarized the community, the distinction between driven and spontaneous reconnection, has to a good deal lost its significance. It is true that in many cases an external driving agent can be identified. In a laboratory plasma device, reconnection may be forced by a voltage applied in the external circuit, or a flux tube emerging from the photosphere may push against the preexisting coronal magnetic loop system enforcing reconnection. But if in the latter example both interacting flux tube systems are quasi-stationary, being shoved around and energized by photospheric motions, a rapid reconnection process between them would clearly be called spontaneous. In the substorm problem, the distinction between driven and spontaneous processes depends on whether the reconnection at a near-Earth X-line is the origin of the tail dynamics or only the consequence of some different primary process. More importantly, the reconnection process itself does not seem to depend on the global driving dynamics, but only on the rate at which the field is locally pushed toward the reconnection site. In this sense, fast reconnection is always a driven process. If, as in the single tearing mode, the dynamics is slow, the process should properly be called magnetic diffusion.

With the problem of fast reconnection *rates* essentially solved, interest in reconnection theory is shifting to more specific questions, in particular that of the energy partition. What fraction of the magnetic energy released is converted into plasma flows, into ion and electron temperatures, and into superthermal particles? As a rule of thumb, in high-β systems with a local region of weak magnetic field, efficient heating is produced by the electric field $E_\perp$, when particles are nonadiabatic. Since this happens more readily for ions, T_i may be significantly higher than T_e, as observed in the magnetotail, and also in the laboratory experiments such as the reversed field pinch. In a low-β system only the parallel field $E_\parallel$ can energize particles, and electrons are more readily accelerated than ions, such that T_e exceeds T_i as observed in ohmically heated tokamaks. The question of energy partition is also important observationally to identify the presence of magnetic reconnection, since energy spectra are more easily accessible than local changes of magnetic field connectivity.

In view of the progress in understanding the physical processes in the reconnection region, reduced equations or fully phenomenological models should allow more realistic global simulations in full three-dimensional geometry, which will be the main task in the future. Here the magnetic activity in the solar corona is a particularly attractive "playground", where one can use, or at least be stimulated by, the stunning observational data from recent satellite missions such as TRACE.

Finally, a word of caution. In spite of the recent advances, the feeling of mastering, after so many years, this scintillating subject called magnetic reconnection might again turn out to be elusive.

...

As the images unwind
Like the circles that you find
In the windmills of your mind.
(A. & M. Bergman)

Bibliography

Adler, E.A., Kulsrud, R.M. & White, R.B. (1980). Magnetic driving energy of the collisional tearing mode, *Phys. Fluids* **23**, 1375–9.

Akasofu, S.I. (1964). The development of the auroral substorm, *Planet. Space Sci.* **12**, 273–82.

Ara, G. Basu, B., Coppi, B., Rosenbluth, M.N. & Waddell, B.V. (1978). Magnetic reconnection and $m = 1$ oscillation in current carrying plasmas, *Ann. Phys.* **112**, 443–76.

Axford, W. I. (1984). Magnetic reconnection, In *Magnetic Reconnection in Space and Laboratory Plasmas*, Geophys. Monogr. **30**, ed. E. W. Hones, Jr (American Geophysical Union, Washington, D.C.), pp. 1–8.

Axford, W.I. & Hines, C.O. (1961). A unifying theory of high-latitude geophysical phenomena and geomagnetic storms, *Can. J. Phys.* **39**, 1433–64.

Aydemir, A.Y. (1991). Linear studies of $m = 1$ modes in high-temperature plasmas with a four-field model, *Phys. Fluids B* **3**, 3025–32.

Aydemir, A.Y. (1992). Nonlinear studies of $m = 1$ modes in high-temperature plasmas, *Phys. Fluids B* **4**, 3469–72.

Aydemir, A.Y. (1997). Nonlinear $m = 1$ mode and fast reconnection in collisional plasmas, *Phys. Rev. Lett.* **78**, 4406–9.

Aydemir, A.Y., Wiley, J.C. & Ross, D.W. (1989). Toroidal studies of sawtooth oscillations in tokamaks, *Phys. Fluids B* **1**, 774–87.

Baker, D.N., Pulkkinen, T.I., Angelopoulos, V., Baumjohann, W. & McPherron, R.L. (1996). Neutral line model of substorms: past results and present view, *J. Geophys. Res.* **101**, 12975–3010.

Balbus, S.A. & Hawley, J.F. (1991). A powerful local instability in weakly magnetized disks. I. Linear analysis, *Astrophys. J.* **376**, 214–22.

Basu, B & Coppi, B. (1981). Theory of $m = 1$ modes in collisionless plasmas, *Phys. Fluids* **24**, 465–71.

Batchelor, G.K. (1969). Computation of the energy spectrum in homogeneous two-dimensional turbulence, *Phys. Fluids* **12**, II 233–9.

Batchelor, D.B. & Davidson, R.C. (1976). Nonlocal analysis of the lower-hybrid-drift instability in theta-pinch plasmas, *Phys. Fluids* **19**, 882–8.

Baty, H., Luciani, J.F. & Bussac, M.N. (1992). Asymmetric reconnection and stochasticity in tokamaks, *Nucl. Fusion* **32**, 1217–23.

Bauer, F., Betancourt, O. & Garabedian, P. (1978). *A Computational Method in Plasma Physics* (Springer, New York).

Baumjohann, W. & Paschmann, G. (1987). Solar wind–magnetosphere coupling: processes and observations, *Physica Scripta* **T18**, 61–72.

Baumjohann, W. & Treumann, R.A. (1996). *Basic Space Plasma Physics* (Imperial College Press, London).

Bayly, B. & Childress, S. (1987). Fast dynamo action in unsteady flows and maps in three dimensions, *Phys. Rev. Lett.* **59**, 1573–6.

Bayly, B. & Childress, S. (1988). Construction of fast dynamos using unsteady flows and maps in three dimensions, *Geophys. Astrophys. Fluid Dyn.* **44**, 207–40.

Beer, M.A., Cowley, S.C. & Hammett, G.W. (1995). Field-aligned coordinates for nonlinear simulations of tokamak turbulence, *Phys. Plasmas* **2**, 2687–700.

Benzi, R., Ciliberto, S., Trippicione, R. *et al.* (1993). Extended self-similarity in turbulent flows, *Phys. Rev. E* **48**, R29–32.

Berger, M.A. (1984). Rigorous new limits on magnetic helicity dissipation in the solar corona, *Geophys. Astrophys. Fluid Dyn.* **30**, 79–104.

Berger, M. A. & Field, G. B. (1984). The topological properties of magnetic helicity, *J. Fluid Mech.* **147**, 133–48.

Bertin, G. (1982). Effects of local current gradients on magnetic reconnection, *Phys. Rev. A* **25**, 1786–9.

Bickley, W.G. (1937). The plane jet. *Phil. Mag.* **23**, 727–31.

Biermann, L. (1951). Kometenschweife und solare Korpuskularstrahlung, *Z. Astrophys.* **29**, 274–86.

Birdsall, C.K. & Langdon, A.B. (1985). *Plasma Physics via Computer Simulation* (McGraw-Hill, New York).

Birn, J. & Hesse, M. (1996). Details of current disruption and diversion in simulations of magnetotail dynamics, *J. Geophys. Res.* **101**, 15345–58.

Birn, J., Thomsen, M.F., Borovsky, J.E., Reeves, G.D., McComas, D.J. & Belian, R.D. (1997). Substorm ion injections: geosynchronous observations and test particle orbits in three-dimensional dynamic MHD fields, *J. Geophys. Res.* **102**, 2325–41.

Birn, J., Hesse, M., Haerendel, G., Baumjohann, W. & Shiokawa, K. (1999). Flow braking and the substorm current wedge, *J. Geophys. Res.* **104**, 19895–903.

Birn, J., Drake, J.F., Shay, M.A., Rogers, B.N., Denton, R.E., Hesse, M., Kuznetsova, M., Ma, Z.W., Bhattacharjee, A., Otto, A. & Pritchett, P.L. (2000). GEM magnetic reconnection challenge, *J. Geophys. Res.*, to be published.

Biskamp, D. (1968). Diagram approach to the theory of collisionless plasma turbulence, *Z. Naturforsch.* **23a**, 1362–72.

Biskamp, D. (1973). Collisionless shock waves in plasmas, *Nucl. Fusion* **13**, 719–40.

Biskamp, D. (1982). Dynamics of a resistive sheet pinch, *Z. Naturforsch.* **37a**, 840–7.

Biskamp, D. (1986). Magnetic reconnection via current sheets, *Phys. Fluids* **29**, 1520–31.

Biskamp, D. (1991). Algebraic nonlinear growth of the resistive kink instability, *Phys. Fluids B* **3**, 3353–6.

Biskamp, D. (1993a). *Nonlinear Magnetohydrodynamics* (Cambridge University Press, Cambridge).

Biskamp, D. (1993b). Current sheet profiles in two-dimensional magnetohydrodynamics, *Phys. Fluids B* **5**, 3893–96.

Biskamp, D. (1994). Magnetic reconnection, *Phys. Rep.* **237**, 179–247.

Biskamp, D., Sagdeev, R.Z. & Schindler, K. (1970). Nonlinear evolution of the tearing instability in the geomagnetic tail, *Cosmic Electrodyn.* **1**, 297–310.

Biskamp, D. & Chodura, R. (1971). Computer simulation of anomalous dc resistivity, *Phys. Rev. Lett.* **27**, 1553–6.

Biskamp, D. & Schindler, K. (1971). Instability of two-dimensional collisionless plasmas with neutral points, *Plasma Phys.* **13**, 1013–26.

Biskamp, D. & Chodura, R. (1973a). Asymptotic behavior of the two-stream instability, *Phys. Fluids* **16**, 888–92.

Biskamp, D. & Chodura R. (1973b). Collisionless dissipation of a cross-field electric current, *Phys. Fluids* **16**, 893–901.

Biskamp, D., Chodura, R. & Dum, C.T. (1975). Ion-sound spectrum and wave–electron interaction in perpendicular shocks, *Phys. Rev. Lett.* **34**, 131–4.

Biskamp, D. & Welter, H. (1980). Coalescence of magnetic islands, *Phys. Rev. Lett.* **44**, 1069–72.

Biskamp, D. & Welter, H. (1989). Dynamics of decaying two-dimensional magnetohydrodynamic turbulence, *Phys. Fluids B* **1**, 1964–79.

Biskamp, D & Drake, J.F. (1994). Dynamics of the sawtooth collapse in tokamak plasmas, *Phys. Rev. Lett.* **73**, 971–4.

Biskamp, D., Schwarz, E. & Drake, J.F. (1995). Ion-controlled collisionless magnetic reconnection, *Phys. Rev. Lett.* **75**, 3850–3.

Biskamp, D. & Drake, J.F. (1996). Dynamics of the sawtooth collapse in tokamak plasmas, *Proceedings of the Fifteenth International Conference on Plasma Physics and Controlled Nuclear Fusion Research* (IAEA, Vienna), Vol. III, pp. 261–72.

Biskamp, D., Schwarz, E. & Drake, J.F. (1996). Two-dimensional electron magnetohydrodynamic turbulence, *Phys. Rev. Lett.* **76**, 1264–7.

Biskamp, D. & Sato, T. (1997). Partial reconnection in the nonlinear internal kink mode, *Phys. Plasmas* **4**, 1326–9.

Biskamp, D., Schwarz, E. & Drake, J.F. (1997). Two-fluid theory of collisionless magnetic reconnection, *Phys. Plasmas* **4**, 1002–9.

Biskamp, D., Schwarz, E. & Zeiler, A. (1998a). Instability of a magnetized plasma jet, *Phys. Plasmas* **5**, 2485–8.

Biskamp, D., Schwarz, E. & Celani, A. (1998b). Nonlocal bottleneck effect in two-dimensional turbulence, *Phys. Rev. Lett.* **81**, 4855–8.

Biskamp, D., Schwarz, E., Zeiler A., Celani, A. & Drake, J.F. (1999). Electron magnetohydrodynamic turbulence, *Phys. Plasma* **6**, 751–8.

Biskamp, D. & Müller, W.-C. (1999). Decay laws for three-dimensional magnetohydrodynamic turbulence, *Phys. Rev. Lett.* **83**, 2195–8.

Biskamp, D. & Schwarz, E. (2000). Scaling properties of two-dimensional magnetohydrodynamic turbulence, to be published.

Boris, J.P., Dawson, J.M., Orens, J.H. & Roberts, K.V. (1970). Computations of anomalous resistance, *Phys. Rev. Lett.* **25**, 706–10.

Brachet, M.E., Meneguzzi, M., Vincent, A., Politano, H. & Sulem P.L. (1992). Numerical evidence of smooth self-similar dynamics and possibility of subsequent collapse for three-dimensional ideal flow, *Phys. Fluids A* **4**, 2845–54.

Brackbill, J.U. & Forslund, D.W. (1982). An implicit method for electromagnetic plasma simulation in two dimensions, *J. Comput. Phys.* **46**, 271–308.

Brackbill, J.U., Forslund, D.W., Quest, K.B. & Winske, D. (1984). Nonlinear evolution of the lower-hybrid-drift instability, *Phys. Fluids* **27**, 2682–93.

Braginskii, S. I. (1965). Transport processes in a plasma, In *Reviews of Plasma Physics*, ed. M. A. Leontovich (Consultants Bureau, New York), Vol. I, pp. 205–311.

Braginskii, S.I. (1963). Structure of the F layer and reasons for convection in the Earth's core, *Sov. Phys. Dokl.* **149**, 8–10.

Brandenburg, A., Moss, D., Rüdiger, G. & Tuominen, I. (1990). The nonlinear solar dynamo and differential rotation: a Taylor number puzzle?, *Solar Phys.* **128**, 243–51.

Brandenburg, A., Moss, D., Rüdiger, G. & Tuominen, I. (1991). Hydromagnetic $\alpha\Omega$-type dynamos with feedback from large-scale motions, *Geophys. Astrophys. Fluid Dyn.* **61**, 179–98.

Brandenburg, A. (1994). Solar dynamos: computational background, In *Lectures on Solar and Planetary Dynamos*, eds. M. R. E. Proctor & A. D. Gilbert (Cambridge University Press, Cambridge), pp. 117–59.

Brittnacher, M., Quest, K.H. & Karimabadi, H. (1994). On the energy principle and ion tearing in the magnetotail, *Geophys. Res. Lett.* **21**, 1591–4.

Büchner, J. & Zelenyi, L.M. (1987). Chaotization of the electron motion as a cause of an internal magnetotail instability and substorm onset, *J. Geophys. Res.* **92**, 13456–66.

Bulanov, S.V., Sakai, J. & Syrovatskii, S.I. (1979). Tearing mode instability in approximately steady MHD configurations, *Sov. J. Plasma Phys.* **5**, 157–63.

Bulanov, S.V., Pegoraro, F. & Sakharov, A.S. (1992). Magentic reconnection in electron magnetohydrodynamics, *Phys. Fluids B* **4**, 2499–508.

Bullard, E.C. (1955). The instability of a homopolar dynamo, *Proc. Camb. Phil. Soc.* **51**, 744–50.

Bullard, E.C. (1968). Reversals of the Earth's magnetic field, *Phil. Trans. Roy. Soc. A* **263**, 481–524.

Buneman, O. (1959). Dissipation of currents in ionized media, *Phys. Rev.* **115**, 503–17.

Busnardo-Neto, J., Pritchett, P.L., Lin, A.T. & Dawson, J.M. (1977). A self-consistent magnetostatic particle code for numerical simulation of plasmas, *J. Comput. Phys.* **23**, 300–12.

Bussac, M.N., Pellat, R., Edery, R. & Soulé, J.L. (1975). Internal kink modes in a toroidal plasma with circular cross-section, *Phys. Rev. Lett.* **35**, 1638–41.

Busse, F.H. (1970). Thermal instabilities in rapidly rotating systems, *J. Fluid Mech.* **44**, 441–60.

Busse, F.H. (1986). Asymptotic theory of convection in a rotating cylindrical annulus, *J. Fluid Mech.* **173**, 545–56.

Campbell, D.J., *et al.* (1987). Sawteeth and disruptions in JET, In *Proceedings of the Eleventh International Conference on Plasma Physics and Controlled Nuclear Fusion Research* (IAEA, Vienna), Vol. I, pp. 433–45.

Carrera, R., Hazeltine, R.D. & Kotschenreuther, M. (1986). Island bootstrap current modification of the nonlinear dynamics of the tearing mode, *Phys. Fluids* **29**, 899–902.

Carreras, B., Waddell, B.V. & Hicks, H.R. (1979). Poloidal magnetic field fluctuations, *Nucl. Fusion* **19**, 1423–30.

Chandrasekhar, S. (1961). *Hydrodynamic and Hydromagnetic Stability* (Oxford University Press, Oxford).

Chapman, S. & Ferraro, V.C.A. (1931). A new theory of magnetic storms, *J. Geophys. Res.* **36**, 77.

Chapman, S. & Kendall, P. G. (1963). Liquid instability and energy transformation near a magnetic neutral line: a soluble non-linear hydromagnetic problem, *Proc. Roy. Soc. London A* **271**, 435–48.

Cheng, C.S. & Lui, A.T.Y. (1998). Kinetic ballooning instability for substorm onset and current disruption observed by AMPTE/CCE, *Geophys. Res. Lett.* **25**, 4091–94.

Connor, J.W., Hastie, R.J. & Taylor, J.B. (1991). Resonant magnetohydrodynamic modes with toroidal coupling. Part 1: Tearing modes, *Phys. Fluids B* **3**, 1532–8.

Cook, A.E. & Roberts, P.H. (1970). The Rikitake two-disc dynamo system, *Proc. Camb. Phil. Soc.* **68**, 547–69.

Coppi, B., Galvao, R., Pellat, R., Rosenbluth, M.N. & Rutherford, P.H. (1976). Resistive internal kink modes, *Sov. J. Plasma Phys.* **2**, 533–5.

Coppi, B. & Detragiache, P. (1992). Magnetic topology transition in collisionless plasmas, *Phys. Lett. A* **168**, 59–64.

Correa-Restrepo, D. (1982). Resistive ballooning modes in three-dimensional configurations, *Z. Naturforsch.* **37a**, 848–58.

Cowley, S.C. & Artun, M. (1997). Explosive instabilities and detonation in magnetohydrodynamics, *Phys. Rep.* **283**, 185–211.

Cowley, S.W.H. (1975). Magnetic field line reconnection in a highly conducting incompressible fluid, *J. Plasma Phys.* **14**, 475–90.

Cowling, T.G. (1934). The magnetic field of sunspots, *Mon. Not. Roy. Astr. Soc.* **94**, 39–48.

Cox, A. (1969). Geomagnetic reversals, *Science* **163**, 237–45.

Dahlberg, R.B., Bocinelli, P. & Einaudi, G. (1998). The evolution of a plane jet in a neutral sheet, *Phys. Plasmas* **5**, 79–93.

D'Angelo, N. (1965). Kelvin–Helmholtz instability in a fully-ionized plasma in a magnetic field, *Phys. Fluids* **8**, 1748–50.

Davidson, R.C. (1972). *Methods in Nonlinear Plasma Theory* (Academic Press, New York).

Davidson, R.C. (1976). Vlasov equilibrium and nonlocal stability properties of an inhomogeneous plasma column, *Phys. Fluids* **19**, 1189–202.

Davidson, R.C., Hammer, D.A., Haber, I. & Wagner, C.E. (1972). Nonlinear development of electromagnetic instabilities in anisotropic plasmas, *Phys. Fluids* **15**, 317–33.

Davidson, R.C. & Gladd, N.T. (1975). Anomalous transport properties associated with the lower-hybrid-drift instability, *Phys. Fluids* **18**, 1327–35.

Davidson, R.C. & Krall, N.A. (1977). Anomalous transport in high-temperature plasmas with applications to solenoidal fusion systems, *Nucl. Fusion* **17**, 1313–72.

Davidson, R.C., Gladd, N.T., Wu, C.S. & Huba, J.D. (1977). Effects of finite plasma beta on the lower-hybrid-drift instability, *Phys. Fluids* **20**, 301–10.

Dawson, J. (1962). One-dimensional plasma model, *Phys. Fluids* **5**, 445–59.

Dobrowolny, M. (1968). Instability of a neutral sheet, *Nuovo Cimento* **B55**, 427.

Drake, J.F. (1978). Kinetic theory of $m = 1$ internal instabilities, *Phys. Fluids* **21**, 1777–89.

Drake, J.F. & Lee, Y.C. (1977). Nonlinear evolution of collisionless and semicollisional tearing modes, *Phys. Rev. Lett.* **39**, 453–6.

Drake, J. F. & Antonsen, Jr, T. M. (1984). Nonlinear reduced fluid equations for toroidal plasmas, *Phys. Fluids* **27**, 898–908.

Drake, J.F., Kleva, R.G. & Mandt, M.E. (1994). Structure of thin current layers: implications for magnetic reconnection, *Phys. Rev. Lett.* **73**, 1251–4.

Drake, J.F., Biskamp, D. & Zeiler, A. (1997). Break-up of the electron current layer during 3D collisionless magnetic reconnection, *Geophys. Res. Lett.* **24**, 2921–4.

Drazin, P.G. & Reid, W.H. (1981). *Hydrodynamic Stability* (Cambridge University Press, Cambridge).

Drummond, W.E. & Pines, D. (1962) Nonlinear stability of plasma oscillations, *Nucl. Fusion Suppl.*, Pt. 3, 1049.

Drummond, W.E. & Rosenbluth, M.N. (1962). Anomalous diffusion arising from microinstabilities in a plasma, *Phys. Fluids* **5**, 1507–13.

Dubois, M.A., Pecquet, A.L. & Reverdin, C. (1983). Internal disruptions in the TFR tokamak: a phenomenological analysis, *Nucl. Fusion* **23**, 147–62.

Dum, C.T. & Dupree, T.H. (1970). Nonlinear stabilization of high-frequency instabilities in a magnetic field, *Phys. Fluids* **13**, 2064–81.

Dum, C.T., Chodura, R. & Biskamp, D. (1974). Turbulent heating and quenching of the ion-sound instability, *Phys. Rev. Lett.* **32**, 1231–4.

Dungey, J.W. (1961). Interplanetary magnetic field and auroral zones, *Phys. Rev. Lett.* **6**, 47–8.

Dupree, T.H. (1966). A perturbation theory for strong plasma turbulence, *Phys. Fluids* **9**, 1773–82.

Dupree, T.H. (1967). Nonlinear theory of drift-wave turbulence and enhanced diffusion, *Phys. Fluids* **10**, 1049–55.

Edenstrasser, J.W. (1980a). Unified treatment of symmetric MHD equilibria, *J. Plasma Phys.* **24**, 299–313.

Edenstrasser, J.W. (1980b). The only three classes of symmetric MHD equilibria, *J. Plasma Phys.* **24**, 515–18.

Edwards, A.W. *et al.* (1986). Rapid collapse of a plasma sawtooth oscillation in the JET tokamak, *Phys. Rev. Lett.* **57**, 210–13.

Elsässer, W.M. (1950). The hydromagnetic equations, *Phys. Rev.* **79**, 183.

Fadeev, V.M., Kvartskhava, I.F. & Komarov, N.N. (1965). Self-focusing of local plasma currents, *Nucl. Fusion* **5**, 202–9.

Finn, J.M. & Antonsen, T.M. (1985). Magnetic helicity: what is it and what is it good for? *Comments Plasma Phys. Controlled Fusion* **26**, 111–26.

Finn, J.M. & Ott, E. (1988). Chaotic flows and fast magnetic dynamos, *Phys. Fluids* **31**, 2992–3011.

Fitzpatrick, R. (1995). Helical temperature perturbations associated with tearing modes in tokamak plasmas, *Phys. Plasmas* **2**, 825–38.

Forbes, T.G. & Priest, E.R. (1987). A comparison of analytical and numerical models for steadily driven magnetic reconnection, *Rev. Geophys.* **25**, 1587–607.

Forslund, D.W., Morse, R.L. & Nielson, C.W. (1970). Electron-cyclotron drift instability, *Phys. Rev. Lett.* **25**, 1266–70.

Forslund, D.W., Morse, R.L. & Nielson, C.W. (1971). Nonlinear electron-cyclotron drift instability and turbulence, *Phys. Rev. Lett.* **27**, 1424–8.

Frank, A., Jones, T.W., Ryu, D. & Gaalaas, J.B. (1996). The magnetohydrodynamic Kelvin–Helmholtz instability: a two-dimensional numerical study, *Astrophys. J.* **460**, 777–93.

Freidberg, J.P. (1987). *Ideal Magnetohydrodynamics* (Plenum Press, New York).

Fried, B.D. & Conte, S.D. (1961). *The Plasma Dispersion Function* (Academic Press, New York).

Fried, B.D. & Gould, R.W. (1961). Longitudinal ion oscillations in a hot plasma, *Phys. Fluids* **4**, 139–47.

Friedel, H., Grauer, R. & Marliani, C. (1997). Adaptive mesh refinement for singular current sheets in incompressible magnetohydrodynamic flows, *J. Comp. Phys.* **134**, 190–8.

Frisch, U. (1995). *Turbulence* (Cambridge University Press, Cambridge)

Frisch, U., Pouquet, A., Sulem, P. L. & Meneguzzi, M. (1983). Dynamics of two-dimensional ideal MHD, *J. Méc. Théor. Appl.* Special issue on two-dimensional turbulence, pp. 191–216.

Fukao, S., Masayuki, U. & Takao, T. (1975). Topological study of magnetic field near a neutral point, *Rep. Ionosph. Res. Jpn.* **29**, 133–9.

Furth, H.P., Killeen, J. & Rosenbluth, M.N. (1963). Finite resistivity instabilities of a sheet pinch, *Phys. Fluids* **6**, 459–84.

Furth, H.P., Rutherford, P.H. & Selberg, H. (1973). Tearing mode in a cylindrical tokamak, *Phys. Fluids* **16**, 1054–63.

Fyfe, D. & Montgomery, D. (1976). High-beta turbulence in two-dimensional magnetohydrodynamics, *J. Plasma Phys.* **16**, 181–91.

Galeev, A.A. & Sagdeev, R.Z. (1973). Nonlinear plasma theory, In *Reviews of Plasma Physics*, ed. M.A. Leontovich (Consultants Bureau, New York), Vol. 7, pp. 1–180.

Galeev, A.A. & Zelenyi, L.M. (1976). Nonlinear instability theory of a diffusive neutral layer, *Sov. Phys. JETP* **42**, 450–6.

Galeev, A.A. & Sagdeev, R.Z. (1983). Theory of weakly turbulent plasma, In *Basic Plasma Physics* I, ed. A.A. Galeev & R.N. Sudan (North-Holland Publishing Company, Amsterdam), pp. 677–731.

Galtier, S., Politano, H. & Pouquet, A. (1997). Self-similar energy decay in magnetohydrodynamic turbulence, *Phys. Rev. Lett.* **79**, 2807–10.

Galtier, S., Nazarenko, S.V., Newell, A.C. & Pouquet, A. (1999). A weak turbulence theory for incompressible MHD, submitted to *J. Plasma Phys.*

Gary, S.P. (1993). *Theory of Space Plasma Microinstabilities* (Cambridge University Press, Cambridge).

Gekelman, W., Stenzel, R.L. & Wild, N. (1982). Magnetic field line reconnection experiments 3. Ion acceleration, flows, and anomalous scattering, *J. Geophys. Res.* **87**, 101–10.

Gilman, P.A. (1983). Dynamically consistent nonlinear dynamos driven by convection in a rotating spherical shell, *Astrophys. J. Suppl.* **53**, 243–68.

Gilman, P.A. (1992). What can we learn about the solar cycle mechanisms from observed velocity fields? In *The Solar Cycle*, ed. K.L. Harvey, ASP Conference Series **27**, pp. 241–55.

Gilman, P.A. & Glatzmaier, G.A. (1981). Compressible convection in a rotating spherical shell. I. Anelastic equations, *Astrophys. J. Suppl.* **45**, 335–49.

Gilman, P.A. & Miller, J. (1981). Dynamically consistent nonlinear dynamos driven by convection in a rotating spherical shell, *Astrophys. J. Suppl.* **46**, 211–38.

Gladd, N.T. (1976). The lower-hybrid-drift instability and the modified two-stream instability in high-density theta-pinch environments, *Plasma Phys.* **18**, 27–40.

Glasser, A.H., Greene, J.M. & Johnson, J.L. (1975a). Resistive instabilities in general toroidal plasma configurations, *Phys. Fluids* **18**, 875–88.

Glasser, A.H., Furth, H.P. & Rutherford, P.H. (1977). Stabilization of resistive kink modes in the tokamak, *Phys. Rev. Lett.* **38**, 234–7.

Glatzmaier, G.A. (1985a). Numerical simulations of stellar convective dynamos. II. Field propagation in the convection zone, *Astrophys. J.* **291**, 300–7.

Glatzmaier, G.A. (1985b). Numerical simulations of stellar convective dynamos. III. At the base of the convection zone, *Geophys. Astrophys. Fluid Dyn.* **31**, 137–50.

Glatzmaier, G.A. & Roberts, P.H. (1995). A three-dimensional dynamo solution with rotating and finitely conducting inner core and mantle, *Phys. Earth Planet. Inter.* **91**, 63–75.

Glatzmaier, G.A. & Roberts, P.H. (1996). An anelastic evolutionary geodynamo simulation driven by compositional and thermal convection, *Physica D* **97**, 81–94.

Glatzmaier, G.A. & Roberts, P.H. (1997). Simulating the geodynamo, *Contemp. Phys.* **38**, 269–88.

Goeler, von, S., Stodiek, W. & Sauthoff, N. (1974). Studies of internal disruptions and $m = 1$ oscillations in tokamak discharges with soft X-ray techniques, *Phys. Rev. Lett.* **33**, 1201–3.

Gordeev, A.V., Kingsep, A.S. & Rudakov, L.I. (1994). Electron magnetohydrodynamics, *Phys. Rep.* **243**, 215–315.

Gould, R.W., O'Neil, T.M., & Malmberg, J.H. (1967). Plasma wave echo, *Phys. Rev. Lett.* **19**, 219–22.

Grad, H., Hu, P.N. & Stevens, D.C. (1975). Adiabatic evolution of plasma equilibrium, *Proc. Natl. Acad. Sci. (USA)* **72**, 3789–93.

Grauer. R. & Sideris, T.C. (1991). Numerical computation of 3D incompressible ideal fluids with swirl, *Phys. Rev. Lett.* **67**, 3511–4.

Grauer, R., Krug, J. & Marliani, C. (1994). Scaling of high-order structure functions in magnetohydrodynamic turbulence, *Phys. Lett.* A **195**, 335–8.

Grauer, R. & Marliani, C. (1995). Numerical and analytical estimates for the structure functions in two-dimensional magnetohydrodynamic flows, *Phys. Plasmas* **2**, 41–7.

Grauer, R., Marliani, C. & Germaschewski, K. (1998). Adaptive mesh refinement for singular solutions of the incompressible Euler equations, *Phys. Rev. Lett.* **80**, 4177–80.

Grauer, R. & Marliani, C. (2000). Current sheet formation in 3D ideal incompressible magnetohydrodynamics, submitted to *Phys. Rev. Lett.*

Green, J. M. (1988). Geometric properties of three-dimensional reconnecting magnetic fields with nulls, *J. Geophys. Res.* **93**, 8583–90.

Hallatschek, K., Gude, A., Biskamp, D., Günter, S. & the ASDEX Upgrade Team, High frequency mode cascades in the ASDEX Upgrade tokamak, *Phys. Rev. Lett.* **80**, 293–6.

Hamasaki, S. & Krall, N.A. (1973). Relaxation of anisotropic collisionless plasma, *Phys. Fluids* **16**, 145–9.

Harris, E.G. (1962). *Nuovo Cimento* **23**, 115.

Hatori, T. (1984). Kolmogorov-style argument for decaying homogeneous MHD turbulence, *J. Phys. Soc. Jpn.* **53**, 2539–45.

Hazeltine, R.D. & Strauss, H.R. (1978). Kinetic and finite-β effects on the $m = 1$ tearing instability, *Phys. Fluids* **21**, 1007–12.

Hazeltine, R.D., Kotschenreuther, M. & Morrison, P.J. (1985). A four-field model for tokamak plasma dynamics, *Phys. Fluids* **28**, 2466–77.

Helmholtz, H. von (1868). Über diskontinuierliche Flüssigkeitsbewegungen, *Monats. Königl. Preuss. Akad. Wiss. Berlin* **23**, 215–28.

Hender, T.C., Hastie, R.J. & Robinson, D.C. (1987). Finite-β effects on tearing modes in the tokamak, *Nucl. Fusion* **27**, 1389–400.

Herzenberg, A. (1958). Geomagnetic dynamos, *Phil. Trans. Roy. Soc.* **A250**, 543–83.

Hesse, M. & Schindler, K. (1988). A theoretical foundation of general magnetic reconnection, *J. Geophys. Res.* **93**, 5559–67.

Hesse, M. & Winske, D. (1994). Hybrid simulations of collisionless reconnection in current sheets, *J. Geophys. Res.* **99**, 11177–91.

Hesse, M. & Winske, D. (1998). Electron dissipation in collisionless magnetic reconnection, *J. Geophys. Res.* **103**, 26479–86.

Hesse, M., Winske, D., Birn, J. & Kuznetsova, M. (1998). Predictions and explanations of plasma sheet dissipation processes: current sheet kinking, In *Substorms IV*, eds. S. Kokubun & Y. Kamide (Kluver Academic, Boston), pp. 437–42.

Hesse, M., Schindler, K., Birn, J. & Kuznetsova, M. (1999). The diffusion region in collisionless magnetic reconnection, *Phys. Plasmas* **6**, 1781–95.

Hewett, D.W. & Langdon, A.B. (1987). Electromagnetic direct implicit plasma simulation, *J. Comput. Phys.* **72**, 121–55.

Hinton, F.L. & Horton, C.W. (1971). Amplitude limitation of a collisional drift wave instability, *Phys. Fluids* **14**, 116–23.

Hinton, F.L. & Hazeltine, R.D. (1976). Theory of plasma transport, *Rev. Mod. Phys.* **48**, 239–308.

Hirschman, S.P. & Molvig, K. (1979). Turbulent destabilization and saturation of the universal drift mode in a sheared magnetic field, *Phys. Rev. Lett.* **42**, 648–51.

Hockney, R.W. & Eastwood, J.W. (1988). *Computer Simulation Using Particles* (Adam Hilger, Bristol).

Hoh, F.C. (1966). Stability of sheet pinch, *Phys. Fluids* **9**, 277–84.

Hones, E.W., Jr (1979). Transient phenomena in the magnetotail and their relation to substorms, *Space Sci. Rev.* **23**, 393–410.

Horiuchi, R. & Sato, T. (1997). Particle simulation study of collisionless driven reconnection in a sheared magnetic field, *Phys. Plasmas* **4**, 277–89.

Hornig, G. & Schindler, K. (1996). Magnetic topology and the problem of its invariant definition, *Phys. Plasmas* **3**, 781–91.

Hoshino, M., Mukai, T., Yamamoto, T, & Kokubun, S. (1998). Ion dynamics in magnetic reconnection: comparison between numerical simulation and Geotail observations, *J. Geophys. Res.* **103**, 4509–30.

Hossain, M., Gray, P.C., Pontius, D.H., Matthaeus, W.H. & Oughton, S. (1995). Phenomenology for the decay of energy-containing eddies in homogeneous MHD turbulence, *Phys. Fluids* **7**, 2886–904.

Hsu, C.T., Hazeltine, R.D. & Morrison, P.J. (1986). A generalized reduced fluid model with finite ion-gyroradius effects, *Phys. Fluids* **29**, 1480–7.

Huba, J.D. & Papadopoulos, K. (1978). Nonlinear stabilization of the lower-hybrid-drift instability by electron resonance broadening, *Phys. Fluids* **21**, 121–3.

Huba, J.D., Drake, J.F. & Gladd, N.T. (1980). Lower-hybrid-drift instability in field-reversed plasmas, *Phys. Fluids* **23**, 552–61.

Imshennik, V.S. & Syrovatskii, S.I. (1967). Two-dimensional flow of an ideally conducting gas in the vicinity of a zero line of a magnetic field, *Sov. Phys. JETP* **25**, 656–64.

Iroshnikov, P.S. (1964). Turbulence in a conducting fluid in a strong magnetic field, *Sov. Astron.* **7**, 566–71.

Ivanova, T.S. & Ruzmaikin, A.A. (1977). A nonlinear magnetohydrodynamic model of the solar dynamo, *Sov. Astron.* **21**, 479–85.

Ji, H., Yamada, M., Hsu, S., Kulsrud, R. Carter, T. & Zaharia, S. (1999). Magnetic reconnection with Sweet–Parker characteristics in two-dimensional laboratory plasmas, *Phys. Plasma* **6**, 1743–50.

Johnson, F.S. (1960). The gross character of the geomagnetic field in the solar wind, *J. Geophys. Res.* **65**, 3049–51.

Jones, T.W., Gaalaas, J.B., Ryu, D. & Frank, A. (1997). The MHD Kelvin–Helmholtz instability II: The roles of weak and oblique fields in planar flows, *Astrophys. J.* **482**, 230–44.

Kadomtsev, B.B. (1965). *Plasma Turbulence* (Academic Press, London).

Kadomtsev, B.B. (1975). Disruptive instability in tokamaks, *Sov. J. Plasma Phys.* **1**, 389–91.

Kadomtsev, B.B. & Pogutse, O.P. (1974). Nonlinear helical perturbations of a plasma in a tokamak, *Sov. Phys. JETP* **38**, 283–90.

Kageyama, A., Watanabe, K. & Sato, T. (1992). A global simulation of the magnetosphere with a long tail: no interplanetary magnetic field, *J. Geophys. Res.* **97**, 3929–43.

Kageyama, A., Watanabe, K. & Sato, T. (1993). Simulation study of a magnetohydrodynamic dynamo: convection in a rotating shell, *Phys. Fluids B* **5**, 2793–805.

Kageyama, A., Sato, T. & the complexity simulation group (1995). Simulation study of a magnetohydrodynamic dynamo. II, *Phys. Plasmas* **2**, 1421–31.

Kageyama, A. & Sato, T. (1997). Generation mechanism of a dipole field by a magnetohydrodynamic dynamo, *Phys. Rev. E* **55**, 4617–26.

Kampen, van, N.G. (1955). On the theory of stationary waves in plasmas, *Physica* **21**, 949–63.

Kaw, P.K., Valeo, E.J. & Rutherford, P.H. (1979). Tearing modes in a plasma with magnetic braiding, *Phys. Rev. Lett.* **43**, 1398–401.

Kelvin, Lord (1871). Hydrokinetic solutions and observations, *Phil. Mag.* **42**, 362–77.

Kennel, C.F. (1995). *Convection and Substorms* (Oxford University Press, New York).

Kennel, C.F. & Engelmann, F. (1966). Velocity space diffusion from weak plasma turbulence in a magnetic field, *Phys. Fluids* **9**, 2377–88.

Kennel, C.F. & Petschek, H.E. (1966). Limit on stably trapped particle fluxes, *J. Geophys. Res.* **71**, 1–28.

Kerr, R.M. & Hussain, H. (1989). Simulation of vortex reconnection, *Physica D* **37**, 474–84.

Kida, S., Yanase, S. & Mizushima, J. (1991). Statistical properties of MHD turbulence and turbulent dynamo, *Phys. Fluids A* **3**, 457–65.

Kida, S. & Takaoka, M. (1994). Vortex reconnection, *Ann. Rev. Fluid Dyn.* **26**, 169–89.

Kikuchi, M. & Azumi, M. (1995). Experimental evidence for the bootstrap current in a tokamak, *Plasma Phys. Controlled Fusion* **37**, 1215–38.

Kingsep, A.S., Chukbar, K.V. & Yan'kov, V.V. (1990). Electron magnetohydrodynamics, In *Reviews of Plasma Physics*, ed. B. B. Kadomtsev (Consultants Bureau, New York), Vol. 16, pp. 243–88.

Kinney, R., McWilliams, J.C. & Tajima, T. (1995). Coherent structures and turbulent cascades in two-dimensional incompressible magnetohydrodynamic turbulence, *Phys. Plasmas* **2**, 3623–39.

Kitauchi, H., Araki, K. & Kida, S. (1997). Flow structure of thermal convection in a rotating spherical shell, *Nonlinearity* **10**, 885–904.

Kleva, R.G., Drake, J.F. & Waelbroeck, F.L. (1995). Fast reconnection in high-temperature plasmas, *Phys. Plasmas* **2**, 23–34.

Kolmogorov, A.N. (1941). Local structure of turbulence in an incompressible fluid at very large Reynolds numbers, *Dokl. Akad. Nauk SSSR* **30**, 299–303.

Kotschenreuther, M., Hazeltine, R.D. & Morrison, P.J. (1985). Nonlinear dynamics of magnetic islands with curvature and pressure, *Phys. Fluids* **28**, 294–302.

Kraichnan, R.H. (1959). The structure of isotropic turbulence at very high Reynolds numbers, *J. Fluid Mech.* **5**, 497–543.

Kraichnan, R.H. (1965). Inertial range spectrum in hydromagnetic turbulence, *Phys. Fluids* **8**, 1385–7.

Krall, N.A. (1968). Drift waves, In *Advances in Plasma Physics*, eds. A. Simon & W. B. Thomson, (Interscience Publishers, New York) Vol. 1, pp. 153–99.

Krall, N.A. & Liewer, P.C. (1971). Low-frequency instabilities in magnetic pulses, *Phys. Rev. A* **4**, 2094–103.

Krause, F.K. & Rädler, K.H. (1980). *Mean-field Magnetohydrodynamics and Dynamo Theory* (Pergamon Press, Oxford).

Krauss-Varban, D. & Omidi, N. (1995). Large-scale hybrid simulations of the magnetotail during reconnection, *Geophys. Res. Lett.* **22**, 3271–4.

Kuang, W. & Bloxham, J. (1997). An Earth-like numerical dynamo model, *Nature* **389**, 371–4.

Küker, M., Rüdiger, G. & Kitchatinov, L.L. (1993). An $\alpha\Omega$-model of the solar differential rotations, *Astron. Astrophys.* **279**, L1–4.

Kuznetsova, M., Hesse, M. & Winske, D. (2000). Toward a transport model of collisionless magnetic reconnection, to be published in *J. Geophys. Res.*

Lampe, M., Manheimer, W.M., McBride, J.B., Orens, J.H., Shanny, R. & Sudan, R.N. (1971). Nonlinear development of the beam-cyclotron instability, *Phys. Rev. Lett.* **26**, 1221–5.

Lapenta, G. & Brackbill, J.U. (1997). A kinetic theory for the drift-kink instability, *J. Geophys. Res.* **102**, 27009–108.

Lapenta, G. & Brackbill, J.U. (2000). An exploration of the connection between tearing and kink instabilities, to be published in *J. Geophys. Res.*

Lau, Y.T. & Finn, J.M. (1990). Three-dimensional kinematic reconnection in the presence of magnetic nulls and closed field lines, *Astrophys. J.* **350**, 672–91.

Lau, Y.T. & Finn, J.M. (1996). Magnetic reconnection and topology of interacting twisted flux tubes, *Phys. Plasmas* **3**, 3983–97.

Lau, Y.Y. & Liu, C.S. (1980). Stability of shear flow in a magnetized plasma, *Phys. Fluids* **23**, 939–41.

Laval, G., Pellat, R. & Vuillemin, M. (1966). Instabilités électromagnetiques des plasmas sans collisions, *Proceedings of the Conference on Plasma Physics and Controlled Nuclear Fusion Research* (IAEA, Vienna), Vol. II, 259–77.

Laval, G. & Pesme, D. (1983). Breakdown of quasilinear theory for incoherent 1D Langmuir waves, *Phys. Fluids* **26**, 52–65; Inconsistency of quasilinear theory, ibid. 66–7.

Lee, L.C. & Fu, Z.F. (1985). A theory of magnetic flux transfer at the Earth's magnetopause, *Geophys. Res. Lett.* **12**, 105–8.

Lee, L.C., Wang, S., Wei, C.Q. & Tsurutani, B.T. (1988). Streaming sausage, kink and tearing instabilities in a current sheet with applications to the Earth's magnetotail, *J. Geophys. Res.* **93**, 7354–65.

Lembège, B. & Pellat, R. (1982). Stability of a thick quasineutral sheet, *Phys. Fluids* **25**, 1995–2004.

Lesieur, M. & Schertzer, D. (1978). Amortissement autosimilaire d'une turbulence à grand nombre de Reynolds, *J. Méc* **17**, 609–48.

Levinton, F.M., Batha, S.H., Yamada, M. & Zarnstorff, M.C. (1993). q-profile measurements in the Tokamak Fusion Test Reactor, *Phys. Fluids B* **5**, 2554–61.

Levinton, F.M., Zakharov, L., Batha, S.H., Manickam, J. & Zarnstorff, M.C. (1994). Stabilization and onset of sawteeth in TFTR, *Phys. Rev. Lett.* **72**, 2895–8.

Libbrecht, K.G. (1988). Solar p-mode frequency splitting, In *Seismology of the Sun and Sun-like Stars*, ed. E.J. Rolfe, ESA SP-286, pp. 131–6.

Lifshitz, A.E. (1989). *Magnetohydrodynamics and Spectral Theory* (Kluwer Academic Publishers, Dortrecht).

Lohse, D. & Müller-Groeling, A. (1995). Bottleneck effects in turbulence: scaling phenomena in r versus p space, *Phys. Rev. Lett.* **74**, 1747–50.

Longcope, D.W. & Cowley, S.C. (1996). Current sheet formation along three-dimensional magnetic separators, *Phys. Plasmas* **3**, 2885–97.

Lorenz, E.N. (1963). Deterministic nonperiodic flow, *J. Atmospheric Sci.* **20**, 130–41.

Lortz, D. (1968). Exact solutions of the hydromagnetic dynamo problem, *Plasma Phys.* **10**, 967–72.

Lottermoser, R.F., Scholer, M. & Matthews, A.P. (1998). Ion kinetic effects in magnetic reconnection: hybrid simulations, *J. Geophys. Res.* **103**, 4547–59.

Lui, A.T.Y. (1996). Current disruption in the Earth's magnetosphere: observations and models, *J. Geophys. Res.* **101**, 13067–88.

Malagoli, A., Bodo, G. & Rosner, R. (1996). On the nonlinear evolution of magnetohydrodynamic Kelvin–Helmholtz instabilities, *Astrophys. J.* **456**, 708–16.

Malkus, W.V.R. (1972). Reversing Bullard's dynamo, *EOS, Trans. Am. Geophys. Union* **53**, 617.

Malkus, W.V.R. & Proctor, M.R.E. (1975). The macrodynamics of the α-effect dynamos in rotating fluids, *J. Fluid Mech.* **67**, 417–43.

Mandt, M.E., Denton, R.E. & Drake, J.F. (1994). Transition to whistler-mediated magnetic reconnection, *Geophys. Res. Lett.* **21**, 73–6.

Manheimer, W.M. & Flynn, R.W. (1974). Formation of nonthermal ion tails in the ion acoustic instability, *Phys. Fluids* **17**, 409–15.

Mason, R.J. (1981). Implicit moment particle simulation of plasmas, *J. Comput. Phys.* **41**, 233–44.

Maunder, E.W. (1913). Distribution of sunspots in heliographic latitude, 1874–913, *Mon. Not. Roy. Astron. Soc.* **74**, 112–16.

McCarthy, D.R., Booth, A.E., Drake, J.F. & Guzdar, P.N. (1997). Three-dimensional simulations of the parallel-shear velocity instability, *Phys. Plasmas* **4**, 300–9.

McPherron, R.L. (1991). Physical processes producing magnetospheric substorms and magnetic storms, In *Geomagnetism*, ed. J.A. Jacobs (Academic press, London), Vol. 4, p. 593.

McPherron, R.L., Russell, C.T. & Aubry, M. (1973). Phenomenological model for substorms, *J. Geophys. Res.* **78**, 3131–49.

Mercier, C. (1960). Un critère nécessaire de stabilité hydromagnétique pour un plasma en symétrie de révolution, *Nucl. Fusion* **1**, 47–53.

Miura, A. & Pritchett, P.L. (1982). Nonlocal stability analysis of the MHD Kelvin–Helmholtz instability in a compressible plasma, *J. Geophys. Res.* **87**, 7431–44.

Moffatt, H.K. (1978). *Magnetic Field Generation in Electrically Conducting Fluids* (Cambridge University Press, Cambridge).

Moffatt, H.K. & Proctor, M.R.E. (1985). Topological constraints associated with fast dynamo action, *J. Fluid Mech.* **154**, 493–507.

Moffatt, H.K. & Ricca, R.L. (1992). Helicity and the Calugareanu invariant, *Proc. Roy. Soc. London A* **439**, 411–29.

Monin, A.S. & Yaglom, A.M. (1975). *Statistical Fluid Mechanics: Mechanics of Turbulence* (The MIT Press, Cambridge, Massachusetts), Vols. I and II.

Montgomery, D., Turner, L. & Vahala, G. (1979). Most probable states in magnetohydrodynamics, *J. Plasma Phys.* **21**, 239–51.

Montgomery, M.D., Asbridge, J.R. & Bame, S.J. (1970). Vela 4 plasma observations near the Earth's bow shock, *J. Geophys. Res.* **75**, 1217–31.

Morrison, P.J. & Hazeltine, R.P. (1984). Hamiltonian formulation of reduced magnetohydrodynamics, *Phys. Fluids* **27**, 886–97.

Müller, W.-C. & Biskamp, D. (2000). Scaling properties for three-dimensional magnetohydrodynamic turbulence, *Phys. Rev. Lett.* **84**, 475–8.

Nagayama, Y. *et al.* (1991). Analysis of sawtooth oscillations using simultaneous measurements of electron-cyclotron emission imaging and X-ray tomography on TFTR, *Phys. Rev. Lett.* **25**, 3527–30.

Nagayama, Y. *et al.* (1996). Tomography of full sawtooth crashes on the Tokamak Fusion Test Reactor, *Phys. Plasmas* **3**, 1647–55.

Newcomb, W.A. (1958). Motion of magnetic lines of force, *Ann. Phys.* **3**, 347–85.

Nishimura, Y., Callen, J.D. & Hegna, C.C. (1999). Onset of high-*n* ballooning modes during tokamak sawtooth crashes, *Phys. Plasmas* **6**, 4685–92.

Ono, Y., Yamada, M., Tajima, T. & Matsumoto, R. (1996). Ion acceleration and direct ion heating in three-component magnetic reconnection, *Phys. Rev. Lett.* **76**, 3328–31.

Ono, Y., Inomoto, M., Okazaki, T. & Ueda, Y. (1997). Experimental investigation of three-component magnetic reconnection by use of merging spheromaks and tokamaks, *Phys. Plasmas* **4**, 1953–63.

Orr, W.M.F. (1907). The stability or instability of steady motions of a perfect liquid and of a viscous liquid, *Proc. Roy. Irish Acad.* **A27**, 9–68 and 69–138.

Ortolani, S. & Schnack, D.D. (1993). *Magnetohydrodynamics of Plasma Relaxation* (World Scientific, Singapore).

Orszag, S.A. & Kraichnan, R.H. (1967). Model equations for strong turbulence in a Vlasov plasma, *Phys. Fluids* **10**, 1720–36.

Orszag, S.A. & Tang, C.-M. (1979). Small-scale structure of two-dimensional magnetohydrodynamic turbulence, *J. Fluid Mech.* **90**, 129–43.

Ossakow, S.L., Ott, E. & Haber, I. (1972). Nonlinear evolution of whistler instabilities, *Phys. Fluids* **15**, 2314–26.

Ottaviani, M. & Porcelli, F. (1993). Nonlinear collisionless magnetic reconnection, *Phys. Rev. Lett.* **71**, 3802–5.

Ozaki, M., Sato, T., Horiuchi, R., & the complexity simulation group (1996). Electromagnetic instability and anomalous resistivity in a magnetic neutral sheet, *Phys. Plasmas* **3**, 2265–74.

Park, W., Fredrickson, E.D., Janos, A., Manickam, J. & Tang, W.M. (1995). High-β disruptions in tokamaks, *Phys. Rev. Lett.* **75**, 1763–6.

Parker, E.N. (1955a). Hydromagnetic dynamo models, *Astrophys. J.* **122**, 293–314.

Parker, E.N. (1955b). The formation of sunspots from the solar toroidal field, *Astrophys. J.* **121**, 491–507.

Parker, E.N. (1957). Terrestrial magnetism and atmospheric electricity, *J. Geophys. Res.* **62**, 509–20.

Parker, E.N. (1958). Dynamics of interplanetary gas and magnetic fields, *Astrophys. J.* **128**, 664–76.

Parker, E.N. (1960). The hydrodynamic theory of solar corpuscular radiation and stellar winds, *Astrophys. J.* **132**, 821–66.

Parker, E. N. (1963). The solar flare phenomenon and the theory of reconnection and annihilation of magnetic fields, *Astrophys. J. Suppl. Ser.* **8**, 177–211.

Parker, E.N. (1980). *Cosmical Magnetic Fields* (Clarendon Press, Oxford).

Parker, E.N. (1994). *Spontaneous Current Sheets in Magnetic Fields* (Oxford University Press, New York).

Paschmann, G., Sonnerup, B.U.O., Papamastorakis, I., Sckopcke, N., Haerendel, G., Bame, S.J., Asbridge, J.B., Gosling, J.T., Russel, C.T. & Elphic, R.C. (1979). Plama acceleration at the Earth's magnetopause: evidence for reconnection, *Nature* **282**, 243–6.

Pellat, R., Coroniti, F.V. & Pritchett, P.L. (1991). Does ion tearing exist?, *Geophys. Res. Lett.* **18**, 143–6.

Petschek, H.E (1964). Magnetic field annihilation, In *AAS/NASA Symposium on the Physics of Solar Flares*, ed. W. N. Hess (NASA, Washington, DC), pp. 425–37.

Pfister, H. & Gekelman, W. (1991). Demonstration of helicity conservation during magnetic reconnection using Christmas ribbons, *Am. J. Phys.* **59**, 497–502.

Phan, T.D. & Sonnerup, B.U.O. (1991). Resistive tearing mode instability in a current sheet with equilibrium viscous stagnation flow, *J. Plasma Phys.* **46**, 407–21.

Politano, H. & Pouquet, A. (1995). Model of intermittency in magnetohydrodynamic turbulence, *Phys. Rev. E* **52**, 636–41.

Politano, H., Pouquet, A. & Sulem, P.L. (1995). Current and vorticity dynamics in three-dimensional magnetohydrodynamic turbulence, *Phys. Plasmas* **2**, 2931–9.

Politano, H. & Pouquet, A. (1998a). Von Karman–Howarth equation for magnetohydrodynamics and its consequences on third-order longitudinal structure and correlation functions, *Phys. Rev. E* **57**, R21–4.

Politano, H. & Pouquet, A. (1998b). Dynamic length scales for turbulent magnetized flows, *Geophys. Res. Lett.* **25**, 273–6.

Politano, H., Pouquet, A. & Carbone, V. (1998). Determination of anomalous exponents of structure functions in two-dimensional magnetohydrodynamic turbulence, *Europhys. Lett.* **43**, 516–21.

Porcelli, F. (1991). Collisionless $m = 1$ tearing mode, *Phys. Rev. Lett.* **66**, 425–8.

Priest, E.R. (1984). *Solar Magnetohydrodynamics* (D. Reidel Publishing Company, Dordrecht).

Priest, E.R. & Forbes, T.G. (1986). New models for fast steady state magnetic reconnection, *J. Geophys. Res.* **91**, 5579–88.

Pritchett, P.L. (1994). Effect of electron dynamics on collisionless reconnection in two-dimensional magnetotail equilibria, *J. Geophys. Res.* **99**, 5935–41.

Pritchett, P.L. (1997). Collisionless reconnection in the magnetotail, *Adv. Space Res.* **19**, 1807–16.

Pritchett, P.L. & Wu, C.C. (1979). Coalescence of magnetic islands, *Phys. Fluids* **22**, 2140–6.

Pritchett, P.L., Lee, Y.C. & Drake, J.F. (1980). Linear analysis of the double-tearing mode, *Phys. Fluids* **23**, 1368–74.

Pritchett, P.L. & Coronoti, F.V. (1996). The role of the drift-kink mode in destabilizing thin current sheets, *J. Geomag. Geoelectr.* **48**, 833–44.

Pumir, A. & Kerr, R.M. (1987). Numerical simulation of interacting vortex tubes, *Phys. Rev. Lett.* **58**, 1636–9.

Quest, K.B., Karimabadi, H. & Brittnacher, M. (1996). Consequences of particle conservation along a flux surface for magnetotail tearing, *J. Geophys. Res.* **101**, 179–83.

Rädler, K.H. (1990). The solar dynamo, In *Inside the Sun*, eds. G. Berthomieu & M. Cribier (Kluwer Academic Publishers), pp. 385–402.

Rayleigh, Lord (1880). On the stability, or instability, of certain fluid motions, *Proc. London Math. Soc.* **11**, 57–70.

Rikitake, T. (1958). Oscillations of a system of disc dynamos, *Proc. Camb. Phil. Soc.* **54**, 89–105.

Robbins, K.A. (1977). A new approach to subcritical instability and turbulent transitions in a simple dynamo, *Math. Proc. Camb. Phil. Soc.* **82**, 309–25.

Roberts, G.O. (1972). Dynamo action of fluid motions with two-dimensional periodicity, *Phil. Trans. Roy. Soc.* **A271**, 411–54.

Roberts, P.H. (1968). On the thermal instability of a self-gravitating fluid sphere containing heat sources, *Phil. Trans. Roy. Soc.* **A263**, 93–117.

Roberts, P.H. & Soward, A.M. (1992). Dynamo theory, *Ann. Rev. Fluid Mech.*, **24**, 459–512.

Rogers, B. & Zakharov, L. (1995). Nonlinear ω_*-stabilization of the $m = 1$ mode in tokamaks, *Phys. Plasmas* **2**, 3420–8.

Rogers, B.N., Drake, J.F. & Shay, M.A. (2000). The onset of turbulence in 3D collisionless magnetic reconnection, to be published in *Geophys. Res. Lett.*

Rosenbluth, M.N., Dagazian, R.Y. & Rutherford, P.H. (1973). Nonlinear properties of the internal $m = 1$ instability in a cylindrical tokamak, *Phys. Fluids* **16**, 1894–902.

Roux, A. *et al.* (1991). Plasma sheet instability related to the westward traveling surge, *J. Geophys. Res.* **96**, 17697–714.

Rüdiger, G. (1973). Behandlung eines einfachen magnetohydrodynamischen Dynamos mittels Linearisierung, *Astron. Nachr.* **294**, 183–6.

Rüdiger, G. (1989). *Differential Rotation and Stellar Convection. Sun and Solar-type Stars* (Akadmie Verlag, Berlin).

Russell, C.T., editor (1995). Physics of collisionless shocks, *Adv. Space Res.*, Vol. 15.

Russell, C.T. & Elphic, R.C. (1978). Initial ISEE magnetometer results: magnetopause observations, *Space Sci. Rev.* **22**, 681–715.

Rutherford, P.H. (1973). Nonlinear growth of the tearing mode, *Phys. Fluids* **16**, 1903–8.

Sagdeev, R.Z. (1967). On Ohm's law resulting from instability, *Proc. Symposia Appl. Math.* (American Mathematical Society, Providence, Rhode Island), Vol. 18, 281–6.

Sato, T. (1979). Strong plasma acceleration by slow shocks resulting from magnetic reconnection, *J. Geophys. Res.* **84**, 7177–90.

Sato, T. & Hayashi, T. (1979). Externally driven magnetic reconnection and a powerful magnetic energy converter, *Phys. Fluids* **22**, 1189–1202.

Schindler, K. (1974). A theory of the substorm mechanism, *J. Geophys. Res.* **79**, 2813–10.

Schindler, K. & Soop, M. (1968). Stability of plasma sheaths, *Phys. Fluids* **11**, 1192–5.

Schindler, K., Pfirsch, D. & Wobig, H. (1973). Stability of two-dimensional collisionless plasmas, *Plasma Phys.* **15**, 1165–84.

Schindler, K., Hesse, M. & Birn, J. (1988). General magnetic reconnection, parallel electric fields, and helicity, *J. Geophys. Res.* **93**, 5547–57.

Scholer, M. (1988). Magnetic flux transfer at the magnetopause based on single X-line bursty reconnection, *Geophys. Res. Lett.* **15**, 291–4.

Shafranov, V.D. & Yurchenko, E.I. (1968). Condition for flute instability of a toroidal-geometry plasma, *Sov. Phys. JETP* **26**, 682–6.

Shay, M.A. & Drake, J.F. (1998). The role of electron dissipation on the rate of collisionless magnetic reconnection, *Geophys. Res. Lett.* **25**, 3759–62.

Shay, M.A., Drake, J.F., Denton, R.E. & Biskamp, D. (1998). Structure of the dissipation region during collisionless magnetic reconnection, *J. Geophys. Res.* **103**, 9165–76.

Shay, M.A., Drake, J.F., Rogers, B.N. & Denton, R.E. (1999). The scaling of collisionless magnetic reconnection for large systems, *Geophys. Res. Lett.* **26**, 2163–6.

She, Z.S. & Leveque, E. (1994). Universal scaling laws in fully developed turbulence, *Phys. Rev. Lett.* **72**, 336–9.

Shibata, K., Yokoyama, T. & Shimojo, M. (1996). Coronal X-ray jets observed with the Yohkoh soft X-ray telescope, *J. Geomag. Geoelectr.* **48**, 19–28.

Shiokawa, K., Haerendel, G. & Baumjohann, W. (1998). Azimuthal pressure gradient as driving force of the substorm current, *Geophys. Res. Lett.* **25**, 1179–82.

Silcock, G. (1975). On the stability of parallel stratified shear flows. Ph.D. dissertation, University of Bristol.

Soltwisch, H., Stodiek, W., Manickam, J. & Schlüter, J. (1987). Current-density profiles in the TEXTOR tokamak, In *Proceedings of the Eleventh Conference on Plasma Physics and Controlled Nuclear Fusion Research* (IAEA, Vienna), Vol. I, pp. 267–73.

Sommerfeld, A. (1908). Ein Beitrag zur hydrodynamischen Erklärung der turbulenten Flüssigkeitsbewegungen, *Proc. 4th Int. Congress Maths.* Rome, Vol. III, pp. 116–24.

Sonnerup, B.U. (1970). Magnetic field reconnection in a highly conducting incompressible fluid, *J. Plasma Phys.* **4**, 161–74.

Sonnerup, B.U.O. & Priest, E.R. (1975). Resistive MHD stagnation-point flows at a current sheet. *J. Plasma Phys.* **14**, 283–94.

Sonnerup, B.U.O., Paschmann, G., Papamastorakis, I., Sckopcke, N., Haerendel, G., Bama, S.J., Asbridge, J.R., Gosling, J.T. & Russell, C.T. (1981). Evidence for magnetic field line reconnection at the Earth's magnetopause, *J. Geophys. Res.* **86**, 10049–67.

Speiser, T.W. (1965). Particle trajectories in model current sheets. 1. Analytical solutions, *J. Geophys. Res.* **70**, 4219–28.

St Pierre, M.G. (1993). The strong-field branch of the Childress–Soward dynamo, In *Solar and Planetary Dynamos*, eds. M.R.E. Proctor *et al.*, (Cambridge University Press), pp. 295–302.

Steenbeck, M., Krause, F. & Rädler, K.H. (1966). Berechnung der mittleren Lorentz Feldstärke $\langle v \times B \rangle$ für ein elektrisch leitendes Medium in turbulenter, durch Coriolis Kräfte beeinflußter Bewegung, *Z. Naturforsch.* **21a**, 369–76.

Stenzel, R.L. & Gekelman, W. (1981). Magnetic field line reconnection experiments. 1. Field topologies, *J. Geophys. Res.* **86**, 649–58.

Stenzel, R.L., Gekelman, W. & Wild, N. (1983). Magnetic field line reconnection experiments. 5. Current disruptions and double layers, *J. Geophys. Res.* **88**, 4793–804.

Stix, M. (1991). The solar dynamo, *Geophys. Astrophys. Fluid Dyn.* **62**, 211–28.

Stix, T.H. (1962). *Theory of Plasma Waves* (McGraw-Hill, New York).

Strauss, H.R. (1976). Nonlinear three-dimensional magnetohydrodynamics of noncircular tokamaks, *Phys. Fluids* **19**, 134–40.

Strauss, H.R. (1977). Dynamics of high-β tokamaks, *Phys. Fluids* **20**, 1354–60.

Sudan, R.N. (1963). Plasma electromagnetic instabilities, *Phys. Fluids* **6**, 57–61.

Sulem, P.L., Frisch, U., Pouquet, A. & Meneguzzi, M. (1985). On the exponential flattening of current sheets near neutral X-points in two-dimensional ideal MHD flow, *J. Plasma Phys.* **33**, 191–8.

Sun, Z.-P. & Schubert, G. (1995). Numerical simulations of thermal convection in a rotating spherical fluid shell at high Taylor and Rayleigh numbers, *Phys. Fluids* **7**, 2686–99.

Suydam, B.R. (1958). Stability of a linear pinch, In *Proceedings of the Second*

United Nations International Conference on the Peaceful Uses of Atomic Energy (United Nations, Geneva), Vol. 31, pp. 157–9.

Sweet, P.A. (1958). The production of high-energy particles in solar flares, *Nuovo Cimento Suppl.* **8**, Ser. X, 188–96.

Syrovatskii, S.I. (1971). Formation of current sheets in a plasma with a frozen-in strong magnetic field, *Sov. Phys. JETP* **33**, 933–40.

Tanaka, M. (1988). Macro-scale implicit electromagnetic particle simulation of magnetized plasmas, *J. Comput. Phys.* **79**, 209–26.

Tanaka, M. (1993). A simulation of low-frequency electromagnetic phenomena in kinetic plasmas of three dimensions, *J. Comput. Phys.* **107**, 124–45.

Tanaka, M. & Sato, T. (1981). Simulations of lower-hybrid-drift instability and anomalous resistivity in the magnetic neutral sheet, *J. Geophys. Res.* **86**, 5541–52.

Taylor, J.B. (1963). The magnetohydrodynamics of a rotating fluid and the Earth's dynamo problem, *Proc. Roy. Soc. London A* **274**, 274–83.

Taylor, J.B. (1974). Relaxation of toroidal plasmas and generation of reversed magnetic fields, *Phys. Rev. Lett.* **33**, 1139–41.

Taylor, J.B. (1986). Relaxation and magnetic reconnection in plasmas, *Rev. Mod. Phys.* **53**, 741–63.

Taylor, J.B. (1993). Filamentation, current profiles and transport in a tokamak, *Phys. Fluids B* **5**, 4378–83.

Ting, A.C., Matthaeus, W.H. & Montgomery, D. (1986). Turbulent relaxation processes in magnetohydrodynamics, *Phys. Fluids* **29**, 3261–74.

Tu, C.Y. & Marsch, E. (1995). *MHD Structures, Waves and Turbulence in the Solar Wind* (Kluwer Academic Publishers, Dordrecht).

Ugai, M. (1995). Computer studies on powerful magnetic energy conversion by the spontaneous fast reconnection mechanism, *Phys. Plasmas* **2**, 388–97.

Usadi, A., Kageyama, A., Watanabe, K. & Sato, T. (1993). A global simulation of the magnetosphere with a long tail: southward and northward interplanetary magnetic field, *J. Geophys. Res.* **98**, 7503–17.

Uzdensky, D.A. & Kulsrud, R.M. (1998). On the viscous boundary layer near the center of the resistive reconnection region, *Phys. Plasmas* **5**, 3249–56.

Vainshtein, S.I. & Zeldovich, Ya.B. (1972). Origin of magnetic fields in astrophysics, *Sov. Phys. Usp.* **15**, 159–72.

Vasyliunas, V.M. (1975). Theoretical models for magnetic field line merging, *Rev. Geophys. Space Phys.* **13**, 303–36.

Vedenov, A.A., Velikhov, E.P. & Sagdeev, R.Z. (1961). *Nucl. Fusion* **1**, 83–100.

Vedenov, A.A., Velikhov, E.P. & Sagdeev, R.Z. (1962). *Nucl. Fusion Suppl.* **2**, 465–75.

Vekshtein, G.E. & Sagdeev, R.Z. (1970). Anomalous resistance of a plasma in the case of ion-acoustic turbulence, *JETP Lett.* **11**, 194–7.

Verhoogen, J. (1961). *Geophys. J. R. Astron. Soc.* **4**, 276–81.

Waelbroeck, F.L. (1989). Current sheets and the nonlinear growth of the $m = 1$ kink-tearing mode, *Phys. Fluids* **B1**, 2372–80.

Wang, S., Lee. L.C. & Wei, C.Q. (1988). Streaming tearing instability in the current sheet with a super-Alfvénic flow, *Phys. Fluids* **31**, 1544–8.

Weibel, E.S. (1959). Spontaneously growing transverse waves in a plasma due to an anisotropic velocity distribution, *Phys. Rev. Lett.* **2**, 83–4.

Wesson, J.A. (1986). Sawtooth oscillations, *Plasma Phys. Control. Fusion* **28**, 243–8.

Wesson, J.A. (1990). Sawtooth reconnection, *Nucl. Fusion* **30**, 2545–9.

Wesson, J.A. (1991). Energy balance in tearing modes, *Plasma Phys. Control. Fusion* **35**, 1–15.

White, R.B., Monticello, D.A., Rosenbluth, M.N. & Waddell, B.V. (1977). Saturation of the tearing mode, *Phys. Fluids* **20**, 800–5.

Wicht, J. & Busse, F.H. (1997). Magnetohydrodynamic dynamos in rotating spherical shells, *Geophys. Astrophys. Fluid Dyn.* **86**, 103–29.

Wilson, H.R., Connor, J.W., Hastie, R.J. & Hegne, C.C. (1996). Threshold for neoclassical magnetic islands in a low collision frequency tokamak, *Phys. Plasmas* **3**, 248–65.

Winske, D. (1981). Current-driven microinstabilities in a neutral sheet, *Phys. Fluids* **24**, 1069–76.

Winske, D. & Liewer, P.C. (1978). Particle simulation studies of the lower-hybrid-drift instability, *Phys. Fluids* **21**, 1017–25.

Wu, C.C., Walker, R.J. & Dawson, J.M. (1981). A three-dimensional MHD model of the Earth's magnetosphere, *Geophys. Res. Lett.* **8**, 523–6.

Yamada, M. *et al.* (1994). Investigation of magnetic reconnection during a swatooth crash in a high-temperature tokamak plasma, *Phys. Plasmas* **1**, 3269–76.

Yamada, M. *et al.* (1997). Study of driven magnetic reconnection in a laboratory plasma, *Phys. Plasmas* **4**, 1936–44.

Yeh, T. & Axford, W.I. (1970). On the reconnection of magnetic field lines in conducting fluids, *J. Plasma Phys.* **4**, 207–29.

Yoon, P.H., Lui, A.T.J. & Chang, C.L. (1994). Lower-hybrid-drift instability operative in the geomagnetic tail, *Phys. Plasmas* **1**, 3033–43.

Yoshimura, H. (1975). A model of the solar cycle driven by the dynamo action of the global convection in the solar convection zone, *Astrophys. J. Suppl.* **29**, 467–94.

Zabusky, N.J., Boratav, O.N., Pelz, R.B., Gao, M., Silver D. & Cooper, S.P. (1991). Emergence of coherent patterns of vortex stretching during reconnection: a scattering paradigm, *Phys. Rev. Lett.* **67**, 2469–72.

Zakharov, L. & Rogers, B. (1992). Two-fluid magnetohydrodynamic description of the internal kink mode in tokamaks, *Phys. Fluids B* **4**, 3285–301.

Zakharov, L., Rogers, B. & Migliuolo, S. (1993). The theory of the early nonlinear stage of $m = 1$ reconnection in tokamaks, *Phys. Fluids B* **5**, 2498–505.

Zeiler, A., Drake, J.F. & Rogers, B. (1997). Nonlinear reduced Braginskii equations with ion thermal dynamics in toroidal plasmas, *Phys. Plasmas* **4**, 2134–8.

Zeldovich, Ya.B., Ruzmaikin, A.A. & Sokoloff, D.D. (1983). *Magnetic Fields in Astrophysics* (Gordon & Breach Science Publishers, New York).

Zhang, K. (1992). Spiralling columnar convection in rapidly rotating spherical fluid shells, *J. Fluid Mech.* **236**, 535–56.

Zhang, K.K. & Busse, F.H. (1988). Finite amplitude convection and magnetic field generation in a rotating spherical shell, *Geophys. Astrophys. Fluid Dyn.* **44**, 33–53.

Zhang, K.K. & Busse, F.H. (1989). Magnetohydrodynamic dynamos in rotating shperical shells, *Geophys. Astrophys. Fluid Dyn.* **49**, 97–116.

Zhang, K.K. & Busse, F.H. (1990). Generation of magnetic field by convection in a rotating spherical shell of infinite Prandtl number, *Phys. Earth Planet. Inter.* **59**, 208–22.

Zhu, Z. & Winglee, R.M. (1996). Tearing instability, flux ropes, and the kinetic current sheet kink instability in the Earth's magnetotail: a three-dimensional perspective from particle simulations, *J. Geophys. Res.* **101**, 4885–97.

Zweibel, E.G. & Bruhwiler, D.L. (1992). The effect of line tying on Parker's instability, *Astrophys. J.* **399**, 318–24.

Index